Electron-Phonon Interactions and Phase Transitions

NATO ADVANCED STUDY INSTITUTES SERIES

A series of edited volumes comprising multifaceted studies of contemporary scientific issues by some of the best scientific minds in the world, assembled in cooperation with NATO Scientific Affairs Division.

Series B: Physics

RECENT VOLUMES IN THIS SERIES

Volume 19 – Defects and Their Structure in Nonmetallic Solids
edited by B. Henderson and A. E. Hughes

Volume 20 – Physics of Structurally Disordered Solids
edited by Shashanka S. Mitra

Volume 21 – Superconductor Applications: SQUIDs and Machines
edited by Brian B. Schwartz and Simon Foner

Volume 22 – Nuclear Magnetic Resonance in Solids
edited by Lieven Van Gerven

Volume 23 – Photon Correlation Spectroscopy and Velocimetry
edited by H. Z. Cummins and E. R. Pike

Volume 24 – Electrons in Finite and Infinite Structures
edited by P. Phariseau and L. Scheire

Volume 25 – Chemistry and Physics of One-Dimensional Metals
edited by Heimo J. Keller

Volume 26 – New Developments in Quantum Field Theory and Statistical Mechanics–Cargèse 1976
edited by Maurice Lévy and Pronob Mitter

Volume 27 – Topics in Theoretical and Experimental Gravitation Physics
Edited by V. De Sabbata and J. Weber

Volume 28 – Material Characterization Using Ion Beams
Edited by J. P. Thomas and A. Cachard

Volume 29 – Electron–Phonon Interactions and Phase Transitions
Edited by Tormod Riste

The series is published by an international board of publishers in conjunction with NATO Scientific Affairs Division

A	Life Sciences	Plenum Publishing Corporation
B	Physics	New York and London
C	Mathematical and Physical Sciences	D. Reidel Publishing Company Dordrecht and Boston
D	Behavioral and Social Sciences	Sijthoff International Publishing Company Leiden
E	Applied Sciences	Noordhoff International Publishing Leiden

Electron-Phonon Interactions and Phase Transitions

Edited by

Tormod Riste
Institute for Atomic Energy
Kjeller, Norway

PLENUM PRESS • NEW YORK AND LONDON
Published in cooperation with NATO Scientific Affairs Division

Library of Congress Cataloging in Publication Data

Nato Advanced Study Institute on Electron-Phonon Interactions and Phase Transitions, Geilo, Norway, 1977.
Electron–phonon interactions and phase transitions.

(NATO advanced study institutes series: Series B, Physics; v. 29)
Includes index.
1. Phase transformations (Statistical physics)–Addresses, essays, lectures. 2. Electrons–Scattering–Addresses, essays, lectures. 3. Phonons–Scattering–Addresses, essays, lectures. 4. Jahn-Teller effect–Addresses, essays, lectures. I. Riste, Tormod, 1925- II. Title. III. Series: NATO advanced study institutes series: Series B, Physics; v. 29.
QC176.8.P45N37 1977 530.4'1 77-14590
ISBN 0-306-35729-1

Lectures presented at the NATO Advanced Study Institute on Electron–Phonon Interactions and Phase Transitions held in Geilo, Norway, April, 1977

A Division of Plenum Publishing Corporation
227 West 17th Street, New York, N.Y. 10011

Printed in the United States of America

PREFACE

This NATO Advanced Study Institute was the fourth in a series devoted to the subject of phase transitions and instabilities with particular attention to structural phase transformations. Beginning with the first Geilo institute in 1971 we have seen the emphasis evolve from the simple quasiharmonic soft mode description within the Landau theory, through the unexpected spectral structure represented by the "central peak" (1973), to such subjects as melting, turbulence and hydrodynamic instabilities (1975). Sophisticated theoretical techniques such as scaling laws and renormalization group theory developed over the same period have brought to this wide range of subjects a pleasing unity. These institutes have been instrumental in placing structural transformations clearly in the mainstream of statistical physics and critical phenomena.

The present Geilo institute retains some of the counter cultural flavour of the first one by insisting whenever possible upon peeking under the skirts of even the most successful phenomenology to catch a glimpse of the underlying microscopic processes. Of course the soft mode remains a useful concept, but the major emphasis of this institute is the microscopic cause of the mode softening. The discussions given here illustrate that for certain important classes of solids the cause lies in the electron phonon interaction. Three major types of structural transitions are considered. In the case of metals and semimetals, the electron phonon interaction relies heavily on the topology of the Fermi surface. In special situations the electronic energy of the conduction band can be lowered by a greater amount than the energetic cost of a lattice distortion, so that lattice distortion will indeed occur. As Professor Friedel has emphasized, instabilities of this type are relatively more likely in band structures of lower dimensionality because of the less severe requirements for "nesting" of various portions of Fermi surface. Several of the papers herein are devoted to the calculation, observation and consequences of this type of electron phonon interaction.

One of the most striking manifestations of this interaction is the so-called "charge density wave" instability, responsible for a structural transformation to a "$2K_F$" superlattice. Since $2K_F$ is most often not an integral submultiple of the reciprocal lattice vector, the unit cell's new structure is not a simple multiple of the old unit cell. While such "incommensurable" structures have been long known, (the spin density wave in chromium is perhaps the most familiar example), their incorporation into the lore of phase transitions is quite recent and still incomplete. Nor are "incommensurable" structures confined to cases where mobile carriers are present. The lectures of Drs. Axe, Di Salvo, Shirane and Yamada highlight the wealth of anomalous physical properties associated with incommensurate transitions in both metals and insulators. Drs. Bak and Luther treat the formal and microscopic theoretical aspects of these systems. Among the concepts that they and Professor McMillan discuss are solitons, a subject that we undoubtedly will hear more about in the future. And Professor de Wolff presents an important systematic means of classifying the symmetries of incommensurate transitions.

The second important class of electron-phonon driven structural transitions is represented by the Jahn Teller systems. The relationsships among the symmetries of the interacting electronic and vibrational states and the resulting lattice distortions are systematized by Professor Thomas. And a variety of experimental discussion of several examples are detailed in this volume by Drs. Harley and Kjems. A quantitative description of many singular properties may be obtained from the "pseudo spin" model of Jahn Teller systems - as discussed here by Dr. Stinchcombe.

The third major area addressed by this institute concerns structural phase transitions and high temperature superconductivity. This subject perhaps epitomizes the meeting inasmuch as structural transitions in the prototypical A-15's (Nb_3Sn, etc.) have been addressed from both the band Jahn Teller and the charge density wave theoretical viewpoint. Professor McMillan's lectures provide an admirable synthesis of both phenomenological and microscopic theories for the charge density wave, the superconducting and structural phase transitions. Dr. Testardi, in his experimental lectures on the A-15 compounds emphasized the recent striking effects which defects in unstable lattices have upon limiting the realizable superconducting T_c.

The lectures of Dr. Romestain called our attention to another class of transition (the metal insulator) to which he has recently applied very sophisticated optical techniques. The several seminars generally provided an excellent complementarity to the subject areas addressed by the main lecturers. In retrospect two of the most exciting and stimulating topics addressed in this institute were the new class of structural transitions represented by "incommensurate" super structures and the beginning of progress in understanding more fully the role of structural transitions in high T_c superconductors.

As in the past, the atmosphere of the setting provided the opportunity for vigorous intellectual and physical activity, so that the participants departed with the clear impression that ideas had not only been exchanged but that new understanding was generated and the beginnings of some new research directions were established at Geilo.

The programme committee joins the other participants in expressing their sincere thanks to Mr. Eigil Andersen, Mrs. Gerd Jarrett and the staff of the Institutt for atomenergi, Kjeller, Norway for their careful planning and creative assistance, upon which largely rested the success of this conference.

J. D. Axe
J. Feder
P. A. Fleury
A. Luther
T. Riste

June, 1977

CONTENTS

PHASE CHANGES AND ELECTRON PHONON COUPLINGS IN PERFECT CRYSTALS. MODULATED STRUCTURES. AN INTRODUCTION.

J. Friedel

Physique des Solides, Université Paris Sud, Orsay

LA du CNRS

The purpose of this introduction is to place the field of this meeting in the general frame work of phase changes and the nature of the interatomic forces involved,then to discuss the possibility of small amplitude phase changes arising in perfect crystals as a weak modulation of the simpler structure. This modulation can be viewed as due to a phonon mode of the simpler structure becoming soft ; and conditions for this to occur through strong electron-phonon coupling are recalled.

I - THE FRAMEWORK

A. Phase changes in perfect crystals

We consider phases in thermal equilibrium, neglecting all problems of nucleation and growth kinetics involved in most phase changes. Each phase is then defined by a free enthalpy

$$F = H - TS,$$

a continuous function of temperature T, pressure p. It might involve also composition ; but we only consider here perfect crystals with fixed composition. $H = U + pV$ is the enthalpy, and variations of the volume V involved in phase changes of condensed phases only introduce small and trivial effects.

A phase change arises for a transition temperature T_t (or a pressure p_t) where two possible phases have the same free enthalpy. The phase change is said of the first or second order, depending on whether the two curves F_A (T or p) and F_B(T or p) cross with an angle or are tangent at the transition (figure 1a,b). To predict the transition temperature T_t of a phase change, one must

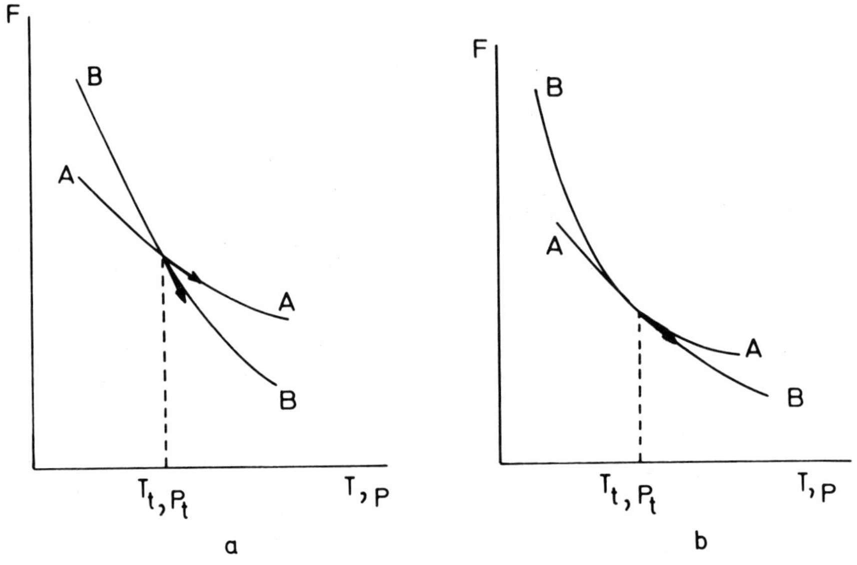

Figure 1 : Phase transitions : a first order ; b second order.

therefore know the difference in enthalpy H and in entropy S between the two phases, and how they vary with temperature. Only the relative stability of phases at zero degree can be discussed neglecting entropy.

In most crystalline phase changes, the difference in internal energy $U_B - U_A$ of the two phases is much less than the latent heat of sublimation, which measures the total stability of the condensed phases. Furthemore in most of these changes, large structural changes occur. As a result, only a limited study of the relative stability of typical simple crystal phases has been made in this general case.

In phase changes by crystal structure modulations however, there is a continuous or nearly continuous transition from phase B to phase A ($U_B - U_A \simeq 0$). Some phonon modes of phase B become soft at or near to the transition temperature T_t or pressure p_t.

Such transitions are usually described 'à la Landau', in terms of the amplitude u and the phase ϕ of the soft modes :

$$F_A - F_B = \alpha u^2 + \beta u^4 + \gamma(\text{grad } u)^2 + \delta(\text{grad } \phi)^2 + \ldots$$

This description assumes u and ϕ to be the relevant order parameters. Even when applicable, it is purely formal in that it hides the computation of the coefficients of α, β, γ, δ in terms of the enthalpy and entropy changes as deduced from a microscopic model.

A special case of crystal modulation is that of a uniform distortion.

B. Interatomic forces

Microscopic models require an analysis of the atomic and electronic structure.

One starts from the two usual fundamental approximations:

- The adiabatic approximation (Born Oppenheimer) uncouples the motion of the atoms from that of the electrons around the atoms, assumed at rest.

- The mean field one electron approximation (Hartree) uncouples the relative motions of the electrons. It takes two extreme selfconsistent forms, between which one nowadays believes there is necessarily a first order cooperative transition at low temperature (Mott transition(1)).

1 - Localized valence electrons

Each electron is assumed to be localized on a small building block of the structure - atom or molecule. The structure is then necessarily an insulator.

This is probably a good description of the stable state in rare gases and most ionic solids or molecular solids.

The building blocks are held together by various forces which correspond to successive refinements in the description of the electronic structure :

- Long range Coulomb interactions. These can be between static monopoles (ionic solids, figure 2.a), static dipoles (ferro or antiferroelectrics, figure 2.b,c) or static multipoles (molecular crystals such as Cl_2, figure 2.d) as described in the stable state of the Hartree approximation.

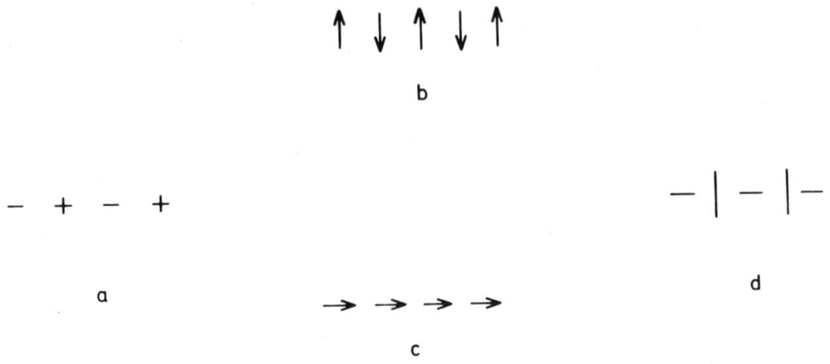

Figure 2 : Three types of Coulomb forces between static mono or multipoles : a monopoles ; b and c dipoles ; d higher multipoles.

- Short range exchange interactions : magnetic couplings between incompletely filled atomic shells or repulsions between filled shells ; medium range magnetic couplings by superexchange. These come in the Hartree Fock scheeme when respectively stable states or virtually excited states of the Hartree scheeme are considered.

- Short range dispersion interactions. These arise through the virtual excitation of Hartree state by long range Coulomb correlation forces between electrons in different building blocks. It can be considered as due to interactions between dynamic dipoles (Van der Waals and more general dispersion forces).

- H bonds, coming from quantum tunnelling of light nuclei.

It is worth pointing out that static Coulomb forces tend to aggregate unlike blocks together (+ with -, up dipole parallel to down dipole in case of figure 2.b, quadrupoles at right angles ... cf figure 2). Contrary wise, dispersion forces tend to aggregate like blocks together (figure 3, cf Appendix A).

One can list a number of phase changes with crystal modulations or small crystal distortions:

- Ferroelectric and antiferroelectric transitions, with the apparition or change of orientation of permanent dipoles.

- Magnetic phases (including cooperative Jahn Teller effects) if spin-orbit coupling and thus magnetostrictive effects are considered.

--- A A A A ---- B B B B---

a

--- B A B A B A B A B A B A B---

b

Figure 3 : Tendency to aggregation by dispersion forces : a preferred to b.

- 'Plastic' phases, where molecules rotate around their center of gravity.

- Insulator metal transitions (Mott transition) ...

Except in the ferroelectric case, the small lattice modulations are but a shadow of a larger change in ordering of an electronic or molecular parameter, and fall therefore really outside the scope of this discussion. The ferro (antiferro) electric cases themselves are made more difficult to study by the ambivalent nature of the interactions of dipoles (cf figures 2.b,c), by the more or less covalent nature of interatomic bondings and, in some cases, by the difficulty of describing correctly the H bonds. Thus if much experimental work has been done in that field, the theoretical analysis remains elementary in many cases. This is especially true for the displacive ferroelectrics where dipoles are created by a small crystal modulation of the non ferromagnetic phase, and which therefore fall most directly within the field of interest here.

It is also worth pointing out that, because electrons are localized, each phonon mode possibly involved in a phase transition interacts necessarily with all the valence electrons. Coupling of individual electrons with phonons are only involved in states excited above or near the conductibility gap, thus in excited conductive electrons or valence holes or in excitons. These couplings are well known both in polar and non polar crystals. They change the effective mass of the carriers, can lead to their self-trapping and to hopping conduction processes. Such processes only involve strongly excited electronic states of no importance in phase changes, except near to a Mott transition.

Two examples of modulated structures with short range interactions

The magnetic helical structures provide historically the first example of modulated structure, with a wave length that can vary continuously and thus can be incommensurate with the period of the lattice. Although actually found in metallic structures, they were first analyzed in terms of short range (magnetic) interactions valid for insulators(2). It is worth recalling this analysis and point out a direct extension to non magnetic modulations.

a) Helical magnetism

One considers a 3 dimensional lattice made of a stacking of parallel and identical planes of magnetic atoms. We assume a ferromagnetic coupling between atoms in each plane, and a strong magnetic anisotropy which forces the atomic moments to lie in the corresponding xy plane. Let $\mathbf{S}_n$ be the magnetic moment per atom in the nth plane. The coupling between atoms of different planes is limited to interactions between first and second neighbouring planes. The coupling energy U per atom is given by

$$N\,U = \sum_{n=1}^{N}\left[A_1 \sum_{i=\pm1} \mathbf{S}_n \mathbf{S}_{n+i} + A_2 \sum_{j=\pm2} \mathbf{S}_n \mathbf{S}_{n+j}\right]$$

If one Fourier analyses the components S_n^x, S_n^y of $\mathbf{S}_n$ in the xy planes (within the first Brillouin zone $0 \leqslant k < K = 2\pi/a$),

$$S_n^x = \sum_k S_k \cos(k R_n + \phi_k)$$

$$S_n^y = \sum_k S_k \sin(k R_n + \phi_k)$$

with $R_n = na$

one finds

$$U = 2 \sum_k S_k^2 (A_1 \cos ka)$$

The different Fourier components are thus additive and their energy is phase independent.

At T = OK, $(\sum_k S_k^2)^{1/2} = S$, length of the magnetic moment. The minimum energy U is obtained for one mode k, which, depending on the values of A_1 and A_2, is (figure 4)

F : ferromagnetic ($ka = 2m\pi$).
A : antiferromagnetic ($ka = (2m+1)\pi$).
H : helical ($\cos ka = -A_1/4A_2$).

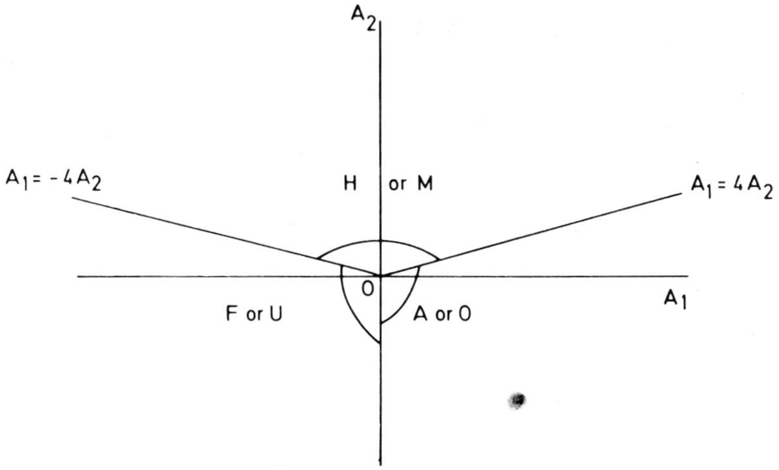

Figure 4 : Magnetic couplings of layered structures with interactions between nearest and second nearest neighbouring planes. Also elastic distortion of layered structures.

At $T \neq 0K$, magnetic entropy coupled with excitation of spin waves introduces terms in S_n for the free energy which reduce the average atomic spin moment $\langle S_n \rangle$ and couple the modes k. At high enough temperatures, it leads $\langle S_n \rangle$ to disappear.

b) Modulated crystal structures

Let a 3 dimensional layered structure have elastic interactions between first and second neighbouring planes so that the energy of distortion is expressed in terms of the displacement u_n of each n plane by

$$N\, U_M = \sum_{n=1}^{N} \left[A_1 \sum_{i=\pm1} (u_n - u_{n+i})^2 + A_2 \sum_{j=\pm2} (u_n - u_{n+j})^2 \right] + O(\delta u_n^{p>2})$$

where $O(\delta u_n^{p>2})$ describes corrective anharmonic terms in relative displacements $\delta u_n = u_n - u_{n'}$.

Taking

$$u_n = \sum_k u_k \cos(k R_n + \phi_k)$$

leads to the same analysis at OK. Depending on the values of A_1, A_2, the most stable state is (figure 4)

U : undistorted ($k = 2\ m\ \pi$)
O : optically distorted ($k = (2m+1)\pi$)
M : modulated sinusoïdally ($\cos ka = -\ A_1/4\ A_2$).

The amplitude u_k of the distortion must now be limited at OK by a supplementary factor : it is the role of the anharmonic terms $O(\delta u_n{}^{p>2})$ in NU_M, which also actually couple somewhat the modes k. At finite temperature, entropy terms must be taken into account, which can make the average $<u_n>$ go to zero.

2 - Delocalized valence electrons

Barring complications such as ionocovalent structures or rare earth metals, which are somewhat on the borderline with the preceeding case, three types of bonding belong to this class :

- covalent structures (figure 5.a), where the atomic structure respects the highly directional conditions for building overlapping atomic s p^n hybrids. This opens a gap between the valence and conductive band, for interatomic distances d smaller than the critical hybridizing distance d_c. Insulators correspond to equilibrium distances $d_o << d_c$; semiconductors and semimetals to $d_o \lesssim d_c$ or $d_o > d_c$.

- normal metals (figure 5.b), where the atomic structure does not respect covalent sp bonding conditions, but is as closepacked as possible. Near d_o, the lower part of the broad sp band looks like that of a nearly free electron gas.

- transition metals : the broad s band overlaps a narrow d band. The atomic structure is again fairly closepacked.

Because in b, c there is no energy gap near the Fermi level, these structures are necessarily conductors.

Linear combinations of atomic orbitals (i.e. tight binding) are a possible starting point on the two extremes a (sp^n) and c (d s) ; nearly free electrons is more adapted to b. More exact computations require the full panapherlia of muffin tin and computers.

General features of cohesion can again be understood using successive approximations for the electronic structure (cf Appendix B).

- In the Hartree approximation, and in a Wigner Seitz approach (3) which uses for computing the energy band the same atomic potential

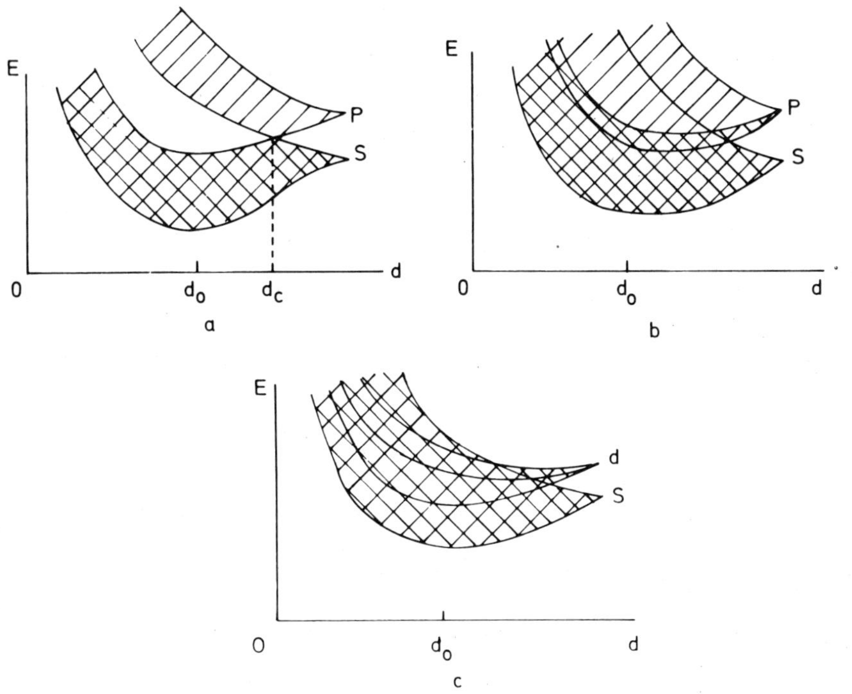

Figure 5 : Schematic band structures : a covalent; b normal metal; c transition metal - Shaded areas : allowed regions; crosshatched areas : regions occupied by valence electrons for a given number of electrons per atom. d interatomic distance.

as seen by an electron in free atoms, cohesion comes from the fact that the broadening of atomic states into bands is essentially symmetrical in energy : for an incompletely filled s, sp or sd shell, the electrons then gain energy by going into the condensed phase ; maximum cohesion occurs for half filled band systems, when the 'bonding' states are full and the 'antibonding' ones empty.

The total band width is related to the frequency of escape of an electron from one atomic orbital, thus both to the strength of the overlap with the orbital on a neighbouring atom and to the number of bonds this orbital takes part in. In covalent structures, bonds are few but very effective (strong overlap, and energy gap lowering the average energy of valence electrons); in metals, cohesion is maximum for maximum number of neighbours : compact and little directional bonding is favoured.

A more refined study in the Hartree scheeme introduces an average electron-electron repulsion which destabilizes somewhat the condensed phase, especially for normal metals[4][5]. Hartree Fock and Coulomb corrections are then essential in normal metals to compute the absolute cohesive energy. They also lead to significant corrections in the middle of transitional series (Appendix C). They should however play only a reduced role in the latent heats of phase changes, as they are mostly sensitive to the atomic volume more than to the crystal structure. This is not quite true of long range Coulomb corrections, which come in when all the atoms do not play equivalent roles in at least one of the phases.

If now crystal modulation is made on one of the structures of figure 5, it will induce 'virtual excitations' or a 'polarization' of the valence electrons. There is therefore a strong relation with the problem of electron-phonon coupling, especially in conductors.

II - ELECTRON-PHONON COUPLINGS (FOR DELOCALIZED ELECTRONS)

A. Soft Modes at OK. Adiabatic and kinematic approximations

1 - General (crystals)

We consider a time independent real perturbation

$$v = \sum_{q} v_{q} \exp(i q r) + \text{C.C.}$$

To first order in (non degenerate) perturbation, each Bloch function $|n,k\rangle$ becomes

$$|n,k\rangle + \sum_{q} v_{q} \sum_{n'} \frac{|n',k+q+K\rangle\langle n',k+q+K| e^{iqr} |n,k\rangle}{E_{k}^{n} - E_{k+q+K}^{n'}}$$

$$+ \sum_{q} v_{q}^{*} \sum_{n'} \frac{|n',k-q+K\rangle\langle n',k-q+K| e^{-iqr} |n,k\rangle}{E_{k}^{n} - E_{k-q+K}^{n'}}$$

K is a period of the reciprocal lattice K_{RL}. In reduced zone scheeme, one chooses K such that k and $K' = k \pm q + K$ are in the first Brillouin zone (figure 6).

If v represents the effect of a (static) phonon mode, q is restricted to

$$q = q_{o} + K'$$

with $K' \in K_{RL}$.

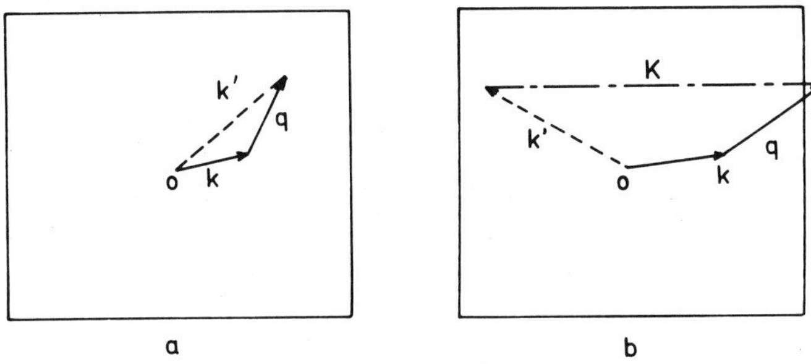

Figure 6 : Convention for K : a normal (K=0); b umklapp (K≠0).

To second order in this non degenerate perturbation scheeme, v introduces then an electron $|n,k\rangle$ phonon q_0 coupling energy

$$\delta E_k^{\,n}(q_0)=\langle n,k|v|n,k\rangle+\sum_{\substack{n'\\K'}}\frac{|\langle n,k|v_{q_0+K'}\,e^{i(q_0+K')r}|n',k+q_0+K\rangle|^2}{E_k^{\,n}-E^{n'}_{k+q_0+K}}$$

$$+\sum_{\substack{n'\\K'}}\frac{|\langle n,k|\,v^{*}_{q+K'}\,e^{-i(q_0+K')r}|n',k-q_0+K\rangle|^2}{E_k^{\,n}-E^{n'}_{k-q_0+K}} \qquad (1)$$

To this order of perturbations, the various Fourier components K' introduce in the energy terms that are additive and independent of the phase in v.

The total energy change for the electron is

$$\delta E_k^n = \sum_{q_0\,occ} \delta E_k^{\,n}(q_0) \quad .$$

The total energy change for the phonon is

$$\delta E_{q_0} = \sum_{n,k\,occ} \delta E_k^n(q_0) \qquad (2)$$

2 - Energy change for the phonon

- First order term. It is zero except if $q_0 = 0$, thus except for a uniform distortion.

Example : if a degenerate strong Van Hove anomaly is lifted by a uniform shear ε without much change in form (figure 7), and if the Fermi level falls initially on the anomaly, the distortion will lower the average (Hartree) energy of the valence electrons by an amount proportional to the lifting of the energy degeneracy, itself proportional to v thus to ε. This negative linear term $\delta E_o = - A\varepsilon$ in the energy must be balanced by an elastic reaction of the lattice $E_o(\varepsilon) = (1/2) E \varepsilon^2$, associated with the rest of the electronic and ionic structure :

$$E_o(\varepsilon) + \delta E_o(\varepsilon) = - A\varepsilon + (1/2) E \varepsilon^2$$

This results necessarily in a spontaneous distortion $\varepsilon = \frac{A}{E}$, which can be termed a band Jahn Teller effect.

- Second order term. Terms where $|n,\underset{\sim}{k}\rangle$ and $|n',\underset{\sim}{k}'\rangle$ are both occupied cancel out two by two. Only unoccupied $|n'\ \underset{\sim}{k}'\rangle$ states need thus be considered.

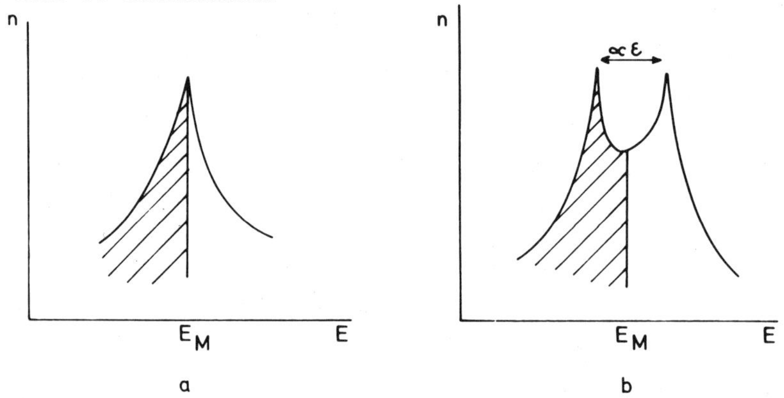

Figure 7 : Band Jahn Teller effect : a $v = 0$; b $v \neq 0$.

They also provide negative contributions : the second order electron-phonon coupling necessarily lowers the phonon energy.

In an insulator , the denominator $E_{\underset{\sim}{k}}^{n} - E_{\underset{\sim}{k}'}^{n'}$ is always finite (larger in amplitude than the energy gap between the valence and conduction band). $\delta E_{\underset{\sim}{q}_o}$ is therefore a correction that varies continuously with $\underset{\sim}{q}_o$. It remains small everywhere, being of the second order in the small phonon amplitude. Explicit estimates of v_q later on will show that the total perturbed energy $E_{\underset{\sim}{q}_o} + \delta E_{\underset{\sim}{q}_o}$ of all phonons are then expected to remain positive : the mode softening due to electron phonon coupling is much too small to become catastrophic. Crystal modulation is therefore expected not to take place normally in covalent crystals. This is after all reasonable: the strongly directional covalent bonds are not expected to lead normally to several structures of comparable energy and differing only little from each other.

In conductors, a large contribution is expected for pairs $|n,k>$, $|n' k'>$ of states both near to the Fermi level. With $k' = \pm q_o + K$, such points are near to lines such as L, figure 8.a where the Fermi Surface S_M cuts a surface S'_M obtained by a translation of S_M by $\pm q_o + K$. When, for a given direction, $\pm q_o + K$ increases in size, L decreases to a point (figure 8.b), then disappears (figure 8.c). One then expects the special contribution from the neighbourhood of L to vary strongly near to the critical values of $\pm q_o + K$ such that S_M and S'_M just touch.

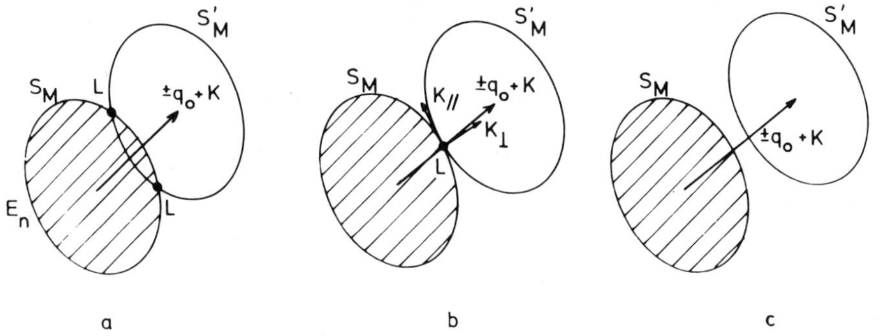

Figure 8 : Critical region L of integration and its variation with $\pm q_o + K$, near a critical value of that vector.

Figure 9, 10 describe schematically four general cases in three dimensions.

α . S_M and S'_M touch on a conical point.

β . S_M and S'_M have in common a point and a tangent plane, but different curvatures.

γ . S_M and S'_M have in common a point, a tangent plane and one common curvature.

δ . S_M and S'_M have in common a point and the two principal curvatures.

Case β is the normal three dimensional one. α arises if the Fermi level happens to fall on a Van Hove anomaly, and q_o is such that S_M and S'_M touch on that point. Cases γ and δ are possible if the Fermi surface has a complex geometry ; they are traditionally called nesting conditions. Inspecting the immediate neighbourhood of L, one sees that the general two dimensional case (where in three dimensions Fermi surfaces are parallel cylinders) corresponds to case γ; the general one dimensional case (in three dimensions, Fermi surfaces are parallel planes) corresponds to case δ.

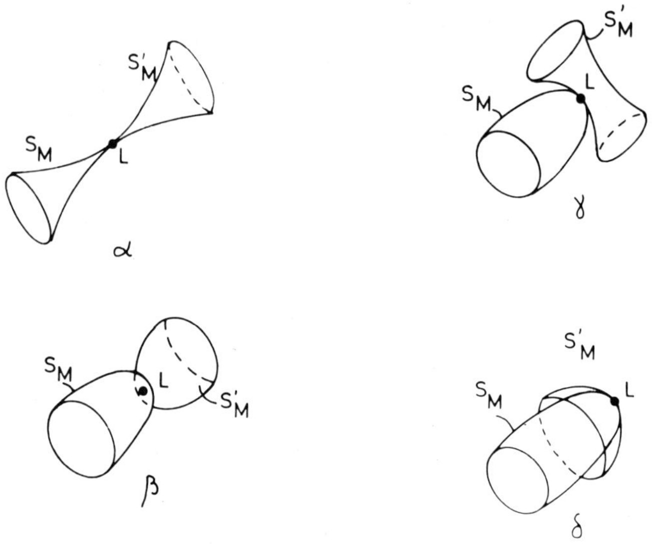

Figure 9 : Four cases of contact between S_M and S'_M.

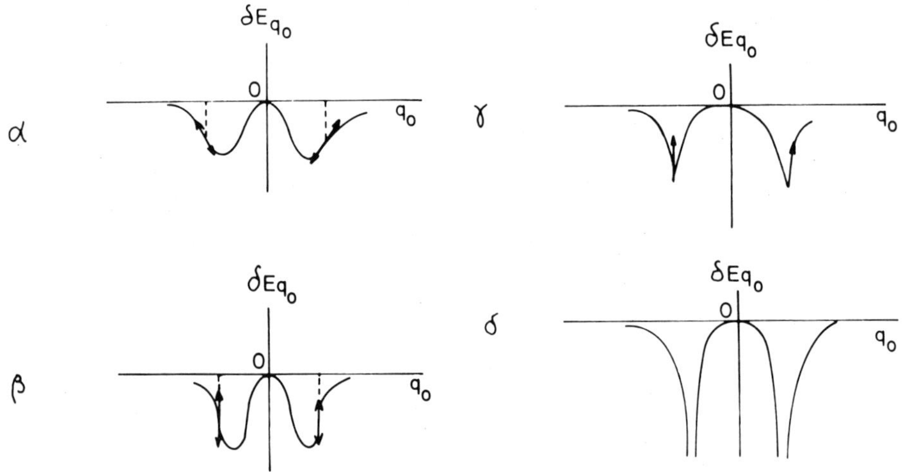

Figure 10 :- Corresponding Kohn anomalies.

Figure 10 gives the corresponding behaviour of δE_{q_0} in the immediate neighbourhood of the critical value for q_0 :

α : inflextion point.

β : inflexion point with infinite tangent.

γ : discontinuous change of slope.

δ : negative infinity.

In the 3 dimensional case β for instance, one can choose suitable axes $K_\perp$, $K_{//}$ (figure 8.b) such that $E^n_{\underset{\sim}{k}}$ and $E^n_{\underset{\sim}{k}-\underset{\sim}{q}_o+\underset{\sim}{K}}$ can be developped near L as

$$E_{\underset{\sim}{k}}^{\,n} = E_1 = E_M + a\,K_\perp^2 + b\,K_{//}^2$$

$$E_{\underset{\sim}{k}-\underset{\sim}{q}_o+\underset{\sim}{K}}^{\,n} = E_2 = E_M - a\,(\underset{\sim}{k}_o - K_\perp)^2 + b\,K_{//}^2$$

The integral in (1), (2) can be written

$$\delta E_{\underset{\sim}{q}_o} = \int \frac{f(\underset{\sim}{k})}{E_1 - E_2}\, d_3k = \text{const} + \frac{1}{b}\, k_o \ln |k_o|$$

where k_o measures the distance or the overlap between S_M ans S'_M, figure 8.

Because of the contribution to integration in (1), (2) of the neighbourhood of the Fermi surface, the negative correction in $\delta E_{\underset{\sim}{q}_o}$ is larger in conductors than in (band) insulators. It has furthermore an anomalous behaviour when this contribution disappears : this is the Kohn anomaly predicted initially for nearly free electrons(6) and observed in the phonon dispersion curves of many normal and transition metals.

The discussion below on v_q will show that $v_{q_o} \propto q_o$, thus $\delta E_{\underset{\sim}{q}_o}$ vanishes with q_o and is usually too small to produce soft phonons. The general form of $\delta E q_o$ pictured figure 10 takes into account the variation in q_o^2 of $\delta E q_o$ near the origin. This form predicts that soft modes

- should always occur at the critical value of q_o in the 3 dimensional case δ of complete nesting (or in general in one dimension).
- might occur at the critical value of q_o in the 3 dimensional case γ of partial nesting (or in general in two dimensions).

- might occur, for q_o near to the critical value, in the general 3 dimensional case β.

- are unlikely to be associated with a Van Hove anomaly at the Fermi surface (case α).

3 - Classical examples

a) Nearly free electrons in 3 dimensions

The first application of these ideas can be said to date from the 1930's. It refers to the explanation by Jones[7] of the regular succession of crystal phases noticed by Hume Rothery[8] in 'normal' metals and alloys when one varies their electron per atom ratio. In such crystals where the valence electrons are only weakly scattered by the atoms, the scattering by the crystal structure can be Fourier analyzed, and each Fourier component acts on the nearly free electrons as a (static) phonon mode. This mode has a stability that varies rapidly when its wave vector $\mathbf{q}_o$ is near to one of the critical values defined figure 8.b, and the whole crystal structure has a stability very sensitive to the electron per atom ratio (or the size of the Fermi sphere) when the critical condition is fullfilled for a maximum number of Fourier components. As the $\mathbf{q}_o$ are here reciprocal lattice vectors, it is the same to say that the structure is such that the Fermi sphere touches a maximum number of Brillouin zone boundaries. Figure 11 pictures the behaviour of the contribution of $\delta E\mathbf{q}_o$ to cohesion versus electron per atom ratio in two such structures. It is here the whole atomic

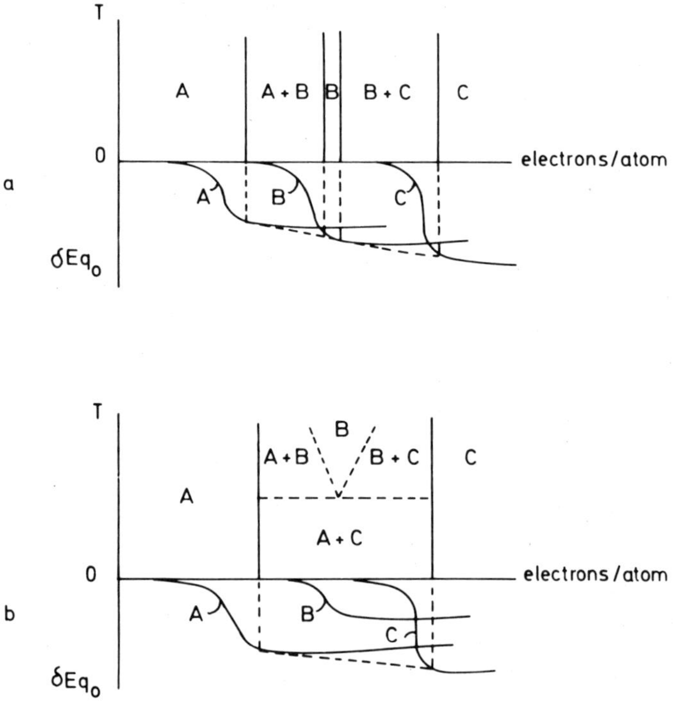

Figure 11 : Two possible cases of Hume Rothery- Jones phases in normal metals and alloys, and the corresponding phases diagrams at OK.

potentials which scatter the electron, and not their displacements from a periodic arrangement. As a result v_{q_o} and δEq_o do not vanish at small values of q_o (for a fixed number of electrons) or at large electron per atom ratio (for fixed q_o and thus crystal structure). As stressed figure 11, the successive Hume Rothery Jones structures met at increasing electron per atom ratio will be actually observed at OK only if their corresponding stabilities are increasing (case a). Even then, one should distinguish two types of conditions for the phase diagram at OK of a solid solution between two elements M_1M_2 : the extremal phases (such as A, C figure 11a) should extend to near the critical electron per atom ratio ; and the intermediary phases (B, figure 11.a) should have a narrow range of stability near the corresponding critical electron per atom ratio. One does not expect much change in the phase boundaries with temperature at low temperatures (because the electron entropy is near to that of free electrons in any case) except that the lattice entropy might stabilize at high temperatures[9] a phase such as B, figure 11.b, which is unstable at OK. A famous example is the BCC β phase of Cu Zn alloys.

The same condition of Brillouin zone boundary tangent to the Fermi sphere has been used to explain more complex alloy structures where superstructures modulate a fundamental simple crystal structure[10]. The modulation can either be a concentration modulation, thus producing a special kind of ordering (Au Cu alloys) or a periodic succession of stacking faults (Ag Cd alloys).

b) Narrow d bands in three dimensions

The conditions for soft modes by critical Kohn anomaly can be expressed explicitely for transitional metals in a tight binding (LCAO) description of the d band[11]. They are usually complex, owing to the complexity of the Fermi surface.

c) Narrow d bands in 2 dimensions (planes)

This case was first seriously considered for the layered 3 dimensional structures of transition dichalcogenides[12], where the transition elements build up close packed planes which are separated by rather insulating layers of S, Se or Te. To first order at least, each plane has a two dimensional d band with an electronic structure which can be treated separately (figure 12.a).

d) Narrow bands in one dimension (chains)

This case was first considered theoretically by Peierls[13] and Fröhlich[14] (figure 12.b) : Kohn anomalies appear very clearly in compounds with parallel conducting chains separated by more or less insulating material :

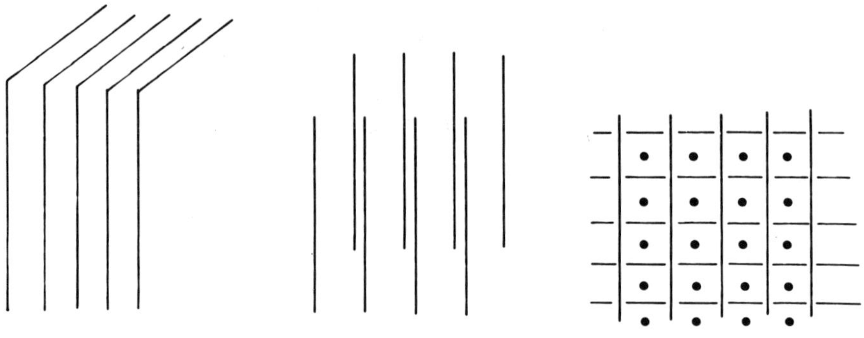

Figure 12 : Three types of two dimensional cases : a parallel planes ; b parallel chains ; c 3 orthogonal sets of parallel chains.

- KCP (chains of Pt atoms)[15].
- TTFTCNQ (organic chains)[16].

e) Band Jahn Teller

The first application of Band Jahn Teller effects was made on A15 transition compounds (V_3 Si, Nb_3 Sn ..), which have three orthogonal sets of parallel chains of transition atoms (figure 12.c). The first studies considered the lifting of degeneracy by shear of the edges of the d bands of the linear chains, assumed independent [17]; further studies considered the lifting of degeneracy of intermediary peaks possibly connected with the interactions between the chains[18][19].

NB : In the historical development of these concepts, there is an intimate relation with metal magnetism. Thus

- ferromagnetism is the analogous of the Band Jahn Teller effect.

- antiferromagnetism and helical magnetism are modulated structures.

Conditions for the stability of these magnetic phases were in all cases expressed in terms of Coulomb and exchange interactions, sometimes before the corresponding conditions for crystal modulation (Stoner's criterion for ferromagnetism[20], Slater for antiferromagnetism [21], Lomer for helical and modulated structures[22]). The quantitative differences involved will be discussed below.

- In all the low dimensional cases, there exist interplane or interchain couplings which actually make the substance a three-dimensional one, at least at low temperature.

B. Discussion of the approximations

1 - The meaning of v

We must define more precisely the meaning of the effective potential v which describes the effect of a phonon on the electrons. We shall show on two extreme simple cases that it is not actually a proper potential. We shall also see that v_q it proportional to q and has necessarily an order of magnitude which makes an instability condition only possible near a Kohn anomaly.

- Nearly free electrons (one atom per unit cell)[23]. If we assume each weak atomic potential $V_i(\mathbf{r}-\mathbf{R}_i)$ to be bodily shifted by the phonon without change of form, we have

$$v_{\mathbf{q}} \exp i\,\mathbf{q}\,\mathbf{r} = \sum_i \left[V_i(\mathbf{r} - \mathbf{R}_i - \mathbf{u}_i) - V_i(\mathbf{r}-\mathbf{R}_i)\right] \simeq \sum_i \mathbf{u}_i\, \nabla R_i\, V_i(\mathbf{r}-\mathbf{R}_i)$$

Assuming

$$\mathbf{u}_i = \mathbf{u}_{\mathbf{q}_o} \exp i\,\mathbf{q}_o\,\mathbf{R}_i \tag{3}$$

and if $V_{\mathbf{K}}$ is the Fourier transform of $V_i(\mathbf{r})$,

$$\langle k|v_{\mathbf{q}_o}\, e^{i\mathbf{q}\mathbf{r}}|k'\rangle = -\, i(\mathbf{q}_o\, \mathbf{u}_{\mathbf{q}_o})\, V_{\mathbf{q}_o}\, \delta(\mathbf{k}' - \mathbf{k} + \mathbf{q}_o).$$

This goes to zero with q_o. For free electrons, there are no umklapp terms, and no coupling of electrons with shear waves.

NB :-The assumption of rigid displacement of V_i will be discussed and somewhat corrected later on.

- V_i is a pseudopotential[24], thus as operator, such that $\langle k|V_i|\mathbf{k}^1 + \mathbf{K}\rangle = V_{\mathbf{K}}$ is a function of K = k'-k but also of k. Thus the continuous variation of V_K with k will distort Eq_o, without changing however the mathematical form of the Kohn anomaly ; and the value of V_K appearing near to Kohn anomaly refers specifically to states k and k' near to the Fermi level.

- In the undistorted crystal structure, the Fermi surface is distorted from a sphere by the scattering by the undisplaced pseudopotentials V_i. However in the approximation so far used, these distortions are of second order in V_i. To take them into account in computing δEq_o would introduce correction of higher order than v_q^2.[25]

- Tight binding (one atomic orbital per unit cell)[25][26]. With the energy of an atomic orbital $|i\rangle$ in a free atom taken as zero of energy, the tight binding hamiltonian for a non degenerate band reads

$$H \simeq \sum_{\substack{i \\ j \text{ near to } i}} \beta_{ij} \, |i\rangle\langle j|$$

if one neglects the electrostatic crystal field term, an approximation valid in pure metals.

In the same spirit as for free electrons, one can assume that the displacement $\underset{\sim}{u}_i$ given by (3) bodily displaces the atoms and their atomic orbitals $|i\rangle$ without otherwise changing the atomic potentials and orbitals. With these shifted orbitals $|i'\rangle$, the change in hamiltonian produced by the displacements $\underset{\sim}{u}_i$ reads

$$\delta H \simeq \sum_{\substack{i \\ j \text{ near to } i}} \delta\beta_{ij} \, |i'\rangle\langle j'|$$

where

$$\delta\beta_{ij} = \beta'_{ij} \, \delta d_{ij} = \beta'_{ij}(u_j - u_i) \cos\theta_{ij} \; .$$

θ_{ij} is the angle made by $\underset{\sim}{R}_j - \underset{\sim}{R}_i = \underset{\sim}{d}_{ij}$ with the direction $\underset{\sim}{u}_{q_o}$ of displacement (figure 13)

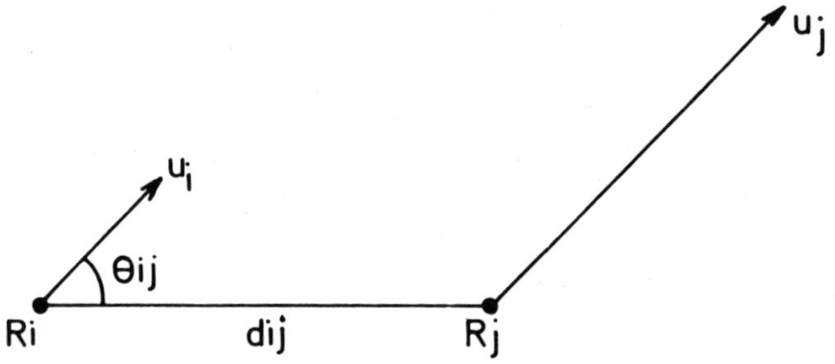

Figure 13 : Definition of θ_{ij}, d_{ij} .

To first order in u, $|i'\rangle$ can be replaced by $|i\rangle$. Hence

$$\langle n, \underset{\sim}{k}|v|n\ \underset{\sim}{k}'\rangle = \sum_{j \text{ near to } i} \frac{1}{N} e^{i(\underset{\sim}{k}'\underset{\sim}{R}_j - \underset{\sim}{k}\ \underset{\sim}{R}_i)} \delta\beta_{ij}$$

$$= \sum_{j \text{ near to } i} \beta'_{ij}\, u_{\underset{\sim}{q}_0} \cos\theta_{ij}\left[1-\exp(-i\underset{\sim}{q}_0\underset{\sim}{d}_{ij})\right] e^{i\underset{\sim}{k}\underset{\sim}{d}_{ij}}$$

$$\times\ \delta(\underset{\sim}{k}'-\underset{\sim}{k}+\underset{\sim}{q}_0+\underset{\sim}{K}_{RL}).$$

This goes again to zero with $\underset{\sim}{q}_0$. In general, shear waves are now coupled with electrons. There are umklapp processes. Finally, as for free electrons, this expression is not only a function of $\underset{\sim}{k}'-\underset{\sim}{k}=\underset{\sim}{q}_0+\underset{\sim}{K}_{RL}$, but also a function of $\underset{\sim}{k}$: v is an operator and not a simple potential function. Only near to k = 0 can one neglect this variation.

– Orders of magnitude for soft modes. Both expressions of v_{q_0} are similar in order of magnitude : $u_{q_0}\, q_0\, V_{q_0}$, where V_{q_0} is the derivative or the Fourier transform of an electronic energy term.

In expression (2), the denominator of $\delta E_k^{\ n}(\underset{\sim}{q}_0)$ is larger than the energy gap g, if one considers an insulator. In that case, the second order correction is

$$\delta E_{q_0} \simeq \chi_{q_0} |v_{q_0}|^2$$

with $\chi_{q_0} = g^{-1}$,

and the mode would be soft only if

$$E_{q_0} + \delta E_{q_0} < 0$$

or

$$\frac{(v_{q_0})^2}{E_{q_0}} > -\frac{1}{\chi_{q_0}} \simeq g$$

with

$$E_{q_0} = \frac{1}{2} M^2\, \omega_{q_0}^2\, u_{q_0}^2$$

M atomic mass, ω_{q_0} pulsation of the phonon without electron-phonon coupling.

Hence the condition :

$$\frac{|q_o V_{q_o}|}{\sqrt{g/2}} > M \omega_{q_o}$$

As $M \omega_{q_o} \gg 1$ atomic unit, and $g \simeq q_o V_{q_o} \simeq$ a fraction of atomic unit, this condition cannot usually be fullfilled, except perhaps for very small gap semiconductors ($g \simeq 0$).

The same applies to metals, with $\chi_{q_o}^{-1}$ of the order of the Fermi energy, except near a Kohn anomaly where the soft mode condition can possible be realized.

In conclusion, although v is not exactly a potential, simple estimates applicable respectively to normal or transition metals show that only modes connected with a Kohn anomaly can usually become soft.

2- Self consistency (to first order)

This point has some general interest, and will therefore be developped in some detail.

In the same approximation used so far, a perturbing potential $v(\mathbf{r})$ produces in the valence electrons a local change in electronic density

$$\delta\rho(\mathbf{r}) = \sum_q v_{\mathbf{q}} \sum_{\substack{n,k_{occ} \\ n'k'_{inocc}}} \frac{\langle n', \mathbf{k}+\mathbf{q}+\mathbf{K} | e^{i\mathbf{q}\mathbf{r}} | n, \mathbf{k} \rangle}{E_{\mathbf{k}}^{n} - E_{\mathbf{k}+\mathbf{q}+\mathbf{K}}^{n'}} u_{\mathbf{k}}^{n*}(\mathbf{r})\, u_{\mathbf{k}+\mathbf{q}+\mathbf{K}}^{n'}(\mathbf{r}) + C \qquad (4)$$

where, as before, C is deduced from the first term by changing $\mathbf{q}$ in $-\mathbf{q}$ and $v_{\mathbf{q}}$ in $v_{\mathbf{q}}^{*}$

a) Asymptotic perturbation, far away from a localized perturbation v.

For insulators, all terms in (4) are regular. The integrals are obviously dominated by the regions where $E_{\mathbf{k}}^{n}$ and $E_{\mathbf{k}'}^{n'}$ are nearly stationary, i.e. the neighbourhood of Van Hove anomalies. If the perturbation v acts near to the origin, the main contribution for $r \to \infty$ will come from the couple of Van Hove singularities for which $E_{\mathbf{k}}^{n} - E_{\mathbf{k}'}^{n'}$ is a minimum, i.e. the top E_V of the valence band and the bottom E_c of the conduction band, figure 14. A regular

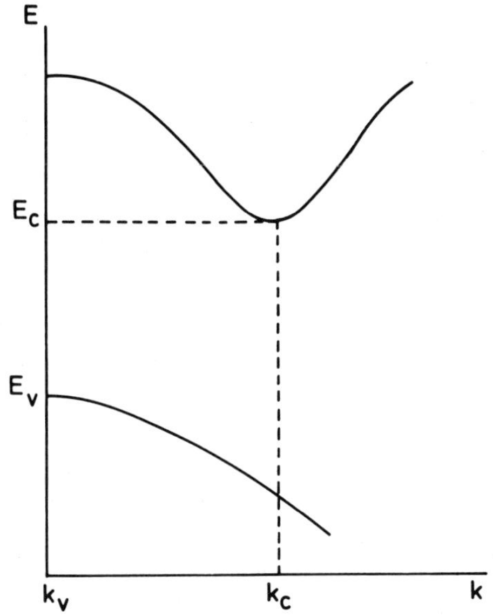

Figure 14 : Energy gap in a band insulator.

development of E_k^n and $E_{k'}^{n'}$ near to the corresponding points $\underset{\sim}{k}_V$, $\underset{\sim}{k}_c$ in the reciprocal space gives easily(27)

$$\delta\rho \underset{r \to \infty}{\to} \sum_{\underset{\sim}{K} \in \underset{\sim}{K}_{RL}} \text{oscill } (\underset{\sim}{K}\,\underset{\sim}{r}) \text{ oscill } (\underset{\sim}{k}_c - \underset{\sim}{k}_V)\underset{\sim}{r} \exp\left(\frac{-r}{a}\right)$$

where oscill (x) means a sinusoïdal function of x. One can show that

$$a \simeq \frac{2h}{[m^*(E_c - E_V)]^{1/2}}$$

where the effective mass m^* is itself proportional to $E_c - E_V$. Hence $a \propto (E_c - E_V)^{-1}$.

The interband terms thus give rise to a rapid exponential degrease of $\delta\rho(\underset{\sim}{r})$ outside the perturbed region where $v \neq 0$.

For conductors, the contributions near the Fermi level give one or several further intraband terms(28)(29). The asymptotic behaviour is dominated by the neighbourhood of lines such as L, figure 8. And the main contributions will arise when the phase of the matrix element of exp i $\underset{\sim}{q}\,\underset{\sim}{r}$ is nearly stationary. This requires L to reduce to a point (figure 8.b and 15) where the normal to the Fermi Surface S_M is parallel to $\underset{\sim}{r}$. If there are only

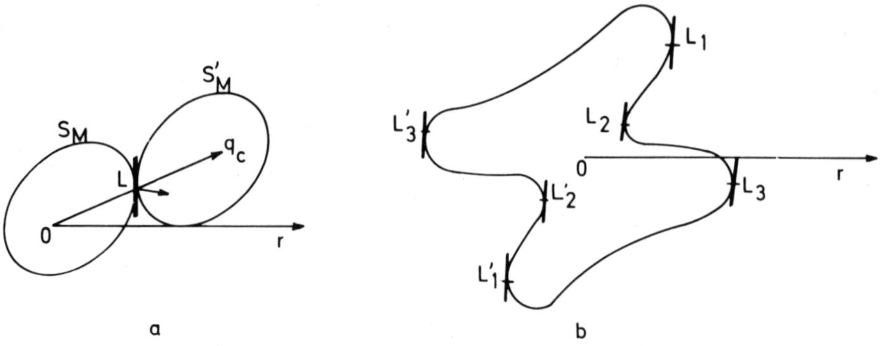

Figure 15 : Value of q dominating the asymptotic behaviour of $\delta\rho(\mathbf{r})$ in a metal : a simple case ; b more complex case.

two such values of $q = \pm\, q_c$ (figure 15.a), an integration near the them gives

$$\delta\rho \underset{r\to\infty}{\to} \Big(\sum_{K\in K_{EL}} \text{oscill}\ \mathbf{K}\mathbf{r}\Big)\ r^{-n}\ \text{oscill}\ (\mathbf{q}_c\ \mathbf{r})$$

with $n = 4, 3, 2, 1$ depending on whether the contact of S_M and S'_M on L is of the type α, β, γ, δ respectively (figure 9). In more complex cases (figure 15.b), there are interference terms between these various points : $\mathbf{q}_c$ = all the possible $L_i\ L_j (L_i = L_1, L_2, L_3, L'_1, L'_2, L'_3)$ (figure 15.b).

b) Screening to an external potential.

Let $v_e(\mathbf{r})$ be an external perturbing potential applied to the system. It will produce a displaced charge $\delta\rho(\mathbf{r})$ which itself produces a supplementary internal screening potential $v_i(\mathbf{r})$, so that the total potential acting on the electrons is

$$v(\mathbf{r}) = v_e(\mathbf{r}) + v_i(\mathbf{r})$$

Poisson's equation gives*

$$\Delta v_i(\mathbf{r}) = -\,4\,\pi\,e^2\,\delta\rho(\mathbf{r})$$

and $\Delta\rho$ is related to v by equation (4).

Fourier transforms give

* As usual, v is the potential energy of an electron.

$$v_{\underset{\sim}{q}} = v_{e\underset{\sim}{q}} + v_{i\underset{\sim}{q}}$$

$$q^2 \; v_{i\underset{\sim}{q}} = 4 \; \pi \; e^2 \; \delta\rho_{\underset{\sim}{q}}$$

$$\delta\rho_q = \sum_{K \in K_{RL}} \chi_{\underset{\sim}{q}}^{\underset{\sim}{K}} \; v_{\underset{\sim}{q}+\underset{\sim}{K}}$$

where the general susceptibility $\chi_q^{\underset{\sim}{K}}$ which relates charge to potential is a tensor. This set of equations allows in principle to compute the selfconsistent perturbing potential v and perturbed electronic density $\delta\rho$ for a given external potential $v_e(\underset{\sim}{r})$.

Thus, introducing the general dielectric tensor

$$\varepsilon_q^K = \left(\delta(K) - \frac{4\pi \; e^2}{q^2} \; \chi_q^K \right)$$

such that

$$\sum_K \varepsilon_q^K \; v_{q+K} = v_{eq}$$

one obtains

$$\sum_K \varepsilon_q^K \; \delta\rho_q = \frac{q^2}{4\pi \; e^2} \sum_K (\delta(K) - \varepsilon_q^K) \; v_{eq}$$

c) Screening of a phonon wave. Charge density wave.

v_e is now the perturbation potential due to a phonon. The preceeding equations show that each Fourier component v_{eq} of the phonon potential produces a charge density wave[30] with various Fourier components $\delta\rho$, at $q' = q + K_{RL}$. This wave screens the external potential v_e into a total potential v with Fourier components $v_{q'}$. Thus in general the charge density wave has a somewhat complex structure ; and it couples the various Fourier components of the total potential v, so that the dielectric constant ε_q^K is a tensor.

The consequence of this screening on the stability of the phonon will only be discussed in the two extreme cases where these formulae simplify.

- Nearly free electrons. The umklapp terms $K \neq 0$ disappear ; and the susceptibility takes a simple form (figure 16):

$$\chi_q = \frac{q^2 (1-\varepsilon_q)}{4 \; \pi} = \sum_{k_{occ}} \frac{1}{E_k - E_{k+q}} \propto - \left[1 + \frac{4k_M^2 - q^2}{4 \; k_M q} \; \ell n \left| \frac{2k_M + q}{2k_M - q} \right| \right] (<0)$$

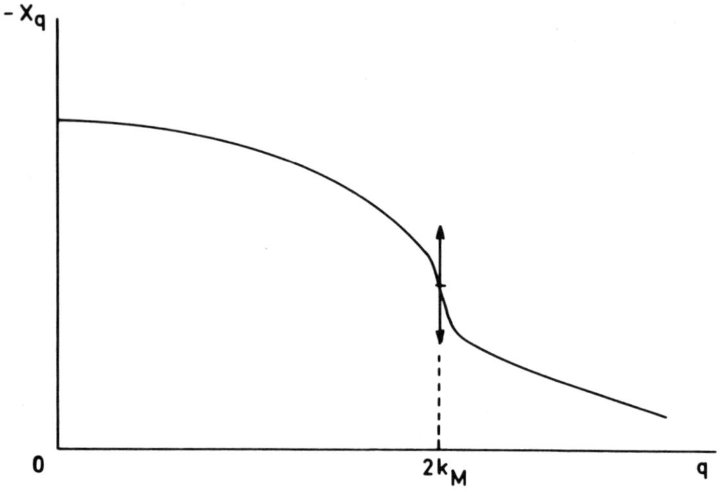

Figure 16: Susceptibility χ_q for free electrons.

In the second order terms in v for δE_{q_0}, equations (1), (2), one must distinguish the contributions of v_e and v_i, and not count twice the Coulomb interactions of $\delta\rho$ with itself. As a result, $|v_q|^2$ should be replaced by $|v_q \, v_{eq}| = \varepsilon_q^{-1} |v_{eq}|^2$. Hence the second order term in δE_q reads(24)(31)

$$\delta E_q = \frac{\chi_q}{\varepsilon_q} |v_{eq}|^2 = \left(\frac{1}{\chi_q} - \frac{4\pi e^2}{q^2}\right)^{-1} |v_{eq}|^2 \cdot$$

Thus the Kohn anomaly due to χ_q is reduced but not suppressed by that in ε_q.

The instability condition of the phonon mode becomes

$$\frac{|v_{eq}|^2}{E_q} \quad \frac{1}{\varepsilon_q} > - \frac{1}{\chi_q}$$

The production of a charge density wave coupled with the phonon and screening its potential thus introduces a reduction factor $\frac{1}{\varepsilon_q} < 1$, for $\chi_q < 0$. The charge density wave stiffens the phonon, and thus makes its instability less easy to produce.

Tight binding

- For half filled bands and elementary alternate structures, there is no charge density wave coupled with a phonon mode[26]. Alternate structures are such that only closed circuits with an even number of interatomic jumps between neighbouring sites are possible (figure 17.a).

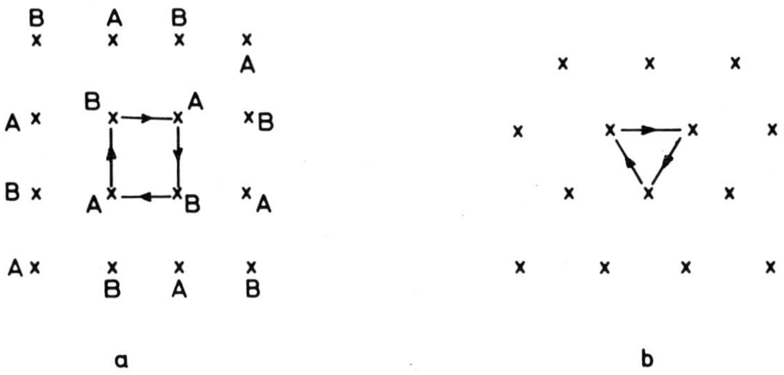

Figure 17 : Two types of crystal structures : a alternate ; b non alternate.

In such structures, one can define two interpenetrating lattices A,B such that each A site has only B nearest neighbours, and vice versa. To each electron function

$$|\psi\rangle = \sum_{nA} a_A^n |n,A\rangle + \sum_{mB} a_B^m |m,B\rangle$$

with energy

$$E = \sum_{\substack{nA \\ mB \text{ near } nA}} (\beta_{AB}^{nm} a_A^{n*} a_B^m + \beta_{BA}^{mn} a_B^{m*} a_A^n)$$

there corresponds a function

$$|\psi'\rangle = \sum_{nA} a_A^n |n,A\rangle - \sum_{mB} a_B^m |m,B\rangle$$

with energy $-E$, symmetrical with respect to the middle of the band ; and these two wave functions give equal electronic densities ($\sum_n |a_A^n|^2$ or $\sum_m |a_B^m|^2$) on each (A or B) site. Thus, whathever the distortion and as long as the alternate topology is preserved, such structures with half filled bands have the same number of electrons on each site, equal to the number of atomic orbitals involved.

For less than half filled bands, such structures have an accumulation of electrons in the compressed regions, where the transfer integrals β_{ij} are larger in amplitude, thus the band effectively larger but still symmetrical ; the converse is true in extended regions ; there should thus be a charge density wave associated with the phonon, providing a repulsive correction in the compressed regions and an attractive one in the extended regions (figure 18.a). The converse should hold for more than half filled bands (figure 18.b), where it is justifiable to talk in terms of positive holes.

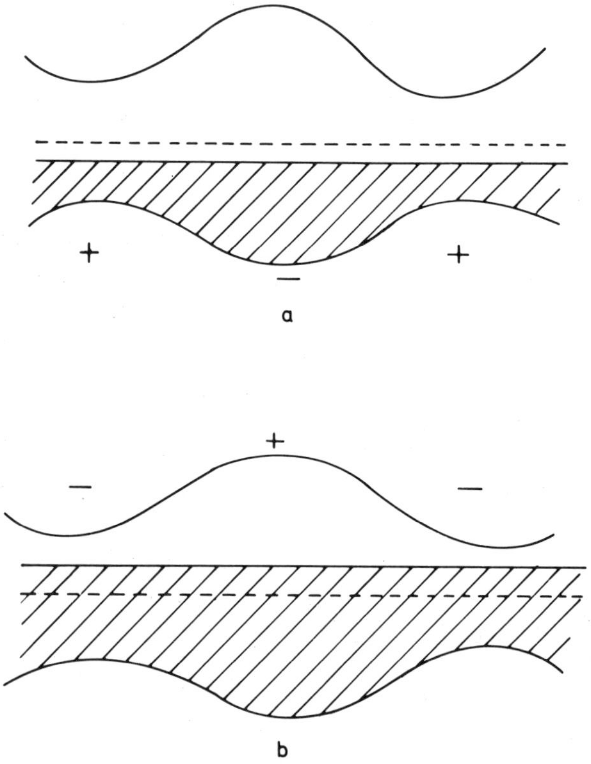

Figure 18 : Longitudinal phonons in alternate structures : a electrons ; b positive holes.

In non alternate structures, the same distinction between electrons and holes respectively for nearly empty and nearly full bands ; but the transition from one type of behaviour to the other, thus the exact filling for which no charge density wave is produced, will depend on the type of phonon considered. The condition for no charge density wave will still occur near the middle of the band.

– If one considers only the intraatomic contribution to screening due to the charge density wave, this again can but stiffen the phonon(32). Let U be the average Coulomb interaction between valence electrons on the same site. To a local change $\delta\rho_i$ in electronic density there would correspond a local change δV_i in the atomic potential of site i such that

$$\delta V_i = U\ \delta\rho_i$$

This replaces Poisson's equation in the preceeding analysis. And, in the extreme case of nearly full or nearly empty non degenerate bands, where a simple geometry of the Fermi surface prevents umklapp processes to occur, the unstability condition reads

$$\frac{|v_{eq}|^2}{E_q} - U > - \frac{1}{\chi_q}$$

The addition of U indeed makes a soft mode less easy to produce.

– One should however also consider interatomic contributions of the screening to the stability of the mode(26). This Madelung energy can lower the energy of suitable modes. A historical example is the special stability of Kohn anomalies along parallel chains when neighbouring chains are in phase opposition, which was

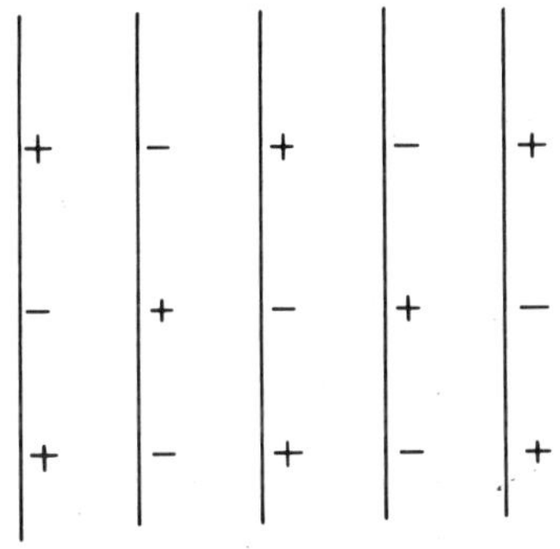

Figure 19 : Coupling in antiphase of Kohn anomalies along parallel chains, stabilized by long range electrostatic interactions.

predicted by Barisic(26) and later observed in linear chain compounds such as KCP(15).

A general study of the tight binding case, combining the intra and interatomic contributions of screening has not been made so far.

It is conceivable that in some cases the overall effect could make the phonon mode softer. It is however very unlikely, because U is a large positive quantity. Indeed the electrostatic correction is most favourable for stabilization if modulation is at wave lengths such that neighbouring atoms have opposite charges ± q. The total electrostatic correction then reads $\frac{1}{2} U_q^2 - \frac{\alpha q^2}{R}$ per atom where R is the interatomic distance and α the Madelung constant. With α/R ≃ 1/4 Atomic unit, and U ≃ 1/4 to 1/2 At. unit, the correction is negative but small. Furthermore we have seen that at such wave lengths, the amplitudes of a charge density wave produced by a phonon is small.

Conclusion

- The screening changes the strength, but not the nature of the Kohn anomaly.

It corresponds usually (but not always) to a charge density wave which is coupled with the phonon ; this usually stiffens the phonon, thus making soft mode conditions less easy.

- Indeed the production of a charge density wave without a phonon(30) would require $\delta\rho \neq 0$ for $v_e = 0$, or, according to the analysis above, the dielectric constant ε_q going to zero. This is clearly impossible for free electrons, where $\chi_q < 0$, and very unlikely in tight binding, except perhaps for modes with short wave lengths which are but weakly coupled with the phonons. It is thus unfortunate, to say the least, that the fashion is nowadays to call a soft phonon mode by the name 'charge density wave' which actually is the part in the mode that fights against its instability!

- The situation would be quite different for a spin density wave, because there the positive U term is replaced by a negative term describing intraatomic exchange effects : a spin density wave can soften into a magnetically modulated stable structure, without necessarily being coupled with a structural distortion. Coupling between spin density waves and phonons by spin orbit coupling can lead indeed to weak but non zero magnetostrictive lattice modulations.

- For nearly free electrons, the second order term in δE_q can be thought of as due to the modulation by the displacement u of the pair interactions between the atomic pseudopotentials, each treated to second order as the interactions of two impurities in a free electron gas. Taking $\varepsilon_q^{-1} |v_{eq}|^2$ is the same as considering the interaction of the screened potential of one atom acting on the maked charge of the other in the pair.

- The (screened) Kohn anomaly can be considered as a resonance between the phonon and the Fermi electrons responsible for the long range oscillations of the screening. It is also an interference effect of the long range oscillating (screened) interatomic forces. Indeed if the Fermi electrons have a long but finite mean free path ℓ, due to say scattering by imperfections, this will dampen exponentially the long range oscillations of the interatomic forces (in $\exp - \frac{r}{\ell}$) and correlatively broaden the Kohn anomaly over a width $\delta k \simeq 2\pi/\ell$. But it will pratically not change the nature and the stability of the possible soft modes.

3 - Other correlation effects

- Magnetism : exchange effects lead to possible ferromagnetism or soft spin density waves, thus various modulated magnetic structures, which will not be considered here.

- Energy : Coulomb correlations introduce corrections in the energy. As stressed above, they are important for evaluating the absolute value of cohesion, but play only a reduced role in the relative stability of crystal phases (cf Appendix D).

- Damping : electron-electron collisions are only active at finite temperatures. They introduce resistive terms which are usually at most comparable with those due to imperfections (impurities, phonons). In all cases, electron phonon coupling seem to dominate on electron-electron scattering ; this last term can be treated as a fairly small correction, analogous to that just mentioned for scattering by imperfections[33].

4 - Degeneracy of electronic states

- First order terms. If the lattice structure has degenerate Van Hove anomalies at different points $\underset{\sim}{k}_r$, $\underset{\sim}{k}_s$ ... in the first Brillouin zone, a phonon mode v_q with $\underset{\sim}{q} = \underset{\sim}{k}_s - \underset{\sim}{k}_r + \underset{\sim}{K}_{RL} \neq 0$ can lift their degenerary to first order :

$$\langle n\, \underset{\sim}{k}_r \mid \underset{\sim}{v}_q \mid n\, k_s \rangle \neq 0$$

This is a natural extension of the effect possibly associated with uniform distortions, and can thus be termed an extended Band Jahn Teller effect.

- Second order terms. The one beam (kinematic) approximation used until now is not valid, strictly speaking, near the lines L, figure 8, where it is used to analyse the Kohn anomaly. In such a region, one Bloch function $|n', \underset{\sim}{k} \pm \underset{\sim}{q}_0 + \underset{\sim}{K}\rangle$ takes an amplitude comparable with $|n, \underset{\sim}{k}\rangle$. It is then better to develop the perturbed function $|\psi\rangle$ in these two wave functions, neglecting the contributions of the other Bloch functions $|n'',k''\rangle$, which will be small and regular.

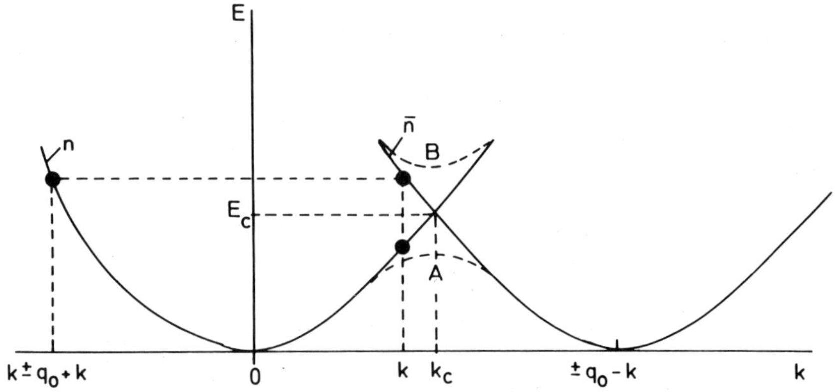

Figure 20 : Two beams approximation (in the case n' = n).

Thus we reduce the hamiltonian

$$H = \sum_{n,k} E_k^n |n,k\rangle\langle n,k| + \sum_{kk'} |n,k\rangle\langle k|v|k'\rangle\langle n'k'|$$

to

$$H = \sum_k \Big[E_k^n |n,k\rangle\langle n,k| + E_k^{\bar{n}} |\bar{n},k\rangle\langle \bar{n},k|$$

$$+ v_{q_o} |n,k\rangle\langle \bar{n},k| + C.C. \Big]$$

where $E_k^{\bar{n}} = E^{n'}_{k \mp q_o + K}$ (figure 20).

The well known solutions of this (nearly) degenerate problem is

$$E_k^n(q_o) \simeq \frac{1}{2} \left[E_k^n + E_k^{\bar{n}} \pm \sqrt{(E_k^n - E_k^{\bar{n}})^2 + 4|v_{q_o}|^2} \right] \qquad (5)$$

with

$$|\psi_k\rangle = u_k |n,k\rangle + v_k |\bar{n},k\rangle$$

$$\frac{v_k}{u_k} = \frac{v_{q_o}}{\frac{1}{2}\left[E_{k}^{n} - E_{k}^{\bar{n}} \pm\sqrt{(E_{k}^{n} - E_{k}^{\bar{n}})^2 + 4|v_{q_o}|^2}\right]}$$

where the sign + refers to the upper continuous curve B, the sign - to the lower one A, which replace the crossing point.

This two beams approximation supresses the divergencies in one electron energy $E_k^n(q_o)$, densities of electronic states n(E), and eventually Kohn anomalies which arise in the one beam approximations.

Thus, in three dimensions, figure 21 pictures schematically the differences arising for n(E) and the corresponding (non self consistent) Kohn anomaly for nearly free electrons. The anomaly is split into two successive anomalies, corresponding to the two

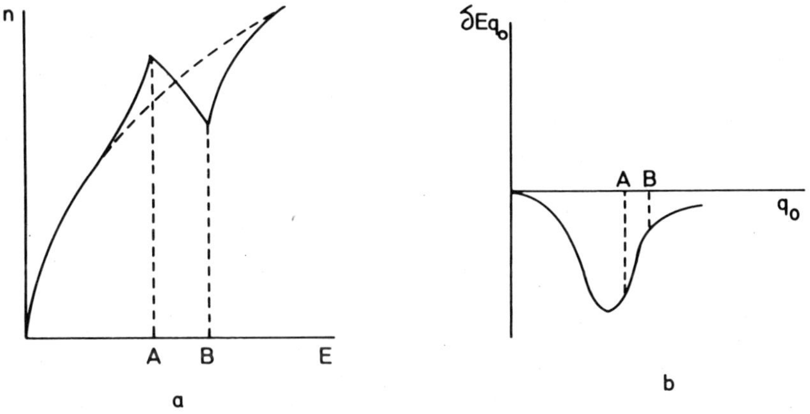

Figure 21 : Three dimensional free electrons scattered by a phonon in the two beams approximation : a density of electronic states ; b Kohn anomaly.

Van Hove anomalies A, B induced by the phonon at the extremities of the energy gap, figure 20. The special stability can thus be viewed as due to the fact that, over a fairly large part of the reciprocal space, the scattering by the phonon lowers the energy of the occupied electronic states and lifts the energy of the empty states. The total effect is clearly less marked than as computed in the one beam approximation. But it can be shown[34] that, when developped to second order in v_q, the results for cohesion of the two beams approximation are identical with those of the one beam approximation, in the general 3 dimensional case (cases α, β of figure 9).

In one or two dimensions, no such equivalence exists between one beam and two beams approximations. This last one systematically gives less marked Kohn anomalies.

Thus, in one dimension, figure 22 gives the density of states and the (non selfconsistent) Kohn anomaly. It is first clear that there is a total energy gap between A and B, figures 20 and

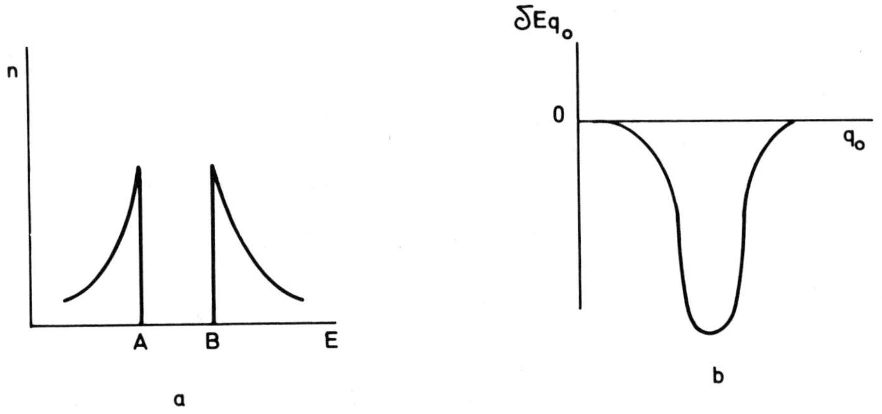

Figure 22 : Effect of a phonon on one dimension electrons : a density of electronic states ; b Kohn anomaly.

22, so that when the Fermi level falls within that gap, one produces an insulator, at least as long as the phonon is static. From equation (5), one deduces, with $n(E) \propto dk/dE''_k(q_o)$,

$$n(E) \simeq \pm\, n_o(E_c) \sqrt{1 + \frac{|v_q|^2}{(E - E_c)^2}} \quad \text{for} \quad E < E_A \text{ or } E > E_B$$

$$= 0 \text{ for } \quad E_A < E < E_B.$$

where $E_c = \frac{1}{2}\,(E_A + E_B)$ is the center of the energy gap, and $n_o(E)$ is the electronic density of states in the absence of the phonon (figure 22.a).

Also

$$\delta E_{q_o} = \frac{1}{\pi}\int^{k_M} (E^n_{\underset{\sim}{k}}(q) - E^n_{\underset{\sim}{k}})\,dk$$

With

$$\frac{d\,E^n_k}{dk} \simeq - \frac{d\,\bar{E}^n_k}{dk} = \frac{k - k_c}{\pi\, n_o(E_c)}$$

where $n_o(E_c)$ is the unperturbed density of electronic states at energy E_c, this gives

$$\delta E_{q_o} = - \frac{1}{\pi}\int_0^{k_M} \left[\sqrt{\frac{(k - k_c)^2}{\pi^2\, n_o^2(E_c)} + |v_q|^2} - \frac{k - k_c}{\pi\, n_o(E_c)}\right] dk$$

$$= - \frac{|v_q|^2}{2\,\pi^2 n_o(E_c)} \left[\frac{q}{|v_q|}\sqrt{1 + \frac{q^2}{|v_q|^2}} - \text{argsh}\,\frac{q}{|v_q|} - \frac{q^2}{|v_q|^2}\right]_{-k_c}^{k_M - k_c}$$

$$\simeq - \frac{|v_q|^2}{2\,\pi^2 n_o}\left[\ell n\,\frac{2\,k_c}{\pi n_o\, v_q} - \left|\frac{k_M - k_c}{\pi\, n_o\, v_q}\right|^2 + \frac{2}{3}\left|\frac{k_M - k_c}{\pi\, n_o\, v_q}\right|^3 + 0_4\left|\frac{k_M - k_c}{\pi\, n_o\, v_q}\right|\right]$$

The anomaly is very peaked (because $|v_q| << k_c/\pi\, n_o$, which is of the order of the electronic band width). But it remains finite, as soon as v_q is finite (figure 22.b).

It must also be stressed that a proper selfconsistent description in the two beams approximations has not yet been developped. It is however not expected to lead to fundamentally different results.

5 - Anharmonic terms

There are various terms higher than second order in the amplitude of the phonons that must be taken into account.

- Terms in u^n with $n > 2$ must be considered to study the equilibrium amplitude of a soft mode. Such terms should be screened selfconsistently, and this has not been studied completely so far.

- Phonon-phonon couplings arise from such anharmonic terms. This leads to various effects :

- Strong attenuation of the nearly soft mode, just above the transition conditions. As a result, such a nearly soft mode with large amplitude might loose its physical meaning near the transition.

- Star of soft modes with equivalent q's : these must be treated together, and their couplings considered[35][36].

- Phase modulation and phase locking[37][38]. Only a few remarks will be made on this topic, which will be treated fully later at this Institute.

Consider a (sinusoidal) longitudinal mode of distortion produced in a crystal, with a wave vector $\underset{\sim}{q}$. This will usually set up a charge density wave which will in turn produce a sinusoidal (or nearly sinusoidal) potential with wave vector $\underset{\sim}{q}$. The lattice will react in turn to this potential : the atomic planes will tend to avoid the top of the sinusoid and concentrate in the lower parts.

- To first order, this reaction can be described by a supplementary sinusoidal distortion which, for nearly free electrons or holes, will certainly be in antiphase with the intial distortion, as stressed above.

- If the initial distortion is large, one must however consider the non linear response of the crystal. This is very similar to the problem of the reaction of a periodic elastic chain of atoms on a rigid periodic substrate, as a model for epitaxy[39][40][41] (fig. 23).

- The first deviation from linearity can be described as if the medium was continuous. It will distort the sinusoidal distortion (or introduce harmonics) in such a way that the phase of the initial simusoid will be modulated. By analogy with a classical hydrodynamical problem, each period in the modulation has been called a soliton.

As the lattice is assimilated to a continuous medium, the initial sinusoidal distortion could be shifted with respect to the lattice without change in energy. In fact, one must distinguish two cases :

- $\underset{\sim}{q}$ is commensurate with a lattice period. When one shifts the sinusoidal wave with respect to the lattice, one necessarily goes through different configurations, such as a, b figure 22, with different energies. One of them must be more stable than the others, and the wave must have its phase locked in that position at OK. The locking is obviously more effective for large amplitude sinusoids, and for short period sinusoids, with wave lengths small multiples of the crystal period.

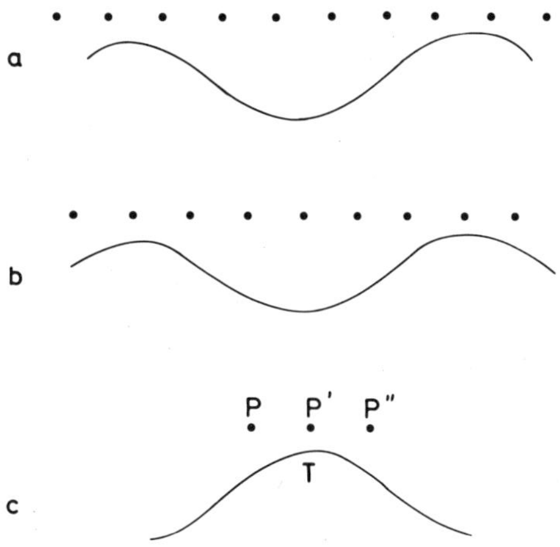

Figure 23: Phase locking. a, b commensurate wave, c uncommensurate wave.

- if q is not commensurate with the lattice, the continuous non linear solution mentioned above leads to atomic planes being distributed all over the top of the sinusoidal potential so that, when the wave is shifted, its energy is not changed : the phase of the distortion wave is not locked ; the wave could in principle be a travelling one, and has thus been called a phason. It is however most probable that it is pinned down by stationary defects or slowed down by anharmonic interactions with normal phonons. But if the amplitude of the initial sinusoidal distortion increases, there will necessarily be a moment when the phase of this uncommensurate wave will also become locked. One is sure this has happened when the negative curvature at the top T of the sinusoidal potential, figure 23.c, becomes larger than the elastic constant that keeps the atomic planes PP'P" more or less equidistant. The plane P' at the top of the sinusoid will then become unstable and be locked on the left or the right of T ; a finite energy will be necessary to move it across T. This is the analogue of the Peierls friction against the 'epitaxial dislocations' whose glide allows the glide of the epitaxial layer. As in epitaxy, and depending on the exact conditions of the problem, the phase locking can lead to commensurate or uncommensurate phase locked structures.

Finally it must be stressed that a selfconsistent description of the phase modulation would require a modulation of the initial sinusoidal screening potential. This is not expected to lead to new physical effects.

6 - Adiabatic approximation

A travelling phonon has a finite frequency ω_q which should be taken into account in the exact description of the electron-phonon coupling(14).

A phonon now couples strongly electronic states with energies E_k^n and $E_{k'}^{n'}$ differing by $\hbar\omega_q$. But as this is usually much smaller than the electronic energies involved, this only rounds off slightly the central part of the Kohn anomaly, without altering its essential characteristics. The effect is furthermore vanishing in the limit of soft modes, where $\hbar\omega_q$ tends to zero.

The only cases where the adiabatic approximation is not sufficient for the study of Kohn anomalies is when the slow electrons near a Van Hove anomaly are considered(42).

The adiabatic approximation is of course also insufficient to compute the indirect attractive electron-electron coupling responsible for supraconductivity, although once the form of this coupling is obtained, a time independent treatment is sufficient.

7 - Entropy at finite temperature

The discussion has centred so far on OK phase changes, induced for instance by a variation of pressure.

If a crystal modulation is stable at OK, one usually observes it to disappear above a sufficient temperature. This is due to entropy effects, which can be complex.

There are indeed a priori several possibilities and only a study case by case can show which of several possible factors is predominant.

The entropy can arise from electrons excited across the (first or second order) energy gap(21). It can also be due to phonons. If it is due to phonons, it can be due mainly to the soft mode(33) renormalized by its anharmonic interaction with other phonons (44); or it can be due to other modes.

In one or two dimensions, critical fluctuations of the soft mode should be dominant at low temperatures(35)(36). They are however usually kept down in that range by three dimensional couplings.

APPENDIX A - SHORT RANGE ORDER DUE TO DISPERSION FORCES

Let A, B be two atoms or molecules interacting at long range by dispersion forces. Let $|0A\rangle$, $|1A\rangle$, $|0B\rangle$, $|1B\rangle$ be the fundamental and main excited states of the two molecules involved in the dispersion forces, with energies E_0^A, E_1^A, E_0^B, E_1^B. The development of the long range Coulomb interaction between A and B gives rise to a pair interaction between electric dipoles on A and B. If A and B have no permanent dipoles, only dipoles induced by mutual polarization interact. Thus if C is the Coulomb correlation term in the hamiltonian describing the pair, its only matrix elements different from zero are $\langle A_0 B_0 | C | A_1 B_1 \rangle = \langle A_1 B_1 | C | A_0 B_0 \rangle$, and it is proportionnal to $u_0 u_1$, with

$$u_i = \langle i_0 | \chi | i_1 \rangle .$$

A perturbation development

$$|\psi\rangle = (|A_0\rangle + \alpha |A_1\rangle)(|B_0\rangle + \beta |B_1\rangle)$$

then gives

$$\begin{vmatrix} E_0^A + E_0^B - E & \langle A_0 B_0 | C | A_1 B_1 \rangle \\ \langle A_1 B_1 | C | A_0 B_0 \rangle & E_1^A + E_1^B - E \end{vmatrix} = 0$$

Hence the dispersion interaction

$$w_{AB} = E - E_0^A - E_0^B = \frac{\langle A_0 B_0 | C | A_1 B_1 \rangle^2}{E_1^A + E_1^B - E_0^A - E_0^B} \propto - \frac{u_A^2 u_B^2}{\delta A + \delta B}$$

where $\delta_i = E_1^i - E_0^i > 0$

Considering now three pairs AB, AA, BB <u>at the same relative distance</u>. One can define a <u>short range order energy as</u>

$$\delta w = \frac{1}{2}(w_{AA} + w_{BB} - 2w_{AB}) \propto \frac{u_A^2 u_B^2}{\delta A + \delta B} - \frac{u_A^4}{2\delta A} - \frac{u_B^4}{2\delta B} \propto - \frac{(u_A - u_B)^2 \delta A \delta B + (u_A \delta B - u_B \delta A)^2}{(\delta A + \delta B)\delta A \delta B} < 0$$

The negative sign of δw for dispersion forces can be constrasted with its positive sign for Coulomb forces between ions, where

$$w_{AB} \propto Z_A Z_B$$

hence $\delta w \propto (Z_A - Z_B)^2 > 0$

A negative sign of δw favours segregation in pure A and B phases ; a positive sign of δw favours ordered compounds (cf figure 3, a and b respectively).

APPENDIX B -- LCAO STUDIES OF THE BAND STRUCTURES OF METALS AND COVALENTS

In the LCAO approximation, the one electron wave functions are analyzed in terms of atomic functions $|i, m>$ (site i, wave function of type m), assumed nearly orthogonal :

$$| \psi > \simeq \sum_{i,m} a_i^m \; | i, m >$$

$$< i, m | j, n > \simeq \delta_{ij} \delta_{mn}$$

With a suitable origin of energies, the hamiltonian reduces to one site and two sites terms :

$$H \simeq \sum_i \Delta_i^{mn} \; |im> <in| + \sum_{\substack{i,j \text{ near to } i \\ m,n}} \beta_{ij}^{mn} \; |im><jn|$$

The transfer integrals β allow interatomic jumps. The intraatomic terms Δ come from electrostatic crystal field interactions (for $m = n$) and from atomic promotion energies if the $|i,m>$ functions are hybrids made with atomic functions of different energies.

1 - H broadens the atomic states into bands without changing the average energy. This can be seen by computing the first moment of the density of states (per unit energy and per atom) in the bands ; using the $|i,m>$ set as a (pseudo) complete set :

$$M_1 = \int n(E)EdE = \text{Trace } H \simeq \frac{1}{N} \sum_{i,m} <i,m| \; H \; |i,m > = 0$$
(N atoms)

A rough estimate of the band width can then be deduced from the second moment[45]

$$M_2 = \int n\,(E)E^2dE = \text{Trace } H^2 \simeq \frac{1}{N} \sum_{\substack{i,m \\ l,n}} <i, m \; |H| \; jn><j,n \; |H| \; i,m >$$

$$\simeq \frac{1}{N} \Big(\sum_{\substack{i,j \text{ near to } i \\ m,\, n}} \beta_{ij}^{mn} \beta_{ji}^{nm} + \sum_{\substack{i \\ m,n}} \Delta_{mn}^{i} \Delta_{nm}^{i} \Big)$$

with usually $|\Delta| << |\beta|$, at least in elementary structures, the effective band width is finally given by an average value of the β'_s multiplied by the square root of the average number of nearest neighbours. Hence large cohesion for strong bands (large β's)or many bonds (large p's).

2 - In sp metals or covalents, the lower limit of the valence band is obtained with pure atomic s states, and coefficients a_i^s of constant sign (minimum number of nodal surfaces). In crystals with one atom per unit cell, the solution is a_i^s = const. and

$$E_{min} = \frac{1}{N} \sum_{\substack{i \\ j \text{ near to } i}} \beta_{ij}^{ss} = p < \beta >$$

As $p > p^{1/2}$, this estimate of the band width is larger than the previous one, but of the same order of magnitude.

3 - In sp covalents, the $|im>$ functions can be taken as sp hybrids such that only functions $|in>$ and $|jn>$ pointing along the same covalent bond between two neighbouring sites i, j have a non vanishing (negative) transfer integral β (46). There are then non vanishing one site terms $\Delta_i^{n \neq m} = \Delta$ associated with the atomic (negative) promotion energy $E_s - E_p$. Thus (figure B.1)

$$H \simeq \beta \sum_{\substack{i,\ j \text{ near to } i \\ n}} |in><jn| + \Delta \sum_{i, m \neq n} |in><im|$$

Writing $H|\psi> = E|\psi>$ and projecting on $|in>$ gives

$$\Delta \sum_{m \neq n} a_i^m + \beta \quad a_j^n = E \ a_i^n$$

or

$$\Delta \sum_m a_i^m + \beta \ a_j^n = (E + \Delta) \ a_i^n$$

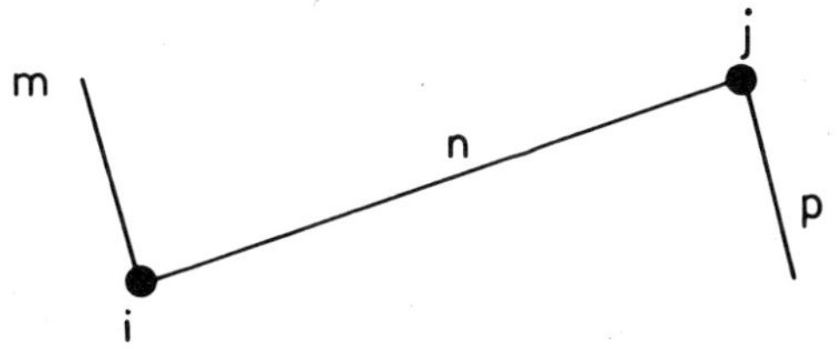

Figure B.1 : Notation of sites and bonds in an sp covalent structure

Similarly

$$\Delta \sum_p a_j^p + \beta \quad a_i^n = (E + \Delta) \ a_j^n$$

Hence $(E + \Delta) \ a_i^n = \Delta\sigma_i + \frac{\beta}{E+\Delta} (\Delta\sigma_j + \beta a_i^n)$

and

$$\left((E + \Delta)^2 - \beta^2 - p\Delta(E + \Delta)\right)\sigma_i - \beta\Delta \sum_j \sigma_j = 0 \qquad \text{(B-1)}$$

with $\sigma_i = \sum_m a_i^m$

Equation (B-1) is the LCAO equation for an s band on the same atomic structure, with transfer integral $\beta\Delta$ and energy $\varepsilon = (E + \Delta)^2 - \beta^2 - p\Delta(E + \Delta)$. General theorems tell us that such an s band has limits ε_1 ε_2 such that

$$- p\ \beta\Delta = \varepsilon_1 \lesssim \varepsilon \lesssim \varepsilon_2 \lesssim + p\beta\Delta$$

The symmetry in energy pointed out in the text shows that ε_2 is equal to $p\beta\Delta$ for alternate structures ; a variational procedure shows that ε_2 is lower than $p\beta\Delta$ for non alternate structures, as defined in the text.

Hence for $|\ \beta\ | > \frac{p}{2}\ |\ \Delta\ |$,

$$\beta + (p - 1)\Delta \lesssim E \lesssim \beta - \Delta \quad \text{or} \quad -\beta + (p - 1)\Delta \lesssim E \lesssim - \beta - \Delta$$

for $|\beta| < \frac{p}{2}\ |\ \Delta\ |$,

$$\beta+(p - 1)\Delta \lesssim E \lesssim - \beta +(p - 1)\Delta \quad \text{or} + \beta - \Delta \lesssim E \lesssim - \beta - \Delta.$$

Figure B.2 schematizes the results, taking into account an exponential variation of β with the interatomic distance d[(47)(48)].

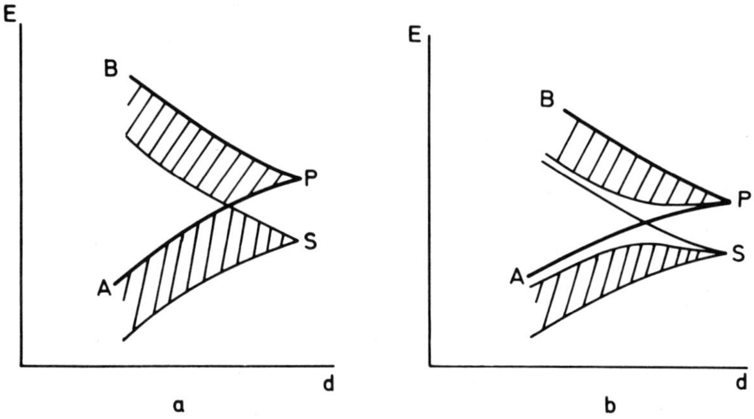

Figure B.2 : Allowed regions of energy for an sp covalent band in the LCAO approximation : a alternate structures ; b non alternate ones (NB : for sp^n bands with $n > 1$, the lines pA, pB are allowed pure p states not discussed here).

3 - More exact details about n(E) are necessary to compute the relative stability of different condensed phases. This can be obtained from a study of higher moments(45)(49).

$$M_q = \int n(E)\, E^q\, dE$$

and from a development of the Hilbert transform of n(E) as a continuous fraction(50)(51)(52)

$$\int \frac{n(E')}{E-E'}\, dE' = \cfrac{a_0}{E-a_1-\cfrac{b_1}{E-a_2-\cfrac{b_2}{\cdots}}}$$

where the a_i's and b_i's can be expressed in terms of the M_q's, and have an asymptotic behaviour dominated in crystals by the van Hove anomalies of n(E).

APPENDIX C - PHASE STABILITY FOR NEARLY FREE ELECTRONS

Treating the scattering of a free electron gas by the atomic pseudopotentials $v_i(r-R_i)$ as small, the total energy of the valence electrons can be written(53)(54)

$$E = E_0 + E_1 + E_2 + 0_3(v)$$

The zero order term E_0 is the cohesive energy of a free electron gas in a uniform background of positive charge that neutralizes the system. It is a function of the electron density or of the Fermi wave number k_M. It contains exchange and Coulomb correlations, essential to stabilize this 'jellium'.

The first order term E_1 corresponds to first order (Born) scattering of the free electrons by the pseudopotentials v_i. Each site scatters independently as an impurity, and the Born approximation is only meaningful if screening of each v_i by the valence electrons is included : in E_1, the v_i are 'clothed', the relation with the 'naked' v_i's being, for each Fourier component K

$$v^c_{iK} = v^n_{iK}/\varepsilon_K$$

(ε_K dielectric constant of the free electron gas).

The second order terms E_2 have two different origins :
- double scatterings on single sites.
- successive scatterings on two different sites.

E_1 and the first terms in E_2 are only a function of the nature and density of atoms. Only the second terms in E_2 depend on the relative arrangement of the atoms, i. e. on the atomic structure. They can be written as a sum of central pair interactions E_{ij}

between two atoms i and j, acting as if they were an isolated pair of impurities in a free electron-gas. General theorems tell us that to this lowest order, E_{ij} is the Coulomb interaction of the clothed potential of one atom on the naked charge of the other. Thus, using Poisson's equations(55)(56)

$$E_{ij} = \iint v^c_{iK} \; e^{i\underset{\sim}{K}(\underset{\sim}{r}-\underset{\sim}{R}_i)} \; d_3K \; \frac{K'^2 v_{jK'}}{4\pi e^2} \; e^{i\underset{\sim}{K}'(\underset{\sim}{r}-\underset{\sim}{R}_j)} \; d_3K'$$

$$= \int \frac{K^2}{4\pi e^2} \; \frac{v^n_{iK} \, v^n_{jK}}{\varepsilon_K} \; e^{i\underset{\sim}{K}\,\underset{\sim}{R}_{ij}} \; d_3K$$

Owing to the logarithmic anomaly of ε_K for $K \simeq 2k_M$, this is an oscillating function of $2k_M R_{ij}$, with an amplitude decreasing as R_{ij}^{-3} at long range. It is the change in $\sum_{i,j\neq i} E_{ij}$ between different lattice structures which gives rise to the latent heat of phase change. In crystals, the integral reduces to the periods K_{RL} of the reciprocal lattice :

$$\sum_{i,j\neq i} E_{ij} = \sum_{K\in K_{RL}} \frac{K^2}{4\pi e^2} \; \frac{v^n_{iK} \, v^n_{jk}}{\varepsilon_K}$$

Owing to the logarithmic decrease of ε_K for $K > 2k_M$, this term increases suddenly, for increasing electron per atom ratios, when $2k_M \simeq K_{RL}$, or for the Fermi sphere tangent to Brillouin zone Boundaries.

Third order terms in v would introduce directional forces and possibly finite mean free path effects. They are however difficult to analyze in a convincing way(25).

APPENDIX D - COHESION IN TRANSITIONAL METALS

Considering only the d band and within the LCAO approximation, this can be written as

$$E = E_0 + E_1 + E_2 + \ldots.$$

The zero order terms is the Hartree one electron term when one neglects interactions between valence electrons (or uses the Wigner Seitz approximation referred to in the text)(57) :

$$E_0 = - \int^{E_M} n(E)E \, dE$$

where the zero of energy is taken as the energy of a d state in the atom. The Fermi level E_M is related to the number z of d electrons per atom :

$$z = \int^{E_M} n(E)\, dE$$

From $dE_0/d_z = - E_n$ and $d^2E_0/d_z^{\,2} = - \frac{1}{n(E_n)}$, one deduces that $E_0(z)$ has a single maximum for a half filled band ($E_n = 0$ or, as the first moment of the band is zero, $z = 5$). Indeed it does not deviate much from the parabola obtained if n(E) was a constant over an energy $-\frac{w}{2} < E < \frac{w}{2}$ (figures D.1)

$$E_0 \simeq \frac{z(10-z)}{20}\, w$$

And the (small) deviations with crystal structures are coherent with the succesion FCC, HCP, BCC, HCP, FCC observed in the transitional series[49].

The first order correction originates from the Coulomb and exchange interactions between valence electrons, as computed in the Hartree Fock approximation i. e. for valence electrons randomly distributed over the lattice sites, taking only into account the exclusion principle[58]. If one only considers the average intraatomic terms $U_{nm} = U$ and $J_{n \neq m} = J$, the differences in population of the 10 d orbitals in atoms and in the paramagnetic metals gives[5]

$$E_1 \simeq - \frac{z(10-z)}{20}\, U - \mu(\mu - 1)\frac{J}{4} - \lambda LS$$

where $\mu = \begin{cases} z & \text{for } z < 5 \\ 10-z & \text{for } z > 5 \end{cases}$

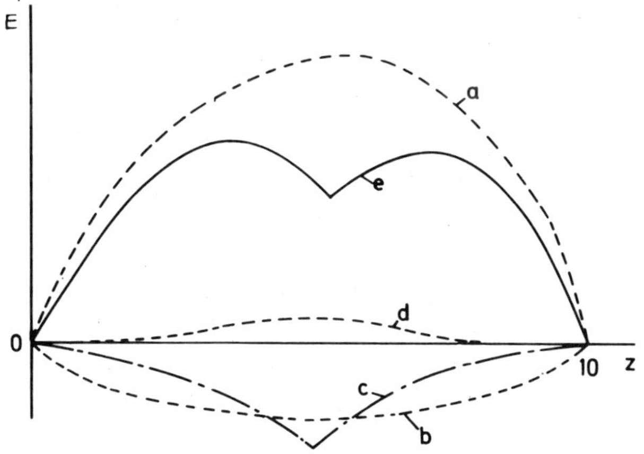

Figure D.1 : Contribution to cohesion in a transitional series : (schematic) :a - E_0;b, c : E_1 ; d - E_2 ; e - Total.

The term in U reduces the amplitude of cohesion ; the term in J, due to the special stability of half filled magnetic d shells, introduces a secondary minimum in cohesion near the middle of the transitional series (figure D.1).

The second order contribution E_2 is the first correction for electron-electron correlations. As U >> J, only the term in V^2 need be considered. As U induces virtual transitions between occupied and unoccupied valence states, with number $\frac{z}{10}$ and $1-\frac{z}{10}$ per atomic orbital, an estimate neglecting conservation of momentum in electron-electron collisions gives

$$E_2 \approx \int_{-\frac{w}{2}}^{E_M} \int_{-\frac{w}{2}}^{E_M} \int_{E_M}^{w/2} \int_{E_M}^{w/2} \frac{Un(E_1)n(E_2)n(E_3)n(E_4)dE_1dE_2dE_3dE_4}{E_1 + E_2 - E_3 - E_4}$$

This is approximately (for rectangular d bands)(5) :

$$E_2 \simeq Z \ A \frac{\left(\frac{3}{10}(1 - \frac{3}{10})U\right)^2}{w}$$

The numerical factor A is of the order of 50.

This is a small positive correction, which does not play a large role in the cohesive energy but explains the secondary minima observed in the surface tension and the elastic constants, near the middle of the transition series(59).

Cohesive energies, elastic constants and surface tensions are coherent with the set of values of the parameters involved as given in the following table(5). As stated in the text, the first order Coulomb interaction (term in U) decreases cohesion by a term independent of crystal structure. The second order correlation correction increases but a little cohesion, by a term somewhat structure sensitive.

TABLE I - Values of parameters in eV

	J	U	w	λ
3d	0.7	(3)	(6)	0.05
4d	0.55	3	9	0.15
5d	0.55	< 3	12	0.40

REFERENCES

(1) Mott N. F., Canad. J. Physics 34, 1356 (1956) ; Phil. Mag. 6, 287 (1961).
(2) Villain J., J. Phys. Chem. Solids 11, 303 (1959) ; Yoshimori A. J. Phys. Soc. Japan 14, 807 (1959)
(3) Wigner E. and Seitz F., Phys. Rev. 43, 804 (1933) ; 46, 509 (1934) ; cf Seitz F., Modern theory of solids Mc Graw Hill New York (1940).
(4) Pines D., Elementary excitations in solids, Benjamin, New York (1963).
(5) Friedel J. and Sayers C. M., J. Physique (1977) under press.
(6) Kohn W., Phys. Rev. Letters 2, 393 (1959).
(7) Jones H. Proc. Roy. Soc. 144, 225 (1934) ; 147, 396 (1934).
(8) Hume Rothery W., J. Inst. Met. 35, 309 (1926) : The structure of metals and alloys, Inst. of Metals Monographs, London(1936).
(9) Friedel J., J. Physique Lettres 35, 159 (1974).
(10) Jehanno G., Thèse Orsay (1965) ; Sato H. and Toth H. S., J. Phys. Chem. Solids 29, 2015 (1968) ; Toth R. S. and Sato H. Acta Met. 16, 413 (1968).
(11) Lomer M., Phase stability in Metals and Alloys (Ed. Rudman P.S.) New York, MacGraw Hill (1966) ; Labbé J., J. Physique 29, 195 (1968) ; cf Friedel J. The Physics of Metals I Electrons (Ed. J. M. Ziman) Cambridge University Press (1969).
(12) Wilson J. A. et al., Adv. Phys. 24, 117 (1975).
(13) Peierls R. E., Quantum theory of solids, Clarendon Press Oxford (1955).
(14) Froehlich H., Proc. Roy. Soc. A 233, 296 (1954).
(15) Comès R., Lambert M. and Zeller H. R., Phys. Stat. Sol.(b) 58 587 (1973) ; Renker B. et al. Phys. Rev. Lett. 30, 1144 (1973).
(16) Comès R. et al., Phys. Rev. B8, 571 (1973) ; Phys. Rev. Lett. 35, 1518 (1975).
(17) Labbé J. and Friedel J., J. Physique 27, 153, 303, 708 (1966).
(18) Weger M., J. Phys. Chem. Solids 31, 1621 (1970).
(19) Gor'kov L. P., J. E. T. P. Lett. 17, 379 (1973) ; Soviet Phys. J. E. T. P. 38, 830 (1974).
(20) Stoner E. C. Proc. Roy. Soc. A 154, 656 (1936).
(21) Slater J. C. Phys. Rev. 82, 535 (1951) ; Lidiard A. B. Proc. Roy. Soc. A 224, 161 (1954).
(22) Lomer M., Proc. Phys. Soc. 80, 489 (1962).
(23) Ziman J. M., Principles of the Theory of Solids, Cambridge University Press, 2d edition (1972).
(24) Harrison W. A., Pseudopotentials in the theory of metals, Benjamin New York (1966) ; Heine V. Solid State Physics 24, 1 (1970)
(25) Heine V. and Weaire D., Solid State Phys. 24, 336 (1970)
(26) Barisic S., Sol. State Comm. 9, 1507 (1971) ; Annales de Physique 7, 23 (1972) ; Saub K., Barisic S. and Friedel J., Phys. Lett. 56A, 302 (1976)
(27) Bloembergen N. and Rowland T. J., Acta Met. 1, 731 (1953)

(28) Roth L.M., Thesis Harvard (1957).
(29) Blandin A., Thèse Orsay (1961).
(30) Overhauser A.W., Phys. Rev. 128, 1437 (1962) ; 167, 691 (1968) ; B 3, 3173 (1971).
(31) Blandin A., Alloying behaviour and effects of concentrated Solid Solutions (ed. Massalsky T.B.) Gordon and Breach 1963.
(32) Chan S.K. and Heine V., J. Phys. F 3, 795 (1973).
(33) Barisic S., Fizika 8, 181 (1976) ; Bjelis A., Saub K. and Barisic S., Nuovohin. 23 B1, 102 (1974).
(34) Pick R. and Blandin A., Phys. Kond. Mat. 3, 1 (1964).
(35) Shapiro S.M. et al., Sol. State Comm. 15, 377 (1974).
(36) Garel A.T., Thèse 3ème Cycle Orsay (1976).
(37) Bak P. and Emery V.J., Phys. Rev. Lett. 36, 978 (1976).
(38) Bjelis A. and Barisic S., Phys. Rev. Lett. 37, 1517 (1976).
(39) Frenkel J. and Kontorova T., Phys. Z. Sowjetunion 13, 1 (1938) ; Fiz. Zh. 1, 137 (1939).
(40) Frank F.C. and van der Merwe J.H., Proc. Roy. Soc. A 200,125 (1950).
(41) Ying S.C., Phys. Rev. B 3, 4160 (1971).
(42) Barisic S., Bjelis A. and Saub K., Sol. State Comm. 13, 1119 (1973) ; Horowitz B., Weger M. and Gutfreund H., Phys. Rev. B 9, 1246 (1974).
(43) de Gennes P.G., J. Phys. Rad. 23, 630 (1962).
(44) Lee P. A., Rice T.M. and Anderson P.W., Phys. Rev. Lett. 31, 462 (1973).
(45) Cyrot-Lackmann F., Adv. Phys. 16, 393 (1967).
(46) Leman G., Annales de Physique 18, 1 (1963).
(47) Thorpe M.F. and Weaire D., Phys. Rev. B 4, 3518 (1971).
(48) Friedel J. and Lannoo M., J. Physique 34, 115, 483 (1973).
(49) Ducastelle F., Thèse Orsay (1972) ; cf Ducastelle F. and Cyrot-Lackmann F., J. Phys. Chem. Sol. 131, 1295 (1971) ; 132, 285 (1972).
(50) Gaspard J.P. and Cyrot Lackmann, J. Phys. C. : Sol. St. Phys. 6, 3077 (1973).
(51) Haydock R. et al, J. Phys. C. Sol.St. Phys. 5, 2845 (1972) ; Desjonquères M.C. and Cyrot Lackmann F., J. Phys. F. Metal Physics 5, 1368 (1975).
(52) Hodges C. H., J. Physique Lettres 38, L187 (1977); Pottier N. Thèse Paris (1976).
(53) Blandin A., Phase Stability in Metals and Alloys (Ed. Rudman JS) Mc Graw Hill (1967).
(54) Harrison W.A., Pseudopotentials in the theory of metals, Benjamin, New York (1966).
(55) Corless G.K. and March N., Phil. Mag. 6, 1285 (1961).
(56) Blandin A. and Déplanté J.L., J. Phys. Rad. 23, 609 (1962) ; Blandin A., Déplanté J.L. and Friedel J., J. Phys. Soc. Japan 18, 1; (1963) ; Déplanté J.L., J. Physique 28, 465 (1967).
(57) Friedel J., Physics of Metals I Electrons (Ed. Ziman J.M.) Cambridge University Press (1969).

(58) Friedel J., J. Phys. Rad. 16, 829 (1955).
(59) Friedel J. and Sayers C.M., J. Physique Lettres (1977) under press.

NEUTRON SCATTERING STUDIES OF ELECTRON-PHONON INTERACTIONS

J.D. Axe

Institute Laue-Langevin, Grenoble, France, and

Brookhaven National Laboratory, Upton, N.Y., U.S.A.

This review is an attempt to summarize the areas in which neutron scattering has been used in studying electron-phonon interactions, and to display some of the key results. The examples chosen reflect the interests of the author and represent in no way a complete survey of the subject.

I. PHONON DISPERSION IN METALS

The most obvious and direct way in which electron phonon interaction manifests itself is in the phonon dispersion of metals. When phonon dispersion curves of simple metals are analyzed by Born-von Karman theory typically force constants between fifth nearest or even more distant neighbors are needed.[1,2] Furthermore the magnitude of the successive force constants is often oscillatory,[2,3] reflecting the rather long ranged oscillatory character of electronic screening. Born-von Karman models are both unweildy and unphysical, and it is now generally recognized that it is more satisfactory to formulate models in which conduction electron -phonon interactions are explicitly dealt with.

In the harmonic approximation phonon frequencies and eigenvectors are obtained by diagonalizing a dynamical matrix $\underset{\sim}{D}(\vec{q})$. For our purposes a sufficiently general form is[4]

$$D_{xy}(\vec{q}) = \frac{1}{v} \sum_{G} \{(\vec{G}+\vec{q})_x \phi(\vec{G}+\vec{q})(\vec{G}+\vec{q})_y - \vec{G}_x \phi(\vec{G})\vec{G}_y\} \quad (1)$$

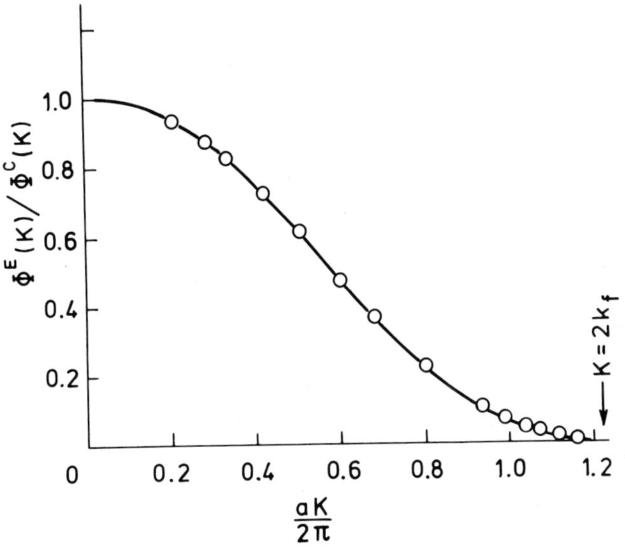

Fig. 1 The effective electron screening potential for sodium metal. The solid curve is deduced from phonon dispersion measurements (Ref. 5). The crosses represent calculations by Toya (Ref. 6).

where $\phi(\vec{K})$ is the Fourier transform of an appropriate ion pair potential, v is the volume of a unit cell and the sum is over all reciprocal vectors $\vec{G}$. For metals it is convenient to write the potential as the sum of three terms,

$$\phi(\vec{K}) = \phi^{C}(\vec{K}) + \phi^{R}(\vec{K}) + \phi^{E}(\vec{K}) \tag{2}$$

representing a) Coulomb interactions between ions, b) core repulsion of ions and c) ion-electron-ion interactions, respectively.

Figure 1 shows a comparison of the potential for sodium derived by Cochran[5] from experimental phonon dispersion curves,[1] compared with an early calculation by Toya[6] which attempted to deal in a fundamental way with the conduction electrons. $\Phi^{E}(K)$ is normalized by the coulomb term $\Phi^{C}(K) = 4\pi e^2/K^2$. Although the agreement is remarkably good, the further comparison of the measured and calculated dispersion curves, while impressive, was not completely satisfactory. This is the result of the near cancellation of the ϕ^C and ϕ^E terms in eq. (2). (ϕ^R is nearly negligible in sodium.) As a result phonon dispersion in metals generally provides a very stringent test of our understanding both of electron-ion potentials and of many body screening effects within the electron gas itself. It is fair to conclude that much of the stimulus for developments in pseudopotential theories[7] over the last decade came with the availability of reliable inelastic neutron scattering measurements of phonons in metals.[8,9]

The additional comment that must be made concerning Figure 1 is that it is not in general possible to uniquely deduce interionic potentials from phonon dispersion curves. There are two distinct difficulties. The first is that both the phonon frequencies and eigenvectors are needed to reconstruct the dynamical matrix $\underset{\sim}{D}(\vec{q})$. Although the eigenvectors can in very favorable cases be deduced from inelastic scattering intensities, in simple structures the problem is best resolved by measuring along those directions in reciprocal space where the eigenvectors are fixed by symmetry. The second problem is that the structure of eq. (1), which in principle involves reciprocal lattice vectors $\vec{G}$ of arbitrarily large value, is such that no unique value of $\phi(\vec{q})$ can be deduced from a knowledge of $D(\vec{q})$. This reflects the fact that in a real space formulation $\underset{\sim}{D}(\vec{q})$ depends only upon $\partial\phi/\partial r$ and $\partial^2\phi/\partial r^2$ evaluated at distances of interatomic separation. Cochran was able to derive $\phi^E(K)$ shown in Figure 1 only by assuming physically plausible constraints on its behavior.

II. KOHN SINGULARITIES

Kohn singularities arise because of the abrupt changes in electronic screening which occur when the phonon wavevector $\vec{q}$, spans the Fermi surface of a metal.[10] The effect is most simply discussed in the case where the electron wavefunctions are sufficiently planewave-like over most of the unit cell to allow $\phi^E(K)$ to be approximately factorized in the form

$$\phi^E(K) = -|v(K)|^2\chi^0(K)/\{1 + v_e(K)\chi^0(K)\} \tag{3}$$

Here v(K) is the effective ion-electron potential and $v_e(K)$ is the electron coulomb plus exchange potential.[11] $\chi^0(K)$ is the familiar one electron susceptibility

$$\chi^0(K) = \sum_K \frac{f_k - f_{k+K}}{\varepsilon_{k+K} - \varepsilon_k} \tag{4}$$

The magnitude of the Kohn singularity depends greatly on how well pieces of Fermi surface separated by the wavevector $\vec{K}$ are matched. For simple ellipsoidal surfaces where the matching is poor $\chi^0(K)$ is regular, but there is a logarithmic singularity in the derivative $\partial\chi^0(K)/\partial K$.[12] Brockhouse, et. al.[13] were the first to find phonon anomalies with these expected properties in a study of lead. Lead is favorable because of the large electron-ion potential, v(K). In most other simple metals the effects are too small to be obvious by direct inspection of the dispersion curves. Nevertheless, Stedman and coworkers, by making very careful measurements (~0.2% precision) and by examining $\Delta\omega/\Delta q$ have identified a large number of other anomalies in Al[14] and Cu[15] as well as Pb[16]. (It is important to realize that while the first moment of the line shape can be determined with such precision, the lines typically have a width ~5% due to instrumental resolution.) A substantial fraction of the anomalies have been assigned to known features of the Fermi surface with reasonable certainty. Ng and Brockhouse[17] have followed the changes in the size of the Fermi surface that occurs when Pb is alloyed with Tl.

In the event that a substantial portion of the Fermi surface "nests" into a matching portion displaced by a wavevector $\vec{K}$, the singularity in $\chi^0(K)$ becomes stronger. In the limit of perfect nesting there is a logarithmic singularity in $\chi^0(K)$ rather than in $\partial\chi^0/\partial K$.[12] When the nesting is less than mathematically perfect it is still possible to have strong cusp-like singularities in $\chi^0(K)$ itself.[12,15] Such cusps are seen in the phonon dispersion in Cr,[18] Mo,[19] and W,[20] and are believed to result from nesting of electron and hole pockets in the rather complex Fermi surfaces of these materials. A particularly strong example occurs in Cr, as shown in Figure 2. Nb and Nb-Mo alloys also have suspected Kohn anomalies at wavevectors that can be reasonably correlated with a rigid band model of the Fermi surface.[21,22] While it is often possible to find qualitative correlations, it is a common observation that anomalies predicted by simple considerations of Fermi surface topology are in some cases too weak to be observed and in other cases relatively strong. There has been little effort to understand the shape and strength of the anomalies in a quantitative way.[18]

It is a very striking fact that many high T_c superconductors exhibit rather broad anomalous dips in their dispersion relations which are not seen in their neighboring low T_c counterparts.[10] For example, (Figure 3) Nb has such features which are not seen in Mo, and similiar relationships are observed in the V,Cr and Ta,W

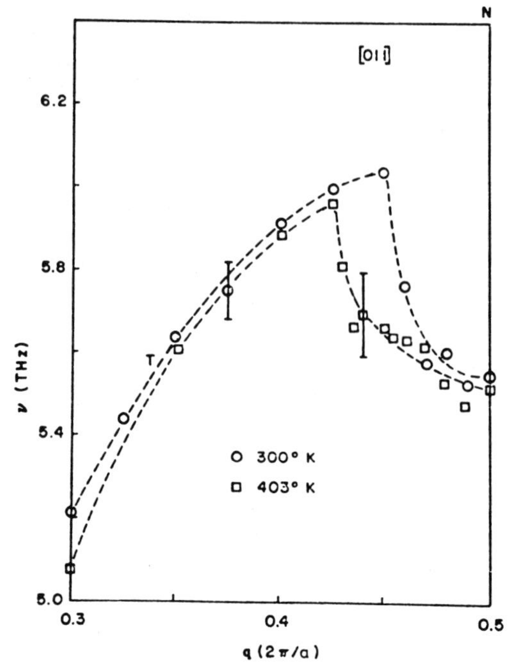

Fig. 2 A portion of the phonon dispersion of chromium near the N symmetry point at two temperatures. (Ref. 19).

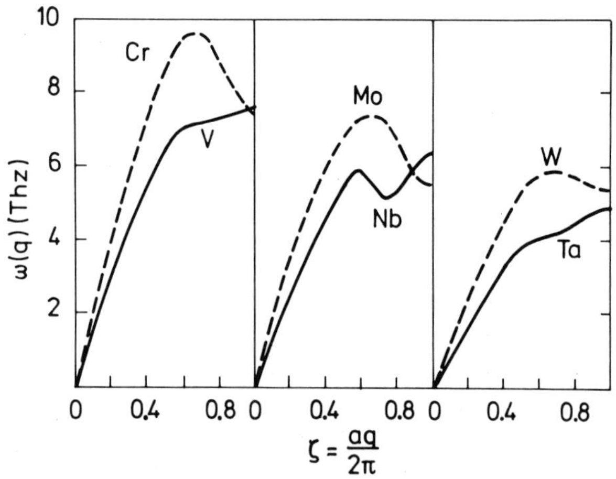

Fig. 3 Longitudinal acoustic phonons in Group V and VI transition metals. (After ref. 23).

pairs. Similar features have been pointed out by Smith and his coworkers[11] for the transition metal carbides. These features are too broad to be Kohn anomalies and furthermore, as has been shown for the Nb-Mo system, do not scale with the size of the Fermi surface. Sinha and Harmon[12] have proposed a model in which collective charge fluctions within the localized d-states softens the lattice response. They suggest that a large density of d-states at the Fermi surface is required. It is not clear whether the correlation with high T_c superconductivity, which undoubtedly exists, is or is not a causal one.

III. NEUTRON SPECTROSCOPY OF SUPERCONDUCTORS

Thus far we have been concerned with the electronic screening effects of the phonon frequencies. These can be adequately described in the adiabatic approximation. In order to discuss the influence of the electrons on phonon linewidths, it is necessary to consider the damping due to excitation of electron-hole pairs. This is accomplished by replacing the static susceptibility $\chi^0(K)$ in eq. (3) by

$$\chi^0(K,\omega + i\eta) = \sum_k \frac{f_k - f_{k+K}}{\varepsilon_{k+K} - \varepsilon_k + \omega + i\eta} \tag{5}$$

The electronic damping is introduced via Im $\chi^0(K,\omega)$, and is in most cases small enough to be completely masked by phonon-phonon scattering. In a neutron scattering experiment this in turn is usually masked by instrumental resolution!

Nevertheless in strong coupling superconductors in the vicinity of T_c there are abrupt changes in electronic damping which are sufficiently strong to be studied by neutron scattering.[26,27] This behavior arises because phonons with energy less than that of the temperature-dependent superconducting energy gap, $2\Delta(T)$, are energitically incapable of decaying by excitation of electron-hole quasiparticle pairs.

Although the theory of this effect dates from the early BCS period[28], the effect was first seen in neutron scattering some ten years later in Nb_3Sn.[26] Recently more refined measurements have been performed in Nb.[27] Figure 4 summarises some of these latter measurements. When $2\Delta(T)$ equals the phonon energy, $\hbar\omega_q$, an abrupt change occurs in the linewidth. Certain qualitative features, such as the displacement of the curves to lower temperature with increasing phonon energy, are obvious from the sketch included in

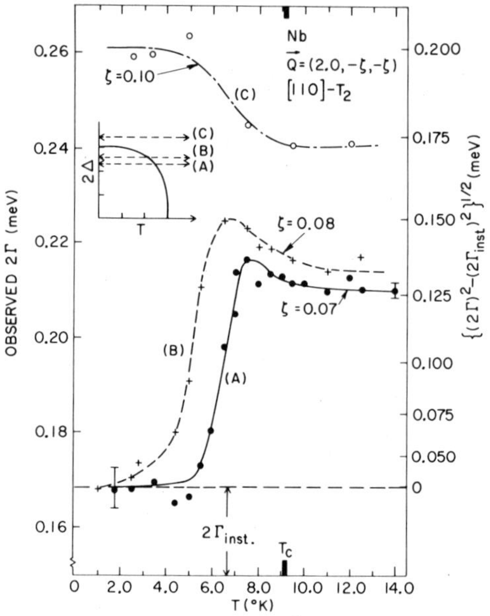

Fig. 4. Temperature dependence of several $[\zeta\zeta 0]T_2$ phonons in Nb, showing the change in width due to the superconducting gap.

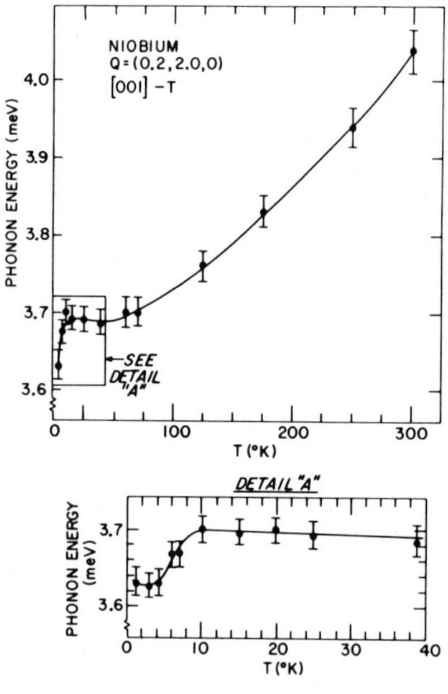

Fig. 5. Temperature dependence of $[00\zeta]T$ phonon frequency in Nb.

the Figure. The rounding of the discontinuity can be partly accounted for by resolution effects. All three sets of data in Figure 4 show that when $\hbar\omega_q$ slightly exceeds $2\Delta(T)$ the phonon linewidths are greater than their values far above T_c. This effect, which is due to an increased density of electron states at the gap energy, is in qualitative agreement with theory. Since the real and imaginary parts of $\chi^0(K,\omega)$ are related by Kramers-Kronig relations, we expect[29] and find anomalies in the phonon frequencies in the vicinity of T_c as well (see Figure 5).

Measurements of this type are of course of interest because they provide an alternate means for direct determination of the temperature dependence and anisotropy of the gap energy. In addition, they measure that part of the phonon linewidth, γ_{ep}, which is due to electron-phonon interaction. Allen[30] has pointed out that it is very closely related to quantities of interest in strong coupling superconductivity by deriving a simple explicit relation between γ_{ep} and the electron-phonon spectral function $\alpha^2F(\omega)$.

Obviously neutron scattering measurements of this sort are successful only if the electron-phonon interaction is sufficiently strong that the quenching of the interaction when $2\Delta(T) \geq \hbar\omega_q$ produces a measureable effect. Given presently available spectrometer resolution, the technique is unfortunately restricted to a small handful of strong coupled superconductors.

IV. MAGNETIC FIELD EFFECTS

In the preceding section we saw how the presence of an energy gap in the conduction electrons can be manifested in the phonon spectrum. Another way of introducing energy gaps in the conduction electrons is by application of an external magnetic field, and under suitable conditions this too may produce interesting effects in the phonon spectrum.

When a magnetic field is applied to a metal the energies of the conduction electrons are quantized into a series of Landau levels. In momentum space this quantization is represented by the condensation of the electron energy states into a series of tubes, each having a constant cross section in a plane perpendicular to the applied field, $\vec{H}$, as shown in Figure 6. The cross sectional area of each tube is proportional to $\vec{H}$. The energy of an electron lying on the n'th tube is

$$\varepsilon_n(k_z) = (n + \tfrac{1}{2})\hbar\omega_c + \frac{\hbar^2 k_z^2}{2m_{||}} \tag{6}$$

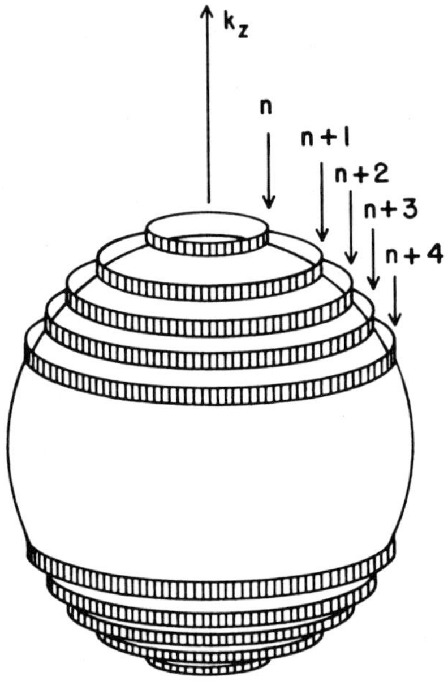

Fig. 6 Free electron Fermi surface and tubes onto which electron states condense in a magnetic field along k_z. (Ref. 34).

where k_z is the component of the electron wavevector parallel to $\vec{H}$, $m_{||}$ is an effective mass and ω_i is the cyclotron frequency.

In the free electron case, as discussed by Cowley,[31] and by Sham,[32] there are two distinct kinds of effects depending upon whether the phonon propagation vector, is parallel or perpendicular to $\vec{H}$. For $\vec{q} \parallel \vec{H}$ there are Kohn-like singularities in $\chi^0(q,\omega,H)$ whenever q equals the length, $2k_n$, of the portion of the n'th tube that lies within the Fermi surface.[33] However even for 100 KG there are $\sim 10^3$ Landau levels below the Fermi energy, so that the tubes are very closely spaced relative to the available momentum resolution of a neutron spectrometer and the strong field induced singularities are greatly smoothed out.

When $\vec{q} \perp \vec{H}$, $\chi^0(q,\omega,H)$ has a different structure[33] reflecting the fact that phonons can now scatter electrons from the n'th to the (n + p)'th tube, subject to the conditions

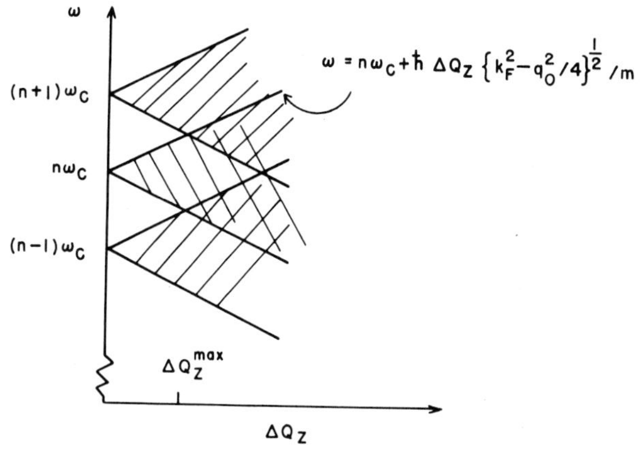

Fig. 7 Im $\chi(Q,\omega)$ is finite in the shaded area with a magnetic field perpendicular to the phonon propagation. (Ref. 34).

$$\varepsilon_{n+p}(k_z) - \varepsilon_n(k_z) \equiv p\hbar\omega_c = \hbar\omega_q \tag{7a}$$

and

$$k_{n+p} - k_n = q \tag{7b}$$

where ω_q and q are the frequency and wavevector of the phonon and k_n is the component of the electron wavevector $\perp$H for the n'th tube. This is just the cyclotron resonance condition, and we consequently expect the electron-phonon interaction to contribute to the lifetimes of only those phonons whose frequencies are integral multiples of the cyclotron frequency.

A more detailed consideration of the possibility of observing the effect of a transverse magnetic field on phonon lifetimes by conventional neutron spectroscopy has been given by Pynn and Axe.[34] Figure 7 shows the effect of a small component of phonon wavevector, ΔQ_z, parallel to $\vec{H}$ on the energy quantization of electrons. $\omega_c = (e\hbar/mc) \sim 1$ meV for H = 100 kG, which is well within the capability of a neutron spectrometer to resolve. The major difficulty that occurs is that typical spectrometer momentum resolution extends well beyond the value ΔQ_z^{max} in this Figure, at least in the case of free electrons. No anomalous effects can then be observed as the smeared value of Im $\chi^0(q,\omega)$ is the same with and without the applied field.

Pynn and Axe suggested that only if there were flat sections of Fermi surface which could be aligned parallel to $\vec{H}$ could the momentum resolution be sufficiently relaxed to make a neutron scattering experiment feasible. They tested these ideas by measuring the effect of a 50 kG field on phonons near the Λ_1 Kohn anomaly in Nb.[25] Instead of measuring the width of the phonons directly, they monitored the peak intensity as a function of $\vec{H}$ and found a small oscillatory component with a period consistent with the cyclotron mass deduced from deHaas-van Alphen measurements.

It is clear that neutrons are potentially very useful to investigate Fermi surfaces, both through Kohn anomalies and cyclotron resonance effects. A distinct advantage, in principle, is that these studies could be extended to impure metals and alloys, which are difficult to study by conventional methods. Similar remarks pertain to the potential of neutron spectroscopy of superconductors. It is fair to conclude, however, that substantial technical improvements will be necessary to make these techniques broadly useful. Order of magnitude increases in reactor fluxes might go a long way toward affecting the necessary resolution, but this is not a likely short term prospect. Unconventional high resolution spectrometers exist,[35,36] but have not as yet been adapted to phonon spectroscopy. It is sobering to recognize that we will often require simultaneous improvements in energy and momentum resolution.

V. CHARGE DENSITY WAVE INSTABILITIES

The charge density wave (CDW) state occurs as the result of a Fermi surface instability, which in the absence of electron phonon coupling would be manifested in a divergent susceptibility, $\chi^0(q_{crit}) \to \infty$, at some critical wavevector. The actual instability is a coupled mode which causes a simultaneous modulation of the electron density as well as a distortion of the lattice, i.e. a structural phase transformation. The neutrons couple to the nuclear distortions only. As might be supposed, simple theories predict that CDW phase transformations are accompanied by a soft phonon mode whose frequency is driven to zero by a "giant" Kohn anomaly.[37] The possibility of such an instability is greatly enhanced in lower dimensional systems because of the possibilities for favorable Fermi surface nesting.

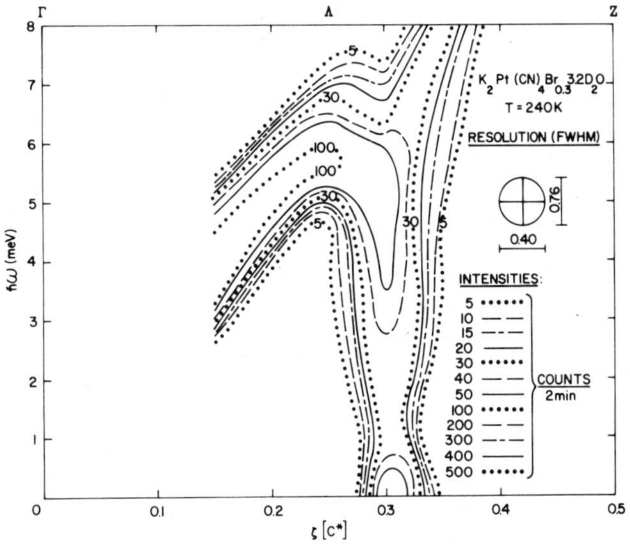

Fig. 8. Inelastic scattering intensity. Contours for KCP showing sharp Kohn anomaly in the acoustic branch near $\zeta \approx 0.3c^*$ (from Ref. 11).

In the past few years very spectacular examples of giant Kohn anomalies have been studied with neutrons in the quasi one-dimensional conductors KCP[38-40] and TTF-TCNQ.[41-42] As is demonstrated in Figure 8 the anomaly in KCP is extremely sharp. It occurs whenever $q_{||}$, the component of the phonon wavevector along the one-dimensional axis, equals $2k_F$. The strength of the anomaly is nearly independent of components of momentum perpendicular to the one-dimensional axis. There is a large quasi-elastic central peak revealing the presence of long lived short ranged correlations over a wide range of temperatures, but no actual transition temperature can be defined. It is possible that impurity pinning rather than the effect of one-dimensional fluctuations is responsible for the lack of long range order.[43]

The layered d^1-metal compounds NbX_2 and TaX_2 (X = S, Se, or Te) show a variety of structural transformations which are ralated to Fermi surface instabilities and CDW formation. Inelastic neutron scattering studies of 2H-$NbSe_2$ and $TaSe_2$ show large Kohn-like anomalies in the LA phonons at wavevectors for which Bragg sattelite peaks occur at the onset of the CDW state.[44,45] However, the softening of the phonon is incomplete near T_o, the divergent behavior occuring instead in a quasi-elastic central peak as $T \rightarrow T_o$. The Fermi surface geometry is rather complex in these materials and alternative models of Fermi surface nesting have been proposed.

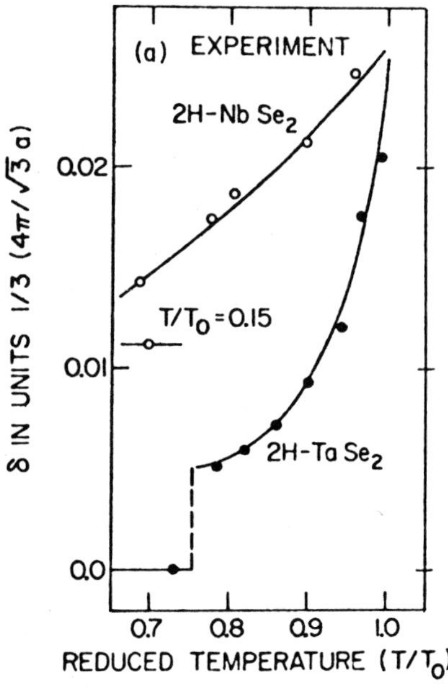

Fig. 9 Temperature dependence of the incommensurate wavevector $\vec{q}_0 = (1-\delta)a^*/3$ in $TaSe_2$ and $NbSe_2$. A "lock-in" transformation (to $\zeta = 0$) occurs in $TaSe_2$ but not $NbSe_2$.

These systems have been extensively reviewed by Wilson et al.[46]

One curious feature of CDW transformations is that the wavevector of the modulation, q_{crit}, is in general not an integral submultiple of a reciprocal lattice vector of the undistorted parent structure. The resulting structures are termed "incommensurate", and since they lack translational periodicity, they are not strictly speaking crystalline phases. However, the periodic potential of the underlying lattice causes non-sinusoidal distortions of the condensed planewave displacements and may lead to subsequent transformations which "lock-in" the period of the displacements with that of the lattice. These effects show up clearly in the neutron scattering results[45] on the layered chalcogenides shown in Figure 9. The most striking feature is the abrupt change of the satellite wavevector from $(1-\delta)a_1^*/3$ to the commensurate value $a_1^*/3$ which occurs at $T \sim 0.76\ T_0$. $NbSe_2$ does not achieve the commensurate state even at the lowest attainable temperatures, but as in $TaSe_2$ the satellite wavevector shows a pronounced temperature dependence, whose origin is closely related to the lock-in phenomenon itself. Moncton et. al.,[45] using a phenomenological Landau theory, showed that this behavior could be understood by allowing additional secondary distortions with wavevectors chosen to take advantage of the periodic lattice potential. They were also able to directly verify the non-sinusoidal nature of the incommensurate state by observing secondary Bragg satellites at the postulated wavevectors. McMillan[47] and Bak and Emery[48] have recently given more detailed discussions of the nature of the incommensurate ground state.

References

1. A.D.B. Woods, B.N. Brockhouse, R.H. March, A.T. Stewart and R. Bowers, Phys. Tev. 128, 1112 (1962).
2. B.N. Brockhouse, T. Arase, G. Caglioti, K.R. Rao and A.D.B. Woods, Phys. Rev. 128, 1099 (1962).
3. S.H. Koenig, Phys. Rev. 135A, 1693 (1964).
4. See, for example, W. Cochran and R. Cowley, Handbuch der Physik, Vol XXV/2a, Springer (1967) p. 83 f.f. More generally $\underset{\sim}{D}(q)$ is not diagonal in G, but in simple metals the off -diagonal terms are thought to be small. (L. Sham, Proc. Roy. Soc. A283, 33 (1965)). For simplicity of notation we consider only one atom/cell.
5. W. Cochran, Proc. Roy. Soc. A276, 308 (1963).
6. T. Toya, J. Res. Inst. Catalysis, 6, 161 (1958).
7. See, for example, articles by Heine et al. in "Solid State Physics", Vol. 24, Academic Press (1970).
8. A useful review of the lattice dynamics of metals has been given by S.K. Joshi and A.K. Rajagopal in "Solid State Physics", Vol. 22, Academic Press (1968) p. 159-312.
9. A complete bibliography of neutron scattering results is contained in "Bibliography for thermal Neutron Scattering", 5'th ed., M. Sakamoto, J. Chihara, Y. Nakahara, H. Kadotani, T. Sekiya and Y. Gotah, Japanese Atomic Research Institute, Tokai-mura, Japan.
10. W. Kohn, Phys. Rev. Letters 2, 393 (1959).
11. Perhaps the simplest but by no means only way to arrive at eq. (3) is by pseudopotential methods (see Ref. 8).
12. See L.M. Roth, H.J. Zeiger and T.A. Kaplan, Phys. Rev. 149, 519 (1966).
13. B.N. Brockhouse, T. Arase, G. Caglioti, K.R. Rao and A.D.B. Woods, Phys. Rev. 128, 1099 (1962).
14. J.W. Weymouth and R. Stedman, Phys. Rev. B2, 4743 (1970).
15. G. Nilsson, "Neutron Inelastic Scattering" (International Atomic Energy Agency, Vienna, 1968) p. 187.
16. R. Stedman, L. Almquist, G. Nilsson and G. Raunio, Phys. Rev. 163, 567 (1967).

17. S.C. Ng and B.N. Brockhouse, "Neutron Inelastic Scattering" (International Atomic Energy Agency, Vienna, 1968) p. 253.
18. T.M. Rice and B.I. Halperin, Phys. Rev. B1, 509 (1970).
19. M.W. Shaw and L.D. Muhlstein, Phys. Rev. B4, 969 (1971); L.D. Muhlstein, E. Gurmen and R.M. Cunningham, "Inelastic Neutron Scattering" (International Atomic Energy Agency, Vienna, 1972) p. 53.
20. S.H. Chen and B.N. Brockhouse, Solid State Commun. 2, 73 (1964). S.H. Chen, Ph.D. Thesis, McMaster Univ. 1964 (unpublished).
21. B.M. Powell, P. Martel and A.D.B. Woods, Phys. Rev. 171, 727 (1968); C.B. Walker and P.A. Egelstaff, ibid 177, 1111 (1969).
22. R.I. Sharpe, J. Phys. C 2, 432 (1969).
23. H.G. Smith, N. Wakabayashi and M. Mostoller in "Superconductivity in d- and f-band metals", D.H. Douglass, ed., Plenum, New York 1976, p. 223.
24. H.G. Smith and W. Gläser, Phys. Rev. Lett. 25, 1611 (1970).
25. S.K. Sinha and B.N. Harmon, Phys. Rev. Lett. 15, 1515 (1975).
26. J.D. Axe and G. Shirane, Phys. Rev. Lett. 30, 214 (1973); Phys. Rev. B8, 1965 (1973).
27. G. Shirane, J.D. Axe and S.M. Shapiro, Solid State Comm. 13, 1893 (1973); S.M. Shapiro, G. Shirane and J.D. Axe, Phys. Rev. B12, 4899 (1975).
28. A. Privorotskii, Sov. Phys. - JETP 16, 945 (1963); V.M. Bobetic, Phys. Rev. A1535 (1964).
29. H.G. Schuster, Solid State Comm. 13, 1559 (1973).
30. P.B. Allen, Phys. Rev. B6, 2577 (1972).
31. R. Cowley, "Neutron Inelastic Scattering" (International Atomic Energy Agency, Vienna, 1968) p. 141.
32. L. Sham, "Modern Solid State Physics", Vol. 2, R.H. Enns and R.R. Haering, ed., Gordon and Breach, New York, 1969, p. 143.
33. A. Ya Blank and E.A. Kaner, Sov. Phys. - JETP 23, 673 (1966).
34. R. Pynn and J.D. Axe, J. Phys. F. 4, 1898 (1974).
35. M. Birr, A. Heidemann and B. Alefield, Nucl. Inst. Methods 95, 435 (1971).
36. F. Mezei, Z. Physik 225, 146 (1972).
37. S.K. Chan and V. Heine, J. Phys. F 3, 795 (1973).
38. B. Renker, H. Rietsch, L. Pintschovius, W. Gläser, P. Brüesch, D. Kuse and M.J. Rice, Phys. Rev. Letters 30, 1144 (1973).
39. K. Carneiro, G. Shirane, S.A. Werner and S. Kaiser, Phys. Rev. B13, 4258 (1976).
40. R. Comes, B. Renker, L. Pintschovius, R. Currat, W. Gläser and G. Schreiber, phys. stat. solidi b71, 171 (1975).
41. H.A. Mook and C.R. Watson, Jr., Phys. Rev. Lett. 36, 801 (1976).
42. G. Shirane, S.M. Shapiro, R. Comes, A.F. Garito and A.J. Heeger, Phys. Rev. B14, 2325 (1976).
43. L.J. Sham and B.R. Patton, "One-Dimensional Conductors", H.G. Schuster, ed., Springer-Verlag, New York, 1975.

44. N. Wakabayashi, H.G. Smith and R. Shanks, Phys. Lett. 50A, 367 (1974).
45. D.E. Moncton, J.D. Axe and F.J. DiSalvo, Phys. Rev. Lett. 34, 734 (1975).
46. J.A. Wilson, F.J. DiSalvo and S. Mahajan, Adv. in Phys. 24, 117 (1975).
47. W.L. McMillan, Physical Review B14, 1496 (1976).
48. P. Bak and V.J. Emery, Phys. Rev. Lett. 36, 978 (1976).

PHASE TRANSITIONS IN QUASI ONE-DIMENSIONAL METALS (TTF-TCNQ AND KCP)

Per Bak

NORDITA, Blegdamsvej 17, Copenhagen, Denmark

1. INTRODUCTION

Quasi one-dimensional conductors are characterized by their very anisotropic conductivity which takes place almost entirely along one particular direction. Most of these systems are organic chain systems. The best one-dimensional organic conductor known is tetrathiafulvalene-tetracyanoquinodimenthane (TTF-TCNQ). The present activity in this field started by the observation of Coleman, Cohen, Sandman, Yamagishi, Garito and Heeger[1] that samples of TTF-TCNQ exhibit an anomalous peak in conductivity at T≃60K but at lower temperatures the conductivity drops sharply and the material becomes an insulator. Although it was widely believed that this behaviour was related to a phase transition it took rather long time before such a transition was actually observed.

Of particular interest in this respect is the coupling between the electron system and the lattice which may lead to at least two different types of phase transformations: either into a superconducting state or into a Peierls state[2] characterized by a static charge density wave (CDW) and an accompanying periodic lattice distortion. Careful theoretical estimates, however, indicate that for an almost one-dimensional system the dielectric Peierls transition will occur before the superconducting phase transition may take place. The physical mechanism behind the Peierls transition is the decay of phonons into electron-hole pair. Because of energy and momentum conservation the phonon wavevector must connect two points on the Fermi surface. The interaction creates a dip in the phonon energy - the Kohn anomaly. For this process to be important, there should be a large density of states at points on the Fermi surface separated by the phonon wavevector. Clearly this condition is fulfilled for one-dimensional, or nearly one-dimensional systems where the entire Fermi surface is two planes separated by $2k_F$. At lower temperatures it may become energetically favourable to have static phonons at this wavevector - the Peierls transition takes place. Energy gaps are formed at the Fermi surface, and the conductivity decreases.

For the theorist, one-dimensional systems are of interest in themselves, since they may exhibit many characteristic and unusual properties, and because one-dimensional models are more amenable to exact theoretical calculations than the three-dimensional counterparts. The most important property of purely one-dimensional systems is that fluctuations prevent them from undergoing phase transitions at all. This leads to an apparant paradox since we have just argued that one-dimensionality is required for a Peierls distortion to take place. Clearly, three dimensional effects, however weak, are bound to be of crucial importance when dealing with "one-dimensional systems". The phase transformation in a real physical 3-d system takes place as a consequence of a subtle interplay between large one-dimensional fluctuations and weak interchain coupling. The transition will set in at a temperature which is significantly lower than the mean field transition temperature

$$T_0 = \varepsilon_F \exp\left(-\frac{\pi v}{g^2}\right) , \qquad 1.1)$$

where v and ε_F are the velocity and energy at the Fermi surface and g is the electron phonon coupling. It is this transition which destroys the possibility of having a very high conductivity in TTF-TCNQ at temperatures below 50K. The Kohn anomaly and the Peierls distortion may be seen by x-ray and neutron scattering experiments which provide the most direct evidence of the phase transition.

In addition to allowing the phase transition to take place, the three dimensional coupling has another important effect. If one ignores the periodic potential from the underlying atomic lattice ("Umklapp terms") the phase of the CDW is free to move along the 1-d direction. However, in the presence of interchain coupling the <u>relative</u> phases on different chains, as specified by $q_\perp$, the component of $\vec{q}$ perpendicular to the chain, are fixed. The value of $q_\perp$ is mainly determined by the Coulomb interaction which attracts regions of opposite excess charge. We shall see that in TTF-TCNQ $q_\perp$ varies with temperature to accomodate different ordering of the two types of chain. If there is coupling to the underlying lattice, the overall phase χ may be "pinned". This coupling can cause $q_\perp$ to lock at a commensurate value to take advantage of the extra free energy. This appears to happen in TTF-TCNQ[3]. If $\vec{q}$ has a component which is almost commensurate -either along or perpendicular to the chain- the CDW state may consist of large commensurate regions with constant χ separated by comparatively narrow regions -"solitons"- where χ changes rapidly.

As the temperature is lowered below T_0, the electrons interact over longer and longer distances, and the effective interaction between chains is enhanced. Ultimately, this brings out the true three dimensional phase transition. In section 2 the effects of weak interchain coupling are calculated by means of a systematic perturbation theory developed in ref. 5. The zeroth order term is not the mean field solution of the system as a whole, but an "exact" solution of the one-dimensional system. The expansion parameter is $\frac{1}{T}$ or, equivalently, $\frac{1}{Z}$ where Z is an effective number of interacting chains. In section 3 the Bak-Emery theory[3] of the phase transformations in TTF-TCNQ will be reviewed. An important feature is that the theory does not depend upon any microscopic model of the system, but makes use of the space group symmetries of the crystal above T_c and the observed symmetry of the ordered state below T_c. The perturbation theory presented in section 2 will provide a prescription of how actually to estimate the numerical coefficients entering the Landau-Ginzburg theory, in addition to providing physical insight into the nature of the phase-transition. We emphasize, however, that the theory of TTF-TCNQ is based upon an expansion of the free energy around the true phase transition, and not the transition temperature as given by any approximative theory. No attempt will be made to actually identifying the details of the microscopic mechanisms involved. The most striking consequence of the theory is the prediction of a new phase transition. This transition has now been confirmed by several experimental groups, and was found to occur at $T \simeq 48K$. This is one of the very rare cases that the existence of a phase transition has actually been predicted before it was observed.

Recently, charge density waves[6,7] and Kohn anomalies[8] with wavevector components $4k_F$ along the chains have been observed, and theoretical explanations have been proposed[9-12]. The possibility to form these waves can be included in the Landau expansion[3]. This does not give rise to additional phase transitions since the symmetry is not lowered further. In section 3.4 an alternative mechanism for the $4k_F$ modes will be suggested, and in section 3.5 the critical properties near the phase transitions are studied from a theoretical point of view.

No one-dimensional system is ideal. Three dimensional coupling is of crucial importance, but also imperfections or impurities are always present. In particular, random impurities are almost necessarily of importance in non-stoichiometric systems such as KCP ($K_2Pt(CN)_4Br_{0.3}$, $3H_2O$) where there is a non-integer average amount of Br atoms in the unit cell. The Br atoms are probably fully ionized and may act as a random potential on the CDW and in turn induce large fluctuations of the phase and possibly destroy long range order. It has been suggested that impurities are responsible for the absense of a CDW transition in KCP[13,14,5].

In section 4 the theory of Bak and Brazovsky[5] on the effects of impurities on quasi-one dimensional systems will be described. The field is assumed to interact linearly with the derivative of the phase of the CDW and not with the phase itself. In the case of a purely one-dimensional system the dynamical structure factor (which can be measured in a neutron scattering experiment) includes a central peak in addition to the phason branches which are also present for the pure system. When the interchain coupling is included, the actual phase transition is determined by an interplay between impurity effects which tend to destroy correlations and prevent long range order and 3-d effects which tend to enhance the susceptibility. It turns out that if the impurity correlation length becomes less than a certain threshold value, then the correlation length is finite at any temperature and no dielectric phase transition can take place. This may explain the saturation of the transverse correlation length in KCP at $T \simeq 100K$[13]. Of course there is the possibility that some other kind of ordering may occur. We suggest that there is a "spin-glass" like ordering where the phases of CDW's on different chains freeze in a random way relative to each other.

2. INTERCHAIN COUPLING

In this section we shall study the effects of weak interchain coupling following closely the systematic perturbation theory derived in ref. 5. A static CDW can be described by the periodic function

$$n(x) = \rho \cos(2k_F x + \chi) \qquad 2.1)$$

where ρ and χ are order parameters. For a one dimensional system the average value of $n(x)$, $\langle n(x) \rangle$, is zero, but its amplitude is sharply peaked around its "mean field" value. At temperatures in the region

$$T \ll T_0 \qquad 2.2)$$

only long wavelength fluctuations are essential. The fluctuations of the whole system are phase fluctuations (phasons). To include the effects of these excitations we shall allow χ to be a space and time dependent function, $\chi(x,t)$. To second order in the derivatives of χ, the phase fluctuations are described by the Lagrangian

$$L_0 = \frac{v}{8\pi} \iint dx d\tau \left(\frac{\dot{\chi}^2}{u^2} + \chi'^2 \right) \qquad 2.3)$$

where τ is the Matsubara time ($0<\tau<\beta$). This Lagrangian does not include terms dependent upon χ itself since the system is invariant to a uniform shift of χ. u is the phason velocity. Brazovsky and Dzyaloshinskij have calculated the dynamic structure factor $S(q,\omega)$ corresponding to L_0 [16,33]. If we are interested in the static integrated correlation function, $S(q)$, only we may restrict ourselves to considering time independent phase fluctuations, with energies

$$F = \frac{v}{8\pi} \int dx\, \chi'^2 \qquad 2.4)$$

Before introducing the interchain coupling, let us show how the phasons destroy long-range order in a one dimensional system. The static charge-charge correlation function, or structure factor, is

$$\begin{aligned} S(x) &= \langle n(0)n(x)\rangle \\ &= \rho^2 \langle \exp i\,\{\chi(0) - \chi(x)\}\rangle \cos 2k_F x \end{aligned} \qquad 2.5)$$

It is essentially this function which is measured in any diffraction experiment. In this paper we shall drop the trivial factor $\cos 2k_F x$, i.e. all momenta will be measured relative to $2k_F$. The correlation function can be calculated using Feynman integral approach:

$$S(x) = \frac{\int D\chi \exp i\,\{\chi(0) - \chi(x)\} \exp(-\beta F_0)}{\int D\chi \exp(-\beta F_0)} \qquad 2.6)$$

where $\int D\chi$ is the functional integral over all possible phase fluctuations. By introducing Fourier transforms χ_q we find

$$\begin{aligned} S(x) &= \frac{\int D\chi_q \exp \frac{1}{2\pi} \int dq\{(1-\exp iqx)i\chi_q - \frac{v\, q^2\chi_q^2}{8\pi T}\}}{\int D\chi_q \exp - \frac{1}{2\pi}\int dq \frac{v\, q^2\chi_q^2}{8\pi T}} \qquad 2.7) \\ &= \exp\; - \int dq \frac{2T\,(1 - \cos(qx))}{v\, q^2} \\ &= \exp\,(- \frac{|x|}{R}) \quad , \; R = \frac{v}{2\pi T} = \frac{1}{\gamma} \end{aligned}$$

We have used a standard formula for Gaussian functional integrals. Hence the correlation length is finite at any finite temperature and no long range order exists.

The three dimensional coupling will be taken to be of the form

$$F_1 = - \sum_{ij} \int dx\, K_{ij} \cos\,(\chi_i(x) - \chi_j(x)) \qquad 2.8)$$

The summation is over interacting chains. This term may represent the Coulomb interaction[17]. The interaction is local, i.e. the phase at position x at chain i interacts with the other chains at the same position only. The thermodynamic average of any quantity, O, is given by the functional integral

$$\langle O\rangle = \frac{\int D\chi\ O(\chi)\ \exp -\beta(F_0+F_1)}{\int D\chi\ \exp -(F_0+F_1)} \tag{2.9}$$

The exponentials in both the denominator and the nominator are now expanded in β:

$$\langle O\rangle = \langle O\rangle_0 - \beta(\langle OF_1\rangle_0 - \langle O\rangle_0 \langle F_1\rangle_0) + \frac{\beta^2}{2}(\langle OF_1^2\rangle_0 - \langle O\rangle_0 \langle F_1^2\rangle_0) + \ldots \tag{2.10}$$

where $\langle\ \rangle_0$ denotes the average with respect to F_0. These averages are all Gaussian functional integrals which can be evaluated. To calculate the structure factor we insert $O = \exp i\,\{\chi_1(0) - \chi_i(x)\}$:

$$\begin{aligned} S_{11}(x) &= \langle \exp i\,\{\chi_1(0) - \chi_1(x)\}\rangle_0 \\ &+ \beta^2 \sum_i K_{1i}^2 \iint dx'dx'' \langle \exp i\,\{\chi_1(0)-\chi_1(x)+\chi_1(x')-\chi_1(x'')\}\rangle_0 \\ &\qquad \times \langle \exp i\,\{\chi_i(x'')-\chi_i(x')\}\rangle_0 \\ &- \beta^2 \sum_i K_{1i}^2 \iint dx'dx'' \langle \exp i\,\{\chi_1(0)-\chi_1(x)\}\rangle_0 \\ &\qquad \times \langle \exp i\,\{\chi_1(x')-\chi_1(x'')\}\rangle_0 \end{aligned} \tag{2.11}$$

The averages are evaluated using the formula

$$\langle \exp i \sum_\nu n_\nu \chi(x_\nu) \rangle \Big|_{\sum_\nu n_\nu = 0} = \exp \tfrac{1}{2} \sum_{\nu\mu} n_\mu n_\nu D_0(x_\nu - x_\mu) \tag{2.12}$$

where n_ν = + or − and $D_0(x) = \frac{|x|}{R}$.

Introducing $K_{1i}^2 = \frac{1}{Z}(\sum_{ij} K_{ij})^2 = \frac{1}{Z}K(0)^2$ and performing the integrations we obtain

$$S_{11}(x) = \exp\left(-\frac{|x|}{R}\right) + \beta^2 \frac{R^2K(0)^2}{Z}\left[\frac{1}{16}\exp\left(-\frac{5|x|}{R}\right) - \exp\left(\frac{-|x|}{R}\right) + \left(\frac{|x|}{4R} + \frac{(x)^2}{2R^2}\right)\exp\left(-\frac{|x|}{R}\right)\right]. \tag{2.13}$$

Z is an effective number of interacting neighbours. The correlation between chains can be calculated in a similar way. The Fourier transform of the correlation function is

$$S(q) = \frac{2\gamma}{\gamma^2+q^2} - \frac{K(0)}{T}\times\left(\frac{2\gamma}{\gamma^2+q^2}\right)^2 + \frac{K^2(0)}{T^2}\times\left(\frac{2\gamma}{\gamma^2+q^2}\right)^3 - \frac{K(0)^2}{ZT^2}\left[\left(\frac{2\gamma}{\gamma^2+q^2}\right)^3 - \frac{R^2}{16}\left(\frac{10\gamma}{25\gamma^2+q^2} - \frac{2\gamma}{\gamma^2+q^2}\right) \frac{(2\gamma^2-12q^2)\gamma}{(\gamma^2+q^2)^3} + \frac{\gamma^2-q^2}{2(\gamma^2+q^2)^2\gamma}\right]. \tag{2.14}$$

where q is the wavevector component along the chain.

To first order in $\beta(=\frac{1}{T})$ this result is equivalent to the mean-field approximation. To second order in β the corrections to the mean-field theory are given by the last term of eq. 2.14). To find the transition temperature, let us find the temperature where the correlation function diverges, i.e. $1/S(0)$ vanishes:

$$\frac{1}{S(0)} \simeq \frac{\gamma}{2}\left(1 - \frac{2K(0)R}{T} + \frac{K(0)^2R^2}{ZT^2}\times\frac{14}{5}\right) \tag{2.15}$$

To first order in $\frac{1}{T}$,

$$T_c = 2K(0)R = \frac{K(0)v}{\pi T_c} \tag{2.16}$$

$$\text{or} \quad T_c^2 = \frac{K(0)v}{\pi} \tag{2.17}$$

This is in agreement with the result of Scalapino, Imry and Pincus[18], but in distinct disagreement with a theory of Lee, Rice and

Anderson[19] which in fact ignores phase fluctuations and leads to a T_c which is only weakly dependent upon K(o). To second order,

$$T_c^2 = \frac{K(o)v}{2\pi} \left[\sqrt{1 - \frac{14}{5Z}} + 1\right] \qquad 2.18)$$

Hence, as could be expected the transition temperature is lowered relative to the "mean-field" theory. The T_c and the susceptibility $\chi_q = 1/T\ S(q)$ as calculated here give an indication of the parameters entering phenomenological calculations, as for example the Bak-Emery theory of TTF-TCNQ[3] which is described in the next section.

3. LANDAU-GINZBURG THEORY OF STRUCTURAL PHASE TRANSFORMATIONS AND CHARGE DENSITY WAVES IN TTF-TCNQ

Since the discovery of the anomalous peak in the resistivity of TTF-TCNQ[1], there has been much speculation about the possibility of a structural "Peierls" transition. However, bulk measurements gave no definite information on this question, and it took surprisingly long time before this very fundamental question was solved by means of obvious diffraction techniques. The first direct evidence of a phase transition in TTF-TCNQ was X-ray diffraction photographs taken at Orsay[20], but recent neutron scattering experiments at Brookhaven have provided a much more detailed picture of the low temperature transitions[21,22]. In these experiments it appeared that there were two transitions, one at 38K, the other at 54K. At low temperatures there is a $4\vec{a} \times 3.4\vec{b} \times \vec{c}$ modulation of the lattice. As the temperature was raised above 38K the modulation period along $\vec{a}$ changes, abruptly at first, then more gradually until it reaches 2a near 51K. These experiments have been explained by Bak and Emery[3] in terms of a Ginzburg-Landau theory. The most striking consequence of this theory is that there should in fact be a third transition between those already observed. The proof of the new transition is based upon symmetry considerations only, whereas the nature of the underlying physical mechanisms plays no role at all for this purpose. The phase transition has now been observed by several experimental groups[23-25] and seems to take place at $T \simeq 48.5$K.

In this section the theory for the three phase transformations will be reviewed, and we shall also study the critical properties at the transitions. In addition, a new mechanism for the $2k_F$ anomaly recently observed will be proposed. Figure 1 indicates schematically the CDW's at the different phases.

3.1 The 54K Transition

The space group of TTF-TCNQ is the monoclinic group P $2_1/c$ with the b axis as the unique axis[26]. The structure consists of chains of TTF molecules and chains of TCNQ molecules arranged in sheets in the bc plane. The x-ray and neutron scattering experiments have shown a second order structural phase transformation at 54K. The ordered phase at this temperature is described by an incommensurate wavevector $\vec{q}_1 = (q_a, q_b, q_c) = (a^*/2, 0.295b^*, 0)$ where a^*, b^* and c are reciprocal lattice constants. Phrased in terms of a Peierls transition, the b^* component is twice the Fermi wavevector, $2k_F$.

According to the theory of Landau and Lifshitz[27] the order parameter describing a second order phase transition should transform as a basis of an irreducible representation of the high symmetry (high temperature) space group of the crystal. The representations are labelled by the wavevector describing the ordered unit cell and by the representation of the group of this wavevector. The wavevector is known from the neutron scattering experiments. The group of $\vec{q}_1$ consists simply of the two-fold axis, which has only two representations described by basis functions which transform into themselves or into minus themselves, respectively. When the order parameter is subject to such a screw axis transformation $(x, y, z) \to (-x, \frac{1}{2}+y, \frac{1}{2}-z)$ it should therefore transform either as

$$\psi_{\vec{q}_1} \to \exp(i\pi 2k_F)\psi_{\vec{q}_1} \qquad 3.1a)$$

$$\text{or} \qquad \psi_{\vec{q}_1} \to -\exp(i\pi 2k_F)\psi_{\vec{q}_1} \qquad 3.1b)$$

We note that the symmetry considerations tells us nothing about which intra-molecular modes are involved. For a detailed discussion of the possible order parameters, see ref. 3. An important point is that it is possible to use optical CDW's on the two chain systems as order parameters. The charge distributions associated with these modes are

$$\rho^i(\vec{r}) = \pm\, \psi^i_{\vec{q}_1} \exp(i\vec{q}_1\cdot\vec{r}),\ i = 1,2 \qquad 3.2)$$

where the + sign applies to one sublattice of type i molecules and the − sign to the other sublattice of type i molecules. Since Coulomb forces tend to favour opposite excess charges on neighbouring chains it is very likely that the optical modes are dominant. In general, however, we are of course free to use any order parameter (as for example the accompanying lattice distortion) which has the correct symmetry properties. The optical CDW's on the two chain systems transform as 3.1a) and 3.1b), respectively. Since the order parameter should transform as one of these basis functions, we may conclude that only one chain system developes an optical CDW at the 54K transition. It is in principle possible that there is an acoustic CDW on the other chain system which transform according to the same representation.

To find the temperature dependence of the order parameters, we expand the free energy close to the phase transitions in terms of the complex order parameters $\psi^1_{\vec{q}_1}$ and $\psi^2_{\vec{q}_1}$. This free energy should be real and invariant under translations, inversion and the two-fold screw-axis transformation, which are the symmetry elements of the high temperature phase. Since the two order parameters transform like two different representations, there are no cross terms of second order.

$$F = r_1|\psi^1_{\vec{q}_1}|^2 + b_1|\psi^1_{\vec{q}_1}|^4 + \ldots \qquad 3.3)$$

$$+ r_2|\psi^2_{\vec{q}_1}|^2 + b_2|\psi^2_{\vec{q}_1}|^4 + \ldots$$

r_1 and r_2 are effective inverse susceptibilities $1/\chi^1_{\vec{q}_1}$ and $1/\chi^2_{\vec{q}_1}$ of the two chain systems, respectively. These quantities may be estimated using the theory of section 2. We found that each set is expected to order at a finite temperature. The phase transition occur when the smallest of the parameters r_1 and r_2 passes through zero. Let us assume that it is r_1, linearize around T_1 = 54K, and minimize F for T not too far from T_1.

$$|\psi^1_{\vec{q}_1}|^2 = \begin{cases} \dfrac{a_1}{2b_1}(T_1 - T), & T < T_1 \\ 0\,, & T > T_1 \end{cases} \qquad 3.4)$$

and $|\psi^2_{\vec{q}_1}|^2$ remains zero at this transition. Here $r_1 = a_1(T-T_1)$. This describes the first phase transition in TTF-TCNQ. Experiments seem to indicate that it is mainly the TCNQ molecules that order[28,24].

3.2 The 47K Transition

Let us investigate the question as to when an optical transition is driven on the <u>second</u> type of chain, following closely the derivation of ref. 3. Since the single chain susceptibility diverges at T=0 on both types of chains, we expect this to happen at finite temperature. It will be shown that this transition is in fact related to the observed shift of the a* component of the wavevector describing the distortions. To study the possibility of forming a charge density wave with $q_a \neq a^*/2$ we expand the free energy in <u>both</u> $\psi^1_{\vec{q}}$ and $\psi^2_{\vec{q}}$. Of course, any possible state with periodicity described by $\vec{q}$ can be formed from linear combinations of these modes. To simplify the notation we introduce $\psi_q{}^i$ = $\psi^i(a^*/2+q,\ 2k_F,\ 0)$. We shall keep $\psi_q{}^1$ exactly since it may not remain small over the whole region of interest. To second order in $\psi_q{}^2$ we have

$$F(\psi_q^1, \psi_q^2, q) = f(|\psi_q^1|^2, q)$$

$$+ A(|\psi_q^1|^2, q)(\psi_q^2\psi_q^{1*} + \psi_q^1\psi_q^{2*}) \qquad 3.5)$$

$$+ B(|\psi_q^1|^2, q)\,|\psi_q^2|^2$$

When the twofold screw axis is applied to the order parameter, $\psi_q^2\psi_q^1 \rightarrow -\psi_{-q}^2\psi_{-q}^1$ and $|\psi_{-q}^i|^2 \rightarrow |\psi_{-q}^i|^2$. Therefore, f and B are even in q and A is odd in q.

We now expand F in powers of q in the neighbourhood of q = 0; and minimize with respect to $|\psi_q^2|$. We find

$$F(\psi_q^2, q) = aq|\psi_q^2| + b|\psi_q^2|^2 + cq^2 + \ldots \qquad 3.6)$$

In principle the coefficient c could vanish at some temperature below 54K indicating a q-deviation within the 1-system only. However, since there is always some coupling between the two systems, a phase transition involving the 2-system would already have taken place before then. Moreover, there is no obvious physical reason for a temperature dependence of c. Near T_1, the coefficients b and c are positive, otherwise the type 2 chains would already have ordered, contrary to assumption, or q = 0 would not be a minimum contrary to experiment. Now, F is minimized with respect to q

$$q = -(a/2c)\,|\psi_q^2| + o|\psi_q^2|^3 \qquad 3.7)$$

and, when this is substituted into 3.6) F becomes

$$F = (b - a^2/4c)|\psi_q^2|^2 + D|\psi_q^2|^4 \qquad 3.8)$$

to fourth order in $|\psi_q^2|$. Equation 3.7) shows how the moving wave-vector is associated with the development of order in type 2 chains. When D > 0, there is a second-order transition at the temperature T_2 for which $b - \frac{a^2}{c}$ vanishes. This is higher than the temperature at which b = 0, because c > 0. Thus the coupling helps to drive the transition of ψ_q^2. Minimizing F in eq. 3.8) with respect to ψ_q^2 gives

$$|\psi_q^2|^2 = -(b - \frac{a^2}{4c})/2D = \alpha(T_2 - T), \quad T < T_2 \qquad 3.9)$$

after expanding about T_2. From Eqs. 3.7) and 3.9) the variation of q is given by

$$q^2 = \begin{cases} 0 \quad , & T_1 < T < T_2 \\ a^2\alpha/4x^2(T_2-T), & T \quad T_2 \end{cases}$$

Hence, q_a stays at $a^*/2$ in a finite temperature interval below T_1, and the deviation of q_a away from this value should occur at a separate phase transition, the "47K transition". Since we first reported our result, several experiments have shown unequivocally the existence of the predicted transition. In a recent neutron scattering experiment, Ellenson et.al.[23] found the transition to occur at $T \simeq 48.5K$ (Fig. 2). Knight shift measurements seem to indicate that the transition is indeed associated with the TTF chains[24]. Specific heat measurements[25] show a clear anomaly at 48K. In the discussion of the 47K transition the existence of the underlying lattice has been ignored. In reality, the CDW is subject to a periodic potential from this lattice. Since $2k_F$ is not close to any simple rational value the effect of such "Umklapp" terms is probably small.

3.3 The 38K Transition

It now remains to understand the transition at 38K and the reason that q_a locks to a value of $a^*/4$ at low temperature. We suggest that this "pinning" is due to extra terms in the free energy, which are allowed only when $q_a = \pm a^*/4$[3]. The simplest possible Umklapp terms are of the form

$$\delta F = K(\psi_{a^*/4} \ \psi_{a^*/4} \ \psi_{-a^*/4} \ \psi_{-a^*/4} + CC) \tag{3.10}$$

as suggested by Bak[29] and by Bjelis and Barisic[30]. δF is dependent upon the phases θ and θ' of $\psi_{a^*/4}$ and $\psi_{-a^*/4}$, respectively

$$\delta F = 2K|\psi_{a^*/4}|^2 \ |\psi_{-a^*/4}|^2 \cos(2\theta-2\theta') \tag{3.11}$$

It is always possible to make this term negative by adjusting the phases. If $|K|$ is large enough, then it becomes favourable for the system to jump to the "symmetric" state consisting of equal amounts of $|\psi_{a^*/4}|$ and $|\psi_{-a^*/4}|$, when the q_a value minimizing the remaining free energy gets close enough to $a^*/4$. This is what happens at the 38K transition[3]. (See fig. 1.) An interesting point can be made here. In principle Umklapp terms may exist at any rational value of the wavevector[31]. This opens a possibility of having phase locking at different wavevectors in the temperature range $38K < T < 47K$. In an experiment, it may be difficult to distinguish such behaviour from a continuous T dependence of q_a.

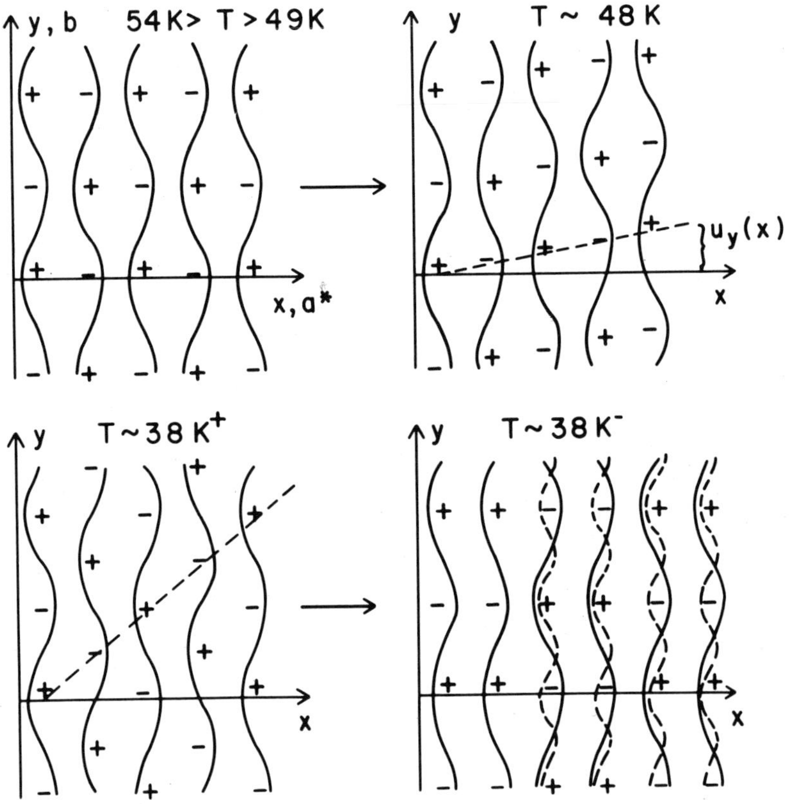

Fig. 1 Schematic diagram of phase transitions in TTF-TCNQ. The full curves indicate the $2k_F$ modulation along the chains. The dotted curves indicate $4k_F$ modulation below 38K. The overall symmetry is monoclinic except for the intermediate phase (48K > T >38K) where the symmetry is triclinic (ref. 29).

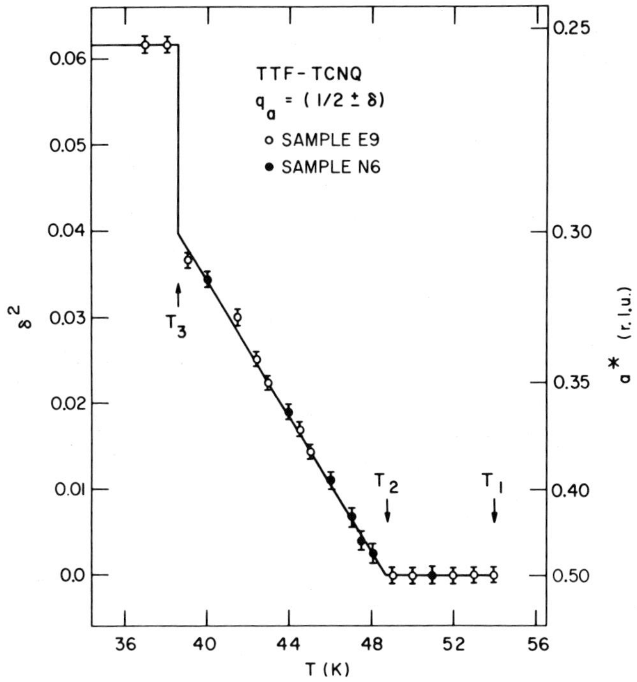

Fig. 2 Plot of observed peak positions $\delta^2(= q^2)$ versus temperature. The three transition temperatures are also indicated (Ellenson et.al., ref. 23).

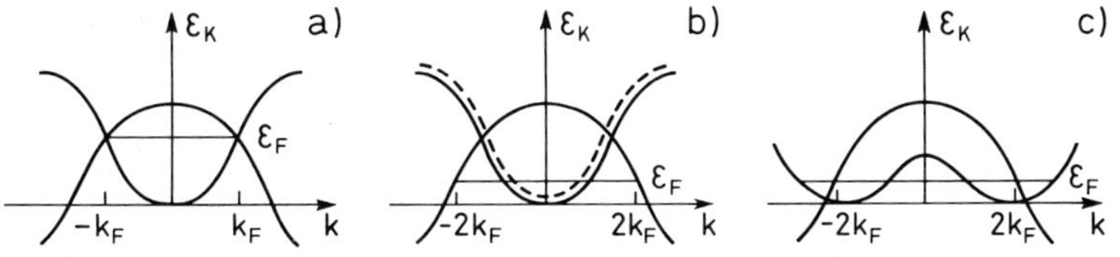

Fig. 3 a) Overlapping valence and conduction bands giving rise to $2k_F$ Peierls distortions. b),c) Band structures which allow both $2k_F$ abd $4k_F$ Kohn anomalies and distortions.

3.4 The $4k_F$ Anomaly

Recently a "Kohn" anomaly[6,8] and charge density waves[7] with wavevector component $4k_F$ along the chain direction have been observed in TTF-TCNQ. The $4k_F$ anomaly is present at room temperatures whereas there is nothing to be seen at $2k_F$ until 150K[6]. This seems to indicate that the $4k_F$ anomaly is not a simple anharmonic effect but is indeed generated by a separate mechanism. Torrance[12] attribute the "$2k_F$" scattering to spinwaves, the "$4k_F$" scattering to the usual Kohn anomaly. Emery[9] explains the $4k_F$ anomaly within the Luttinger model by assuming a large and positive intra molecular Coulomb repulsion, U, Sham proposes that anharmonic phonons are responsible[11], and Weger and Friedel guess that it is simply due to higher order diffraction from libration modes for which the "$2k_F$" scattering (approximately) vanishes[12].

Even if all these explanations are in principle possible, I shall now use this opportunity to confuse the situation further. The usual charge transfer mechanism can be understood by means of figure 3a. A valence (TTF) band and a conduction (TCNQ) band are overlapping, and because of one dimensionality and charge conservation the Fermi level is at the intersection between these bands. This picture is based upon an assumption that there is only one available orbital on each molecule. However, let us assume that there are two (almost) degenerate orbitals on one type of molecule. These orbitals could be associated with the nitrogen atoms situated at each end of the TCNQ molecule. The situation is then described by figure 3b: a charge transfer of $4k_F \times 2/b^*$ electrons from the TTF molecule will fill the two TCNQ bands up to $2k_F$ only. Clearly, the usual Peierls mechanism might then give rise to Kohn anomalies and lattice distortions with wavevector $2k_F$ on the TCNQ chains and $4k_F$ on the TTF chains. Another possibility is that there exist a conduction (valence) band with minimum (maximum) at $q_b = 0, \frac{b^*}{2}$. Then, because of inversion symmetry, there should exist another extremum at $-q_b$, as shown in figure 3c. Again, the bands corresponding to one type of molecules would be filled only up to half the Fermi wavevector of the other type of molecules. Clearly, more theoretical and experimental work on the band structure of TTF-TCNQ is required. The possibility of forming $4k_F$ modes can be included in the Landau Ginzburg theory[3]. Al all temperatures there should exist waves with periodicity given by the wavevector $2\vec{q}=(2q_a, 4k_F, 0)$. Above 47K this mode is acoustic, and below 38K it is associated with one chain system only in analogy with the $2k_F$ mode above 47K. An interesting consequence of our theory of the 38K transition[3] is that there should also exist a CDW with wave vector $(0, 4k_F, 0)$ below 38K. This mode is induced by terms like

$$\delta F = \psi_{(a^*/4,\ 2k_F,\ 0)}\ \psi_{(-a^*/4,\ 2k_F,\ 0)}\ \psi_{(0,\ -4k_F,\ 0)}$$

in the expansion of the free energy. We predict that this mode should be acoustic. According to symmetry this mode is forbidden when measuring around (0,2n+1,0) reflections. This is in agreement with the experiment by Kagoshima et.al.[7]: the "$4k_F$" peak at $\vec{S}$ = (0,3-0.59,0)=(0,$\vec{2}$+0.41,0) vanishes at temperatures bolow 38K whereas the peak at $\vec{S}$ = (0,3-0.59,1) is very sharp. We find that there is no obvious reason that this behaviour is due to an accidental polarization of the mode. The allowed $4k_F$ peaks disappear abruptly at 38K as it should.

3.5 Critical Behaviour

Because of universality, the critical behaviour is believed to depend upon very few properties of the system, namely the dimensionality of the order parameter and the space group of the crystal. For TTF-TCNQ the star of the wave vector associated with the 54K transition consists of the two vectors $\pm\vec{q}_1$. Since the order parameter transforms as a one dimensional representation of the group of $\vec{q}_1$, the dimensionality of the order parameter is two, corresponding for example to the amplitude and phase degrees of freedom, and one should expect three dimensional XY(n=2) exponents. Close to the transition, the temperature dependence $\psi_{q_1}^1$ therefore should be

$$|\psi_q^1|^2 \simeq (T_1-T)^{2\beta}, \ \beta \simeq 0.33 \qquad 3.12)$$

In the case where the phase transition takes place at a non-symmetric $\vec{q}$ value with $q_a \neq \frac{1}{2}$, the star of $\vec{q}$ would consist of four vectors, $\pm(q_a,q_b,0)$ and $\pm(-q_a,q_b,0)$, and one would expect n=4 critical behaviour. At the 47K transition the "CDW lattice" distorts so that the twofold axis is destroyed, as shown in figure 1. It is this distortion that manifest itself as a shift of the $\vec{q}$ vector in a diffraction experiment. A possible order parameter is a homogeneous strain of the CDW:

$$\varepsilon_4' = \frac{d\ u_y(x)}{dx}$$

where $u_y(x)$ is the displacement of the CDW in the b-direction at position $\vec{r}$. The symmetry of this transition is exactly the same as that of a homogeneous q=0 structural phase transition in a monoclinic atomic lattice, where the order parameter is an ε_4 (ε_6) strain. Hence the critical behaviour for these two systems should also be the same. Just as the phase transition in an atomic crystal is triggered by a soft phonon mode the 47K phase transition is triggered by a soft transverse phason mode[29]. Recently, Cowley[32] found that the critical behaviour for the monoclinic system is classical so no corrections to the mean field exponents are expected. Since ε_4 and ε_4' have the same symmetry, a spontaneous ε_4 strain should develope at the phase transition. Again it should be stressed that

Umklapp terms, i.e. the periodic potential from the underlying lattice, has been ignored and could change the critical behaviour at the 47K transition.

4. IMPURITIES

In real physical systems there are always some imperfections or random impurities. In KCP the non-integer amount of Br-ions in the formula unit may act like an external Coulomb potential on the CDW. To investigate the effects of impurities, both in the strictly one-dimensional system and in the quasi three-dimensional case, we shall extend the Feynman integral method of section two, following closely the work by Bak and Brazovsky[5]. The most important contribution to the Coulomb interaction on a CDW is due to backward scattering of electrons and may be written

$$U_i = -\frac{1}{2\pi}\int dx\, V(x)\,\chi'(x) \qquad 4.1)$$

where $V(x) \sim \sum_i u_i(x-l_i) - c\,\bar{u}_i$.

Here $u_i(x-l_i)$ is the Coulomb potential from an impurity ion at site l_i. These ions are randomly distributed with concentration c. $\bar{u}_i$ is the "average" potential of one impurity. Note that the random field acts linearly with the derivative of the order parameter and not with the phase itself as in the model studies by Sham and Patton[15,35].

4.1 One Dimensional Random Systems

We consider a one dimensional conductor with random impurities described by the Lagrangian

$$L = \frac{v}{8\pi}\iint dx d\tau\;(\dot{\chi}^2/u^2 + \chi'^2 - \frac{4V}{v}\chi') \qquad 4.2)$$

It turns out to be convenient to transform the phason field χ to a new unconstrained field ψ through the relation

$$\chi(x) = \bar{\chi}(x) + \psi(x) \qquad 4.3)$$

$$\text{where} \qquad \bar{\chi}(x) = \int_{-\infty}^{x}\frac{2V(x')}{v}\,dx'.$$

The Lagrangian becomes

$$L(\psi) = \frac{v}{8\pi}\iint dx d\tau\,(\dot{\psi}^2/u^2 + \psi'^2) \qquad 4.4)$$

plus a constant term which does not affect thermodynamic averages. To calculate an observable quantity one should first calculate it for any possible impurity configuration and then average over the impurity distribution function. For the structure factor we find, using 4.3):

$$S(x) = \langle\langle \langle \exp i\{\chi(0,0) - \chi(x,t)\} \rangle \rangle\rangle = S_i(x)S_p(x) \qquad 4.5)$$

$$= \langle \exp i\ \{\psi(0,0) - \psi(x,t)\} \rangle \times \langle\langle \exp i\{\ \bar{\chi}(0) - \bar{\chi}(x)\} \rangle\rangle$$

where the double brackets denote averages over impurity configuration. The averages can thus be performed independently. In the small concentration, small potential limit it can be shown that

$$S_i(x) = \langle\langle \exp i\{\ \bar{\chi}(0) - \bar{\chi}(x)\} \rangle\rangle = \exp(-\frac{|x|}{\xi_i}) \qquad 4.6)$$

or in Fourier space,

$$S_i(q) = \frac{2\gamma_i}{q^2+\gamma_i^2}$$

where $\xi_i=1/\gamma_i$ is the impurity correlation length. For KCP, ξ_i is probably comparable to the correlation length $R_p = 1/\gamma_p$ for the pure system. The first factor in 4.5) has been calculated by Brazovsky[33]. In (q,ω) space

$$S_p(q,\omega) = A \frac{2\gamma_p}{\{(\omega+qu)^2+(u\gamma_p)^2\} + \{(\omega-qu)^2+(u\gamma_p)^2\}} \qquad 4.7)$$

The dynamical structure factor is obtained as a convolution of the Fourier transform of the pure structure factor with the impurity correlation function 4.6):

$$S(q,\omega) = \frac{1}{2\pi} \int_k S_i(k)\ S_p(q-k,\omega)dk$$

$$= A \times \Bigg[\frac{\gamma_i}{(\omega^2+\gamma_p^2u^2)(q^2+\gamma_i^2)u^2} \qquad 4.8)$$

$$+ \frac{2q^2(\gamma_p+\gamma_i)-\gamma_i\{\omega^2/u^2+(\gamma_p+\gamma_i)^2\ q^2\}}{\{(\omega+uq)^2+(\gamma_p+\gamma_i)^2u^2\}\ \{(\omega-uq)^2+(\gamma_p+\gamma_i)^2u^2\}\ (q^2+\gamma_i^2)}\Bigg]$$

The second term gives the contribution from phason modes at $\omega=\pm uq$. The first term describes a central peak not present in the pure systems, and the q-width is the inverse impurity correlation length. A very accurate neutron diffraction is required to resolve the spectrum since there is considerable overlap between the various terms. The expression should be valid for KCP at temperatures far above the "three dimensional" phase transition at $T \simeq 100K$[13].

4.2 Combined Effects of Random Impurities and Interchain coupling[5]

The model that we shall investigate is the combination of the random 1-d model and the three dimensional model studied in section 2. The random potentials are assumed to act within each chain and not affect the coupling between chains. The corresponding free energy functional is then

$$F = \int dx \sum_i \{\frac{v}{8\pi}(\chi_i' - \frac{2V_i(x)}{v})^2 - \sum_j K_{ij}\cos(\chi_i - \chi_j)\} \qquad 4.9)$$

Applying the transformation 4.3) we get a new effective free energy

$$F(\psi) = \int dx \, (\sum_i \frac{v}{8\pi}\psi_i'^2 - \sum_{ij} K_{ij} \cos(\bar{\chi}_i(x) - \bar{\chi}_j(x) + \psi_i(x) - \psi_j(x))) \quad 4.10)$$

We note that F is not independent of $\bar{\chi}(x)$, as it was in the case of the one dimensional system, and the averages < > and << >> do not commute any more. Hence, first the thermodynamic average < > should be performed for any possible impurity configuration, and <u>then</u> averaged over the impurity distribution function.

The expansion 2.10), therefore, should be modified in the following way

$$\begin{aligned} \langle\langle \langle O\rangle \rangle\rangle = & \langle\langle \langle O\rangle_0 \rangle\rangle \qquad 4.11) \\ & - \beta\,(\langle\langle \langle OF_1\rangle_0\rangle\rangle - \langle\langle \langle O\rangle_0 \langle F_1\rangle_0\rangle\rangle) \\ & + \frac{\beta^2}{2}(\langle\langle \langle OF_1^2\rangle_0\rangle\rangle - \langle\langle \langle O\rangle_0\langle F_1^2\rangle_0\rangle\rangle) + \dots \end{aligned}$$

where F_1 is the interacting part of 4.10).

To calculate the structure factor we insert

$$\begin{aligned} O &= \exp i\{\chi_i(0) - \chi_j(x)\} \qquad 4.12) \\ &= \exp i\{\bar{\chi}_i(0) - \bar{\chi}_i(x)\} \exp i\{\psi_i(0) - \psi_i(x)\} \end{aligned}$$

The bookkeeping of the various terms can be performed in a diagrammatic way[5]. Each term in the expansion for the pure system is replaced by a product of this term and an impurity term. The Fourier transform of 4.11) has been evaluated. To estimate the transition temperature the inverse correlation function is expanded in $\frac{1}{T}$. To second order

$$\frac{1}{S(0)} = \frac{\Upsilon}{2} \times \{1 - \frac{2K(0)R}{T} + \frac{K^2(0)R^2}{ZT^2} \times \frac{14}{5} + \frac{K(0)^2R^2}{2ZT^2} f(\frac{\xi_i}{R_p})\}. \qquad 4.13)$$

where the function f is dependent upon temperature through R_p. For example, $F(\infty)=0$, $f(1)=0.375$, $f(0.1)=1.74$.

Again, to first order in $\frac{1}{T}$ we find the mean field result

$$\frac{T_c}{R} = 2K(0) \; , \; \frac{1}{R} = \frac{2\pi T_c}{v} + \frac{1}{\xi_i} \qquad 4.14)$$

T_c is lowered by the impurities but remains finite. To second order

$$\frac{T_c}{R} = K(0)\left(1+ \sqrt{1- \frac{1}{Z}\left(\frac{14}{5} + \frac{f}{2}\right)}\right) \qquad 4.15)$$

For small Z, therefore, a large impurity concentration will tend to suppress the phase transition.

At low temperatures the high temperature expansion will break down. In this region, the structure factor can be calculated using a continuum representation in the transverse direction too[5]. For not too large concentrations of impurities the dynamical structure factor will consist of three different features: 1) A Bragg peak which indicates that long range order exists. 2) An elastic impurity induced peak with finite width in momentum space and 3) two phason branches originating from phase fluctuations. However, when the impurity concentration exceeds a certain limit, the continuum representation breaks down, indicating that no long range order can exist due to a "melting" of the CDW lattice. This seems to be the case for KCP. As the temperature is lowered below 100K, the transverse correlation length increases rapidly but saturates rather abrubtly around 100K. This relative sharp saturation indicates that some other ordering may occur. We suggest that there develops a "spin glass"-like ordering, where the CDW's at distant positions freeze in a random way relative to each other, but where near neighbours are correlated. In fact, our model of KCP is very similar to the three dimensional rotator considered by Edwards and Anderson[34], (which is believed to exhibit spin-glass ordering) since both models are formally three dimensional random xy-models.

5. REFERENCES

1. L.B. Coleman, M.J. Cohen, D.J. Sandman, F.G. Yamagishi, A.F. Garito, and A.J. Heeger, Sol.St.Comm. 12, 1125 (1973).
2. R.E. Peierls, Quantum theory of solids, Oxford University Press, p. 108.
3. P. Bak and V.J. Emery, Phys.Rev.Lett. 36, 978(1976), and to be published.
4. W.L. McMillan, Phys.Rev. B14, 1496(1976).

5. P. Bak and S.A. Brazovsky, to be published.
See also A.J. Larkin and V.I. Melnikov, preprint,
Z. Zavadovski, ZhETF 54, 1429(1968).
6. J.P. Pouget, S.K. Khanna, F. Denoyer, R. Comès, A.F. Garito and A.J. Heeger, Phys.Rev.Lett. 37, 437(1976).
7. S. Kagoshima, T. Ishiguro and H. Anzai, J.Phys.Soc. Japan, 41, 2061(1976).
8. H.A. Mook and C.R. Watson, Phys.Rev.Lett. 36, 801(1976).
9. V.J. Emery, Phys.Rev.Lett. 37, 107(1976).
10. L.J. Sham, Sol.St.Comm. 20, 623(1977).
11. M. Weger and J. Friedel, Journal de Physique 38, 241(1977).
12. J.B. Torrance, H.A. Mook and C.R. Watson, Preprint.
13. J.W. Lynn, M. Iizumi, G. Shirane, S.A. Werner and R.B. Saillant, Phys.Rev. B12, 1154(1975).
14. K. Carneiro, G. Shirane, S.A. Werner and S. Kaiser, Phys.Rev. 13, 4258(1976).
15. L.J. Sham and B.R. Patton, Phys.Rev. B13, 315(1976), Phys.Rev. Lett. 36, 733(1976).
16. S. Brazovsky and I. Dzyaloshinsky, ZhETF 71, 2338(1976) and ZhETF 72(1977), (to be published).
17. K. Saub, S. Barisic and J. Friedel, Phys.Lett. A56, 302(1976).
18. D.J. Scalapino, Y. Imry and P. Pincus, Phys.Rev. B11, 2042 (1975).
19. P.A. Lee, T.M. Rice and P.W. Anderson, Phys.Rev.Lett. 31 462 (1973).
20. F. Denoyer, R. Comès, A.F. Garito and A.J. Heeger, Phys.Rev. Lett. 35, 445(1975).
21. R. Comès, S.M. Shapiro ,G. Shirane, A.F. Garito and A.J. Heeger, Phys.Rev.Lett. 35, 1518(1975).
22. R. Comès, G. Shirane, S.M. Shapiro, A.F. Garito and A.J. Heeger, Phys.Rev. B14, 2376(1976).
23. W.D. Ellenson, R. Comès, S.M. Shapiro, G. Shirane, A.F. Garito and A.J. Heeger, Sol.St.Comm. 20, 53(1976).
24. E.F. Rybaczewski, L.S. Smith, A.F. Garito, A.J. Heeger and B.G. Silbernagel, Phys.Rev. B14, 2746(1976).
25. K. Franulović, S. Tomic, M. Prester, D. Djurek, L. Girel and J.M. Fabre, Preprint.
26. B.T. Kistenmacher, T.E. Phillips and D.O. Cowan, Acta Crystallogr. B30, 763(1974).
27. L.D. Landau and M.E. Lifshitz, Statistical Physics, 2nd ed. (Pergamon, New York, 1968) Chap. XIV.
28. S. Etemad, Phys.Rev. B13, 2254(1976).
29. P. Bak, Phys.Rev.Lett. 37, 1071(1976).
30. A. Bjelis and S. Barisic, Phys.Rev.Lett. 37, 1517(1976).
31. S.A. Brazovsky and I. Dzyaloshinsky, Proceedings of the 14th international conference on low temperature physics, 5, 337 (1975).
32. R.A. Cowley, Phys.Rev. B13, 4877(1976).
33. S.A. Brazovsky, NORDITA-preprint.
34. S.F. Edwards and P.W. Anderson, J.Phys. F6, 1927(1976).

35. The term 4.1) may represent forward scattering of conduction electrons by the impurity potential, whereas the coupling considered by Sham and Patton may represent backward scattering across the Fermi surface. We expect this latter coupling to be comparatively weak since it involves the $2k_F$ component of a long range Coulomb potential.

SOLITONS AND CHARGE DENSITY WAVES

Alan Luther

NORDITA
Blegdamsvej 17, 2100
Copenhagen Ø, Denmark.

1. Abstract

The microscopic physics underlying the sine-Gordon equation in quasi-one-dimensional conductors and two dimensional layered structures is explored.

An interesting parallel between these two systems is discussed, and the quantum nature of the former, involving the time variable, is found to be equivalent to the second space dimension of the latter. The soliton spectrum of both are studied and the implications of these discrete states for real physical systems, with three-dimensional couplings, are developed.

2. Introduction

The study of charge density waves in lower dimensional systems has been advanced beyond the stage of speculation due to experiments in both quasi-one [1] and quasi-two [2] dimensional systems. While it is by no means clear that the special properties of one and two dimensional models can explain these phenomena, it seems appropriate to understand the predictions of these models, keeping in mind a healthy skepticism about the ultimate outcome.

There appears to be much in common between the quasi-one-dimensional and some of the quasi-two-dimensional charge density structures. In the former, time appears as an important dynamical variable. Obviously, renaming this variable a second space variable, one arrives at a classical two dimensional problem. To the extent that quantum fluctuations are unimportant in the layered structures, the two are equivalent.

An interesting consequence of this equivalence concerns the existence of soliton states [3]. Propagating wave-type solutions along a one dimensional chain have a direct correspondence to stationary domain-like excitations in the classical two dimensional system. These domain structures, in turn, have consequences for the thermodynamics and static correlation functions. This paper is intended to explore the types of soliton states which can be of interest, and to deduce some observable consequences of these states, if they exist.

If the reader has sensed skepticism about solitons solving all problems of lower dimensional systems, it is not unfounded. There can be no doubt about the importance of these states in the purely one or two dimensional case. But when interactions are turned on between the strings, or between the layers, the effects of these must be included in the original soliton solution. At present, a good understanding of the destruction of solitons, or perhaps only slight damping of solitons due to these interactions, is lacking.

On a heuristic level, the problem could be phrased in terms of energy scales. If the energies of interaction within, say, the string, are much stronger than the interactions between the strings, it is a good approximation to apply the solutions to the string problem, and treat the interactions between them as a perturbation. This will be a good approximation for temperatures larger than the mean interaction between strings, which is roughly the mean field transition temperature for the strings. The same words obviously apply in the layered case as well.

3. Change Density Waves and the Sine-Gordon Equation

With these introductory hedges in mind, let us turn to a study of charge density waves in lower dimensions. Not surprisingly, the close relation between charge density waves and the solitons is one of the first points to be encountered. The mathematics literature is filled with various equations which exhibit nonlinear wave propagation [4]. Of these, the sine-Gordon equation [5] is the most relevant for the physical applications of interest here. Before discussing this equation, it is important to understand where it came from, what assumptions went into it, and why it might be of interest.

An example from the one dimensional conductor problem consists of a filled Fermi sea of electrons moving in a potential capable of causing transitions from one side of the Fermi line, at momentum plus k_F, to minus k_F, the usual charge density wave instability. To analyze this situation, we focus attention on those states at the Fermi energy. These are described by the free particle Hamiltonian,

$$H_0 = v_F \sum_k (k-k_F)\, a^+_{1k}\, a_{1k} - (k+k_F)\, a^+_{2k}\, a_{2k} \tag{1}$$

where v_F is the Fermi velocity, a_{1k} (a_{2k}) describes the electrons moving to the right (left). To this, is added the mixing with the charge density wave potential,

$$H_{CDW} = V_0 \sum_k a^+_{1k}\, a_{2k+k_F} + a^+_{2k}\, a_{1k-2k_F} \tag{2}$$

where V_0 is the strength of the interaction. In a more realistic model, the electron-electron interactions should be included, as well as a description of the phonon degrees of freedom. The present model, however oversimplified, does contain the essential physics, and can be generalized to the more complicated situations without undue hardship.

One of the fundamental simplifications of the one dimensional electron gas starts with the recognition [6] that Eq. (1) can be transformed into a simple harmonic oscillator, or one dimensional phonon problem. While much of our intuition about metals is based on electron Hamiltonians, results for soliton problems are based on phonon-type fields, or fields describing a phase variable. The flexibility to view the problem from either vantage point is exceedingly insightful.

The equivalent phonon problem is given by the Hamiltonian,

$$H_0 = \frac{2\pi v_F}{L} \sum_k [\rho_1(k)\rho_1(-k) + \rho_2(-k)\rho_2(k)] \tag{3}$$

where L is the length of the string of electrons, and the phonon operators satisfy the commutation relations

$$[\rho_1(-k),\rho_1(k')] = [\rho_2(k'),\rho_2(-k) = \delta_{k,k'}\, kL\, (2\pi)^{-1}$$

$$[\rho_1,\rho_2] = 0 \tag{4}$$

In terms of the electron operators, the phonon operators are given by

$$\rho_1(k) = \sum_p a^+_{1k+p}\, a_{1p}$$
$$\rho_2(k) = \sum_p a^+_{2k+p}\, a_{2p} \tag{5}$$

The key to understanding the soliton-charge density wave equivalence is contained in the H_{CDW} written in terms of the phonon variables. It is here necessary to make use of the so-called phase representation [7] of electron operators, given by

$$\sum_k a_{1k}\, e^{ikx} = (L/2\pi s)^{\frac{1}{2}}\, e^{ik_F x + \phi_1(x)}$$

and

$$\sum_k a_{2k}\, e^{ikx} = (L/2\pi s)^{\frac{1}{2}}\, e^{-ik_F x + \phi_2(x)} \tag{6}$$

where s is a cut-off length, equal to a lattice constant, and the phase variables $\phi_1(x)$ and $\phi_2(x)$ are, in turn, related to the phonon operators through

$$\phi_1(x) = 2\pi\, i \int^x dy\, \rho_1(y)$$
$$\phi_2(x) = -2\pi\, i \int^x dy\, \rho_2(y) \tag{7}$$

when substituting these into Eq. (2), one obtains the simple answer,

$$H_{CDW} = \frac{V_0 L}{2\pi s} \int dx\, \cos \Phi(x) \tag{8}$$

where the total phase $\Phi(x) = i(4\pi)^{\frac{1}{2}} (\phi_1(x) + \phi_2(x))$ has been introduced. Now it is possible to write Eq. (3) in a slightly more familiar form, if we use the additional operators, $\Phi(x)$ and $\pi(x) = i\, [\Phi(x), H_0]$, where

$$\Phi(x) = i \sum_p \frac{1}{\sqrt{\pi V_F}} \left[\rho_1(p) + \rho_2(p)\right] \frac{e^{-ipx}}{pL}$$
$$\pi(x) = i \sum_p \frac{-1}{\sqrt{\pi V_F}} \left[\rho_1(p) - \rho_2(p)\right] \frac{e^{-ipx}}{L} \qquad (9)$$

with the simple result that the charge density wave problem can be written in the form

$$H = \int dx \, \frac{1}{2}\left[\pi^2(x) + V_F^2(\nabla\varphi)^2\right] + \frac{V_0}{2\pi s} \cos\sqrt{4\pi}\,\varphi \qquad (10)$$

which is recognized as the quantum mechanical sine-Gordon equation, and, at the same time, from Eqs. (1) and (2), is the equation for an electron moving in the potential of a $2\ k_F$ density wave.

The essential ingredient leading to this result is the phase representation of the charge density wave, which lead to Eq. (8). Any interaction which permits such a representation will lead to the same equation. It is interesting to note the many seemingly complicated interactions which can be written in this form. Including electron-electron interactions leads to two modifications. The first is trivial, a renormalization of the Fermi velocity. The second is a replacement of the $(4\pi)^{\frac{1}{2}}$ by another constant, β. This constant determines the nature of the soliton state solutions, and is itself determined by the strength of the electron-electron interaction. If U represents this interaction strength, then $\beta = \sqrt{4\pi}\,(1-U\rho)^{\frac{1}{2}}\,(1+U\rho)^{-\frac{1}{2}}$. (We might imagine U to be a screened Coulomb interaction, which can be comparable to the reciprocal bandwidth ρ.) These solutions will be discussed below.

It is worth a brief mention of other interactions which reduce to this sine-Gordon form. Obviously, the phase variable was constructed from two electron states - moving in opposite directions in the above example. But any combination of two phases will lead to an identical result. We might imagine a phase associated with the up spin electrons, and another associated with the down. This creates a phase field for the spin density, and we would expect a sine-Gordon equation for these degrees of freedom [8]. A further - but more complicated situation - concerns the electron-phonon interaction. When the temperature is very low, and the electron-phonon system is strongly coupled, the phase of the displacement becomes a good variable [9], and the motion of a charge density wave is governed by the free phonon part of Eq. (10). If there is an impurity present, the interaction with the charge density wave is given by

$$H = \frac{1}{2} \int dx \left[\pi^2 + c^2(\nabla\varphi)^2 \right] + \lambda \sum_i \cos \beta\varphi(x_i) \qquad (11)$$

where c is a phonon velocity, λ the interaction strength, and the sum runs over the impurity sites, x_i. It is this, modified slightly from Eq. (10), which is the sliding conductivity problem. Except for the free particle case, $\beta = \sqrt{4\pi}$, little is known about Eq. (11). This problem will be studied below.

4. Solitons and Charge Density Waves

From a mathematical viewpoint, the problem is now well-posed - namely, what are the eigenvalues and eigenfunctions of the Hamiltonian Eq. (10), for general values of β? But the physics is not so simple. Even with this complete (but nonexistent) information, the questions of transport, impurity scattering, etc. would remain. These are well beyond the scope of this paper.

Rather, a simple physical picture of these states is the goal of this paper, and it is hoped that the mathematical details can be kept to a minimum. It is possible for the reader interested in such questions as rigor and completeness to satisfy herself as to the current state of the art in the relevant literature.

There is substantial information about the eigenvalue spectrum [4,10] of the sine-Gordon equation, but little else. There are several limiting cases which are quite helpful in understanding the additional questions, such as matrix elements, selection rules and the like, but here the situation is far from satisfactory.

The trivial limiting case is one good example to keep in mind. Consider $\beta^2 = 4\pi$. From Eq. (1), Eq. (2), and Eq. (10), it is clear that this case is a one-electron problem, which is immediately solved. The eigenvalue spectrum is

$$E = V_F K_F \pm V_F \left[(K-K_F)^2 + \left(\frac{V_0}{2\pi s}\right)^2 \right]^{\frac{1}{2}} \qquad (12)$$

and the new feature of the problem is the appearance of a gap at the Fermi energy, a gap required to create a particle-hole pair excitation. From the equivalence to the sine-Gordon equation, there comes the identification of the single electron states with a propagating solition. That is, through the miracles of modern science, the highly complicated non-linear equation involving the phase variables becomes a simple free electron problem when viewed from another vantage point.

Another interesting limiting case concerns the region $\beta^2 \to 0$. In this situation, it is permissable to expand the cosine, and retain the lowest order correction. The resulting equation is

$$H \to \int \frac{dx}{2} \left[(\nabla\varphi)^2 \, V_F^2 + \pi^2 + \frac{V_0}{2\pi s} \, \beta^2 \, \varphi^2 \right] \tag{13}$$

which is very similar to the one dimensional optical phonon problem, with the eigenvalue spectrum given by:

$$E_K = V_F \left[K^2 + \frac{V_0 \, \beta^2}{2\pi s \, V_F^2} \right]^{\frac{1}{2}} \tag{14}$$

As found above, the characteristic feature is the gap in the spectrum, of magnitude:

$$E_0 = \beta \left[\frac{V_0}{2\pi s} \right]^{\frac{1}{2}} \tag{15}$$

but the reader should recognize the rather profound difference between the two cases. The former was explicitly a fermion-type gap, while the latter is a boson gap! In a peculiar manner, the Hamiltonian changes continuously from a free fermion problem to a free boson problem, with complications in between.

It is possible to offer a plausible interpretation of this gap, regardless of its fermi-bose schizophrenia. If a static correlation function is calculated, such as the phase-phase correlation function at large distance between the phase measurement points, there will be an exponential fall-off, while at shorter distances the effects of the gap will be negligible. When divided by the Fermi velocity, the gap thus determines a wave number scale of local ordering.

The complicated behaviour of the spectrum has been studied by the WKB approximation [4] and by the device of finding another equivalence to a solved model in statistical mechanics [10]. The results of these investigations have found

$$E_K^2 = \Delta_n^2 + V_F^2 K^2$$
$$\Delta_n^2 = \Delta \sin\left(\frac{n\gamma'}{16}\right) \tag{16}$$
$$\Delta = \left(\frac{V_0}{2\pi s}\right)^{\nu}$$

where $\gamma' = \beta^2(1 - \beta^2/8\pi)^{-1}$, $2\nu = (1 - \frac{\beta^2}{8\pi})^{-1}$, $n = 0,1,2,\cdots$ and n is a new quantum number, which has an interpretation as labelling bound solitons. Again, a heuristic interpretation of these states is possible, and it compares them to excitons forming within the gap of a semiconductor. Since $\beta^2 < 4\pi$, the interaction implied by the result under Eq. (10), is particle-hole attractive when bound solitons appear.

A few words about the rigor of this solution one in order. The WKB results have been demonstrated to be exact for the eigenvalue spectrum, by studying the relationship to the spin $\frac{1}{2}$ x-y-z spin chain in one dimension [10]. This relationship was used not only to study the eigenvalue spectrum, but to provide an intuitive picture for the soliton states. Without an overwhelming mass of mathematics, the full justification of this relationship is not possible - but some observations are of interest.

The x-y-z model is given by the Hamiltonian

$$H_s = -\sum_i J_\alpha S_i^\alpha S_{i+1}^\alpha \tag{17}$$

where $\alpha = x,y,$ or z, S_i^α is a spin $\frac{1}{2}$ operator at site i, and the sum is over N sites of a chain. Again, the study of a few limiting cases helps establish the physics of Eq. (17). When all J_α are equal, the Heisenberg ferromagnetic case, we know that Eq. (17) should have spin waves. Comparison to Eq. (15), for $\beta^2 = 0$, shows the similar behaviour. Now for the case $\beta^2 = 4\pi$. It has been known that the x-y model, with $J_z = 0$, can be transformed into a free electron problem, and indeed, the equivalence between these two cases has been rigorously established.

But more has been established. In the entire region with $J_x = J_y$, and $|J_z| < J_x$, it is intuitively clear that the rotational degeneracy about the z-axis, within the x-y plane can be described by a phase variable. A spin wave, corresponding to precession about the z-axis, corresponds to the phonon propagation of Eq. (3), corresponds to a potential tending to "orient" the phase in the x-y plane, along a particular direction, analogous to orientation of the spin along the x axis, if J_x is larger than J_y. In this view, a soliton is a spin wave propagating along the chain. Bound solitons are spin wave bound states.

In addition to this simple physical picture, some new results can be derived from this equivalence. A peculiar instability in the sine-Gordon equation, at the value $\beta^2 \geq 8\pi$, has led to confusion about the meaning of charge density waves in these circumstances.

It is easy to recognize $\beta^2 = 8\pi$ as a special point, from the result states in Eq. (16). The exponent ν , in the relation between the observed gap and the applied field, goes to infinity at this point, then becomes negative. Has something gone wrong in the solution? It is instructive to consider this question from the spin chain viewpoint.

Recall that $\beta^2 = 0$ corresponds to the Heisenberg ferromagnet, $J_z = J_x = J_y$, and the phase of the sine-Gordon equation, in general, has the interpretation as a phase variable of a basal plane spin. As J_z is decreased to zero, and then downward to $J_z = -J_x = -J_y$, the Heisenberg antiferromagnetic point is reached. For more negative J_z , the spin no longer lies in the basal plane. Indeed this marks the crossover to a spin problem of Ising symmetry, and the phase variable cannot be defined as before.

This spin equivalence suggests that a new type of gap appears when $\beta^2 > 8\pi$, a gap analogous to the appearance of an Ising gap as the J_z exchange becomes large in magnitude. Obviously, in the spin problem, the equilibrium direction of the spin simply rotates to accommodate the new ground state symmetry. Once the spin operators are rotated to recognize the new equilibrium direction, we may again view the soliton as a spin wave. The $\beta^2 \geq 8\pi$ problem is simply the requirement that the phase variable must be defined about the new equilibrium direction.

To state this resolution more precisely, the sine-Gordon equation assumes that the parameters β^2 and V_0 have been defined such that, in the corresponding spin problem, $|J_z|$ is the smallest interaction. If this is not the case, the spins must be rotated to new axes, such that the new variables, denoted by a prime, do have $|J_z'|$ the smallest. If this prescription is followed, $\beta^2 < 8\pi$ is always satisfied.

Collecting these results leads to the following simple picture of charge density waves in the one-dimensional conductor. For $0 < \beta^2 < 8\pi$, that is $-1 < Up < 1/2$, there is a gap in the excitation spectrum caused by the appearance of $2k_F$ charge density waves, and for $-1 < Up < 0$, there are bound soliton states, representative of charge bunching, which may propagate. For Up outside this region, it is necessary to recognize the equivalence to an Ising-like spin problem, with a preferential orientation. Solitons, in this case, are analogous to the Ising-like spin flip excitations of the chain.

5. Solitons in Disordered Systems

A question of recurring interest, both in the quantum one-dimensional systems, and in the classical two-dimensional system, concerns the role of impurities. One hears repeatedly the common view, that since all electron states are localized in a random potential in one dimension, localization dominates the problem. It is helpful to analyze this situation from a slightly broader viewpoint, which recognizes the random potential as simply another type of interaction, in addition to the many others which are present.

This attitude is well-known from studies of phase transitions in disordered compounds. The random variable, for example a local transition temperature which varies randomly throughout a ferromagnetic alloy, can be considered as a new type of "interaction" which competes with the usual exchange, to give new physics, such as a spin glass. But the question about which phase dominates, spin glass or ferromagnetism, can only be answered if both random and exchange fields are considered simultaneously. Statements about localized states in the one-dimensional electron gas have been based on non-interacting electrons, and the question about the relevance of localization for the interacting system, is both interesting and important.

There emerges an interesting picture for this phenomena when considered from the soliton viewpoint, and the problem can be posed in the language of localized versus uniform solitons. Consider first the situation of a one-dimensional electron gas with a single impurity present. If we include an electron-electron interaction, the problem is described by Eq. (11), but for a single impurity only.

Perhaps it is worth emphasizing that, in this view, the impurity is trying to induce a local gap at $2K_F$, while in the original electron variables, the impurity is simply scattering the electron from $+K_F$ to $-K_F$. The equivalence of these two pictures follows from the fermion-boson duality. While it may seem arbitrary to choose either viewpoint, it is possible to use the soliton picture to derive a relation of this problem to another, namely, the Kondo problem of a magnetic impurity in a metal. This problem has been understood, and it is this understanding which can be applied to solve the localization problem.

It might seem surprising at first glance that these two problems are equivalent. But there is a simple plausibility argument that, while not rigorous, is convincing. The rigorous equivalence will be given later. The Kondo problem involves a spin $\frac{1}{2}$ impurity which scatters electrons with a matrix element that depends on the electron spin, in the form:

$$H_{KONDO} = H_e - \bar{J}_z S^z \sigma^z(0) - \bar{J}_\perp (S^+ \sigma^-(0) + h.c.) \quad (18)$$

where $\vec{S}$ is the impurity spin operator and $\vec{\sigma}(0)$ is the spin density of the electrons at the impurity site, the origin $x = 0$, H_e the free electron Hamiltonian, and $\bar{J}_z$ and $\bar{J}_\perp$ refer to the longitudinal (spin non-flip) and transverse (spin flip) processes. But why is this relevant here?

First of all, the Kondo model is also one-dimensional, because we need work with only one partial wave, the s-wave, which depends on a one-dimensional wave vector $|k|$. There are, in fact, two types of one-dimensional electrons in this problem, spin up or spin down. These correspond to the two types of electrons in the one-dimensional conductor problem (remember that these are "spinless" electrons), namely, moving to the right, a "1" state, or moving to the left, a "2" state.

From this basis, we see that a spin flip at the impurity in the Kondo model is the same as a reflection of an electron from a static impurity in the one-dimensional conductor problem. Both drive an electron from one type to the other, do it locally, and repeatedly. There are a few subtle points which complicate the picture, such as the equivalence of the longitudinal part of the Kondo interaction, $\bar{J}_z$, to the forward scattering part of the one dimensional conductor system, which serves only to complicate matters, but can be included satisfactorily.

The important parameters in this problem, are λ of Eq. (11) and the spin-flip amplitude, $\bar{J}_\perp$, in the Kondo problem. The parameter β, arising from electron-electron interactions in Eq. (11), is related to the longitudinal coupling, $\bar{J}_z$. The relations are

$$\lambda = \bar{J}_\perp$$

$$\beta^2 = 2\pi\ (2-\varepsilon)$$

Here ε is the scattering phase shift for $\bar{J}_z$ as in [11]. From the solution of the Kondo impurity problem, it is possible to "solve" this one as well.

The current understanding of the Kondo problem revolves around the issue whether the ground state is a singlet, with a constant magnetic susceptibility as the temperature goes to zero, or a triplet with a Curie law [11]. The former implies a gap of sorts, namely a gap between the singlet and triplet excited states. The latter has degeneracy. The scaling law arguments [11] indicate that the ground state is a singlet when $|\bar{J}_\perp| > -\bar{J}_z$, and a triplet for

$\bar{J}_\perp \lesssim -\bar{J}_z$. The latter case is viewed as being adequately described by ordinary perturbation theory, while the former involves a complicated many body ground state, which is treated by sophisticated renormalization group methods.

Application of the analogy to the one-dimensional conductor says that, for attractive interactions which satisfy $\beta^2 \leq 4\pi(1-\lambda/2)$ (for small λ) , there is no interesting local correlations or behaviour. The scattering is adequately described by perturbation theory in that the scattering rate goes to zero, that is an infinite conductivity, as in the Kondo problem for ferromagnetic coupling.

For the other sign of the inequality, which includes the case of no electron-electron interactions, the situation is more interesting. There the analogy suggests the build-up of a localized highly correlated state which corresponds to a resonance in the scattering. This resonance develops for the temperature less than a characteristic "Kondo" temperature, T_c , which can be determined for the special case $\beta^2 = 2\pi(2-\rho)$ to be $T_c \sim E_F \sqrt{\lambda}\, e^{-1/\lambda}$, a special case which corresponds to the isotropic Kondo problem. For other values of λ and β^2 the estimates of T_c are complicated and will not be discussed here.

The build-up of this localized correlation, corresponding to a local $2K_F$ gap, can be extended to the case of many impurities, the case normally considered. If the separation between impurities is much larger than the correlation length implied by T_c , namely $V_F\,(T_c)^{-1}$ at $T = 0$, the properties will be determined by single impurity characteristics and the analogy here is applicable. In the other limit, the high concentration limit, no conclusion can be drawn, although it is plausibly related to the "Kondo necklace" model [13] studied for high concentrations of Kondo spins in metals.

We complete the discussion by deriving the equivalence between the Kondo impurity model and the one-dimensional conductor problem. It is easy to show the equivalence of the free energy for the system described by a single impurity described by Eq. (11), and the formula for the Kondo problem,

$$e^{-\frac{F}{T}} = \sum_n \left(\frac{\bar{J}_\perp}{2\pi s}\right)^{2n} \int_0^{T^{-1}} dt_1 \ldots \int^{t_{2n-1}} dt_{2n}\; e^{\sum_{i<j}^{2n} (-)^{i-j}(2-\varepsilon)\ln\left(\frac{t_i - t_j}{s}\right)} \tag{20}$$

where T is the temperature in the Kondo problem, ε is related to a scattering phase shift, and for small J_z is $\varepsilon = 2J_z\,\rho$, F the change in free energy due to the presence of an impurity, and s is a cutoff parameter, introduced according to the prescriptions [7] of the Luttinger-Tomanaga models. The same method used to to derive Eq. (20), applied to the single impurity version of Eq. (11) leads to

$$e^{-\frac{F}{T}} = \sum_n \left(\frac{\lambda}{2\pi s}\right)^{2n} \int_0^{T^{-1}} dt_1 \ldots \int_0^{t_{2n-1}} dt_{2n}\, e^{\sum_{i<j}^{2n} (-)^{i-j} \left(\frac{\beta^2}{2\pi}\right) \ln\left(\frac{t_i - t_j}{s}\right)} \qquad (21)$$

where s is the same cutoff parameter. Obviously the identifications $\lambda = J_\perp$ and $\beta^2 = 2\pi\ (2-\varepsilon)$ follow immediately.

6. Classical Solitons in Two Dimension

One of the recurring problems in the theory of charge density waves for the two dimensional layered structures involves the classical solution to the sine-Gordon free energy. It arises whenever a local phase variable can describe the relevant degrees of freedom, and in addition, an interaction with a lattice seeks to impose a particular phase relationship on these variables. While a microscopic derivation of this free energy expression is not intended here, it is helpful to consider a phenomenological motivation for it.

Suppose a charge density wave exists and can be described by a Ginzburg-Landau free energy expression, which fixes its amplitude according to the usual mean field theory. The phase is still left undetermined, and inversion symmetry would require a contribution to the free energy of the form

$$F_0 = C \int d^2x\ (\nabla\phi - 1)^2 \qquad (22)$$

where C is a microscopic parameter characterizing the energy required to cause a phase distortion, and the period of the CDW is taken equal to unity.

If a periodic potential generated by the lattice is included, the CDW would prefer to arrange its phase field to be commensurate with the lattice to minimize this additional interaction, and there results a contribution of the form:

$$F_{CDW} = \int d^2x \quad \alpha(\cos \beta\phi - 1) \qquad (23)$$

where β is an integer, and α is another microscopic parameter characterizing the underlying lattice interaction, typically an Umklapp process. Obviously F_{CDW} prefers a phase multiple of $2\pi/\beta$, while F_0 prefers a phase increase proportional to distance.

In a rough physical sense, the solution might be expected to have phase jumps between the allowed multiples, in such a way as to maintain a stair step approximation to the straight line increase. From another viewpoint the phase increases linearly with distance with the F_{CDW} basically introducing harmonics. As the contributions from F_{CDW} become dominant, there ceases to be a linear increase of the phase with distance, and a type of transition, the commensurate-incommensurate transition occurs. It is believed this transition is continuous, a conclusion based on numerical studies. It is interesting to consider this problem in light of the known solutions to the quantum sine-Gordon equation.

The beginning of our analysis is the total free energy given by Eq. (22) and Eq. (23). But the phase variables which appear in these expressions are just that: variables. They must be integrated out in order to find the observable free energy. The situation here is perfectly analogous to the familiar Ginzburg-Landau free energy applied to the theory of a second order phase transition. In that case, one has an order parameter variable, and two possible interpretations of the meaning of that variable. The first, and historically first as well, simply minimizes the free energy expression to solve for the mean field version of the transition. After Wilson, it became clear that the second, which treated the order parameter as a thermodynamic variable, was correct. In the same manner, by the same arguments, the phase variable of the sine-Gordon free energy must be integrated.

With this understanding about the meaning of the phase variable, the expression for the partition function is:

$$Z = \int \delta\phi \quad e^{-\beta F\{\phi\}} \tag{24}$$

where the $\delta\phi$ stands for the density of states available to the system which is

$$\int \delta\phi = \prod_k \int_{-\infty}^{\infty} d\phi_K \tag{25}$$

and the product is over all wave numbers in the Fourier transform of the phase field. The object in the partition is conventionally called the free energy functional which simply gives the energy of a particular configuration specified by $\phi(x)$. The density of states then insures that all functions are included in the partition function. The observable, or thermodynamic, free energy is then $F = -K_BT \ln Z$, as usual.

The solution of the sine-Gordon classical phase problem is constructed directly from the solution to the corresponding quantum problem. The principles underlying this equivalence have been discussed previously [14], but it is helpful to recall them briefly here again. Consider first the calculation of the angle correlation function for the special case $\alpha = 0$. This correlation function is given by:

$$C(\vec{x}) = \langle e^{i\phi(\vec{x})} \; e^{-i\phi} \rangle \tag{26}$$

where the slanted brackets denote an average in the weighting functional of Eq. (24). Since the integrals are just Gaussian, and the result is simply:

$$\begin{aligned} C(\vec{x}) &= Z^{-1} \int \delta\phi \quad e^{-\beta F\{\phi\} + i\phi(\vec{x}) - i\phi} \\ &= e^{\langle \phi(\vec{x})\phi \; - \; \phi^2 \rangle} \end{aligned} \tag{27}$$

The phase correlation function is given by

$$\langle \phi(\vec{x})\phi \rangle = \frac{1}{L^2} \sum_{k,k'} e^{i\,\vec{K}\cdot\vec{x}} \langle \phi_K \; \phi_{K'} \rangle \tag{28}$$

where L^2 is the area of the layer. This leads to

$$\begin{aligned} \langle \phi(\vec{x})\phi - \phi^2 \rangle &= \frac{K_B \; TC}{4\pi^2} \int \frac{d^2K}{K^2} \left(e^{i\,\vec{K}\cdot\vec{x}} - 1 \right) \\ &= \frac{K_B \; TC}{8\pi^2} \; \ell n \left(\frac{x^2+y^2}{iS} \right) \end{aligned} \tag{29}$$

where S is the same type of cut-off as in Eq. (7). Finally, the angle correlation function is given by

$$C(\vec{x}) = \left(\frac{S^2}{x^2+y^2} \right)^{\frac{K_B \; TC}{8\pi^2}} \tag{30}$$

This result, long known in the literature of the planar model, not only illustrates the meaning of the free energy functional, but provides the clue to use the quantum result if we calculate the quantum phase-phase correlation function using Eq. (10), the answer is essentially the same, with the exception that $t \to iy$, that is, the time variable becomes the other space dimension, and the temperature in this classical problem becomes related to the coupling constant β in the quantum problem.

The corresponding angle correlation function in the quantum problem is calculated by precisely the same methods, for the corresponding case $V = 0$, giving the result:

$$<e^{i\phi(x,t)} \, e^{-i\phi}> = \left(\frac{S^2}{x^2-t^2}\right)^{\frac{\beta^2}{4\pi}} \tag{31}$$

Such an equivalence between the angle correlation functions alone would not be very helpful. But this equivalence holds for all correlation functions, which can be built up from products of operations $e^{i\phi}$ at arbitrary space-time points. In particular, the correlation functions involving an arbitrary product of $\cos\phi$, are identical in the quantum and classical theories. Consequently, we can conclude that all matrix elements of the interaction, H_{CDW} or F_{CDW}, are identical. Therefore the two problems are the same.

This means the solution to the quantum sine-Gordon equation provides the solution to the classical problem, in the sense that the ground state energy of the quantum problem is the thermodynamic free energy in the classical one. Correlation functions are also equivalent, after $t \to iy$.

What features of the quantum solution are interesting here? The first qualitative feature is the gap in the excitation spectrum for the region $0 < \beta^2 < 8\pi$. A gap in the excitation spectrum implies a correlation length in the classical problem, corresponding to this gap. The behaviour of the gap, as a function of coupling constant, describes the behaviour of the correlation length as a function of temperature.

Some of the features of the quantum solution are, unfortunately, not sufficiently clear at present to resolve the obviously interesting question about a "phase transition" at $\beta^2 = 8\pi$, that is $K_B\, CT_c = 16\pi^2$. At this value, the gap vanishes, but the precise manner in which it vanishes as $\beta^2 \to 8\pi$ has not been elucidated yet. There is a scaling argument that this gap, Δ, vanishes as

$$\Delta \sim \sqrt{t}\; e^{-\frac{1}{t}} \tag{32}$$

where t is the temperature in reduced units, $1 - \frac{T}{T_c}$. This is a continuous transition, and it is characterized by an infinite length, as $t \to 0$.

The construction of this argument is roundabout, but it is perhaps of sufficient interest to state the general features. It relies, first of all, on the equivalence of the backward scattering model of 1-d fermions to the sine-Gordon equation. The renormalization group has been applied to the fermion problem, and through that equivalence, it implies the exponential dependence of gap on coupling constant. There is, as yet, no equivalent calculation for the ground state energy in the region $\beta^2 \to 8\pi$, that is $t \to 0$, but it is probably logarithmic, which would imply logarithmic temperature dependences in the specific heat.

Of particular interest, and some puzzlement, is the role played by the soliton bound states, which can occur for small β^2, that is well away from T_c. Presumeably, these correspond to phase jumps, and their juxtaposition must correspond to an excited state. It is tempting to identify these with the discommensuration of McMillan [15], for these appear as excited states of the free energy functional, and a correspondingly more rapid decay of correlations in space.

7. Three Dimensional Ordering

The obvious physics underlying much of the discussion here, is that it does make sense to consider the purely one and two dimensional systems independently of the three dimensional aspects of the problem. There is a limit when this view point makes physical sense, and a few remarks concerning the nature of this limit might help to put this question in perspective.

The characteristic energies of the one (or two) dimensional system in comparison with the interactions between these lower dimensional subunits is the determining consideration. In some circumstances, it is rather obvious which energies determine this ratio, but in those of interest in the real lower dimensional systems, it is not obvious.

Some qualitative remarks are of interest. Consider the problem from the mean field viewpoint. There are several ways to define what we mean by mean field. The obvious one, treating all interactions on an equal footing, predicts an ordering temperature determined by the divergence of a three dimensional susceptibility. For the CDW state, this is given by

$$\chi(\vec{Q}) = \frac{\chi\,(\vec{Q})}{1 - V_Q\,\chi(\vec{Q})} \tag{33}$$

where χ_0 is a 3-dim, non-interacting susceptibility, and V_Q is a 3-d screened Coulomb interaction. Compare this with the equivalent problem viewed from a layered mean field theory viewpoint, given by

$$\chi(\vec{Q}) = \frac{\chi_{2D}(\hat{Q})}{1 - V\cos(Q_Z S)\,\chi_{2D}(\hat{Q})} \tag{34}$$

where $\vec{Q} = (\hat{Q}, Q_Z)$ has been used. Here V is the interaction between layers, and all complications non-essential to the comparison between the two types of mean field theory have been neglected.

In the former result, the ordering is determined by a non-interacting susceptibility. In the latter, it is the interacting susceptibility in the lower dimensional subdivision which appears. Obviously, the latter approach makes physical sense when the correlation length within a layer becomes much longer than the correlation length perpendicular to it. The correlation lengths play an important role in this approach. One length involves essentially one (or two) dimensional considerations. An estimate of this length is easy to obtain, it is just the soliton or charge density correlation length, ξ .

This length is to be compared with the perpendicular correlation length. Phenomenologically, we might suppose the susceptibility for the charge density wave is

$$\chi_d(Q) = \frac{\chi_0}{1 + \hat{Q}^2\xi^2}$$

where χ_0 is the static susceptibility. Expanding the $\cos Q_Z S$ gives rise to the determination of the correlation length ratio to be $(V\chi_0)^{\frac{1}{2}} S/\xi$. To the extent that $\chi_0 \propto \xi^2$, this ratio is unaffected by the specific nature of excitations.

However, the ordering temperature, determined by equating the perpendicular length with the layer separation, is most sensitive to the nature of the excitations. With the expectation that χ_0^{-1} is proportional to the gap in the soliton spectrum, the greater the gap, the lower the ordering temperature. Strong interactions within a layer can therefore lead to a lower temperature than would be expected within a more conventional approximation, since χ_0 is typically divergent at low temperatures when interactions are ignored.

References

1. J. P. Pouget, S. K. Khanna, F. Denoyer, R. Comes, A. F. Garito, and A. J. Heeger, Phys. Rev. Lett. 37, 437 (1976). S. Kagoshima, T. Ishiguro, and H. Anzai (to be published in J. Phys. Soc. Japan)

2. D. E. Moncton, J. D. Axe, and F. J. Di Salvo, Phys. Rev. Lett. 34, 734 (1975)

3. A. Luther, Phys. Rev. B15, 403 (1977)

4. R. Bullough, to be published in Topics in Modern Physics (Springer). L. D. Fadeev, JETP Letters 21, 64 (1975). R. Dashen, B. Hasslacher, and A. Neveu, Phys. Rev. D11, 3424 (1975)

5. S. Coleman, Phys. Rev. D11, 2088 (1975)

6. D. C. Mattis and E. H. Lieb, J. Math. Phys. 6, 304 (1965)

7. A. Luther and I. Peschel, Phys. Rev. B9, 2911 (1974)

8. A. Luther and V. J. Emery, Phys. Rev. Lett. 33, 589 (1975)

9. S. Brazovsky and I. Dzyaloshinsky, Zh EFT 71, 2338 (1976) to be published

10. A. Luther, Phys. Rev. B14, 2153 (1976)

11. G. Yuval and P. W. Anderson, Phys. Rev. B1, 1522 (1970). Here the parameter $\varepsilon = 2J_z\rho$ for small J_z is found.

12. V. J. Emery and A. Luther, Phys. Rev. B9, 215 (1974)

13. S. Doniach (to be published)

14. V. L. Berezinskii, Sov. Phys. JETP 32, 493 (1971) Ibid. 34, 610 (1972). This connection is reviewed by J. Jose (to be published)

15. W. L. McMillan, Phys. Rev. B14, 1496 (1976)

CHARGE DENSITY WAVES IN LAYERED COMPOUNDS

F. J. Di Salvo

Bell Laboratories

Murray Hill, New Jersey 07974 USA

The work I will describe was performed in collaboration with many others including J. D. Axe, S. Mahajan, D. E. Moncton, J. A. Wilson and J. V. Waszczak. While it is impossible to cover all the interesting phenomena observed in layered compounds, I will reference these phenomena where appropriate.

Let me outline what we will talk about. We'll start by studying the structure of those layered compounds which are expected to be metallic conductors. The physical properties, such as electrical transport, show clear anomalies that I will assert are associated with charge density wave (CDW) formation and/or changes in the CDW structure. At this point I will describe what a CDW is and how it comes about using a simple model, thus introducing the parameters that characterize the CDW. Next we'll come back to the layered compounds to get a feel for the magnitude and temperature dependence of the CDW parameters. Then we will discuss some of the effects of impurities on the CDW. We'll see, by comparison to experiments and band calculations, that the simple model first presented for CDW formation is not adequate to explain all the data. While more complicated models have been proposed, detailed microscopic theories are still lacking. A number of review articles have been written concerning the CDW properties of these compounds,(1,2,3) including one similar to this given at the 1976 NATO conference on one dimensional conductors.(4)

1. STRUCTURE AND PROPERTIES

The layered transition metal dichalcogenides (5) have the chemical formula MX_2, where X = S, Se, or Te and M can be any

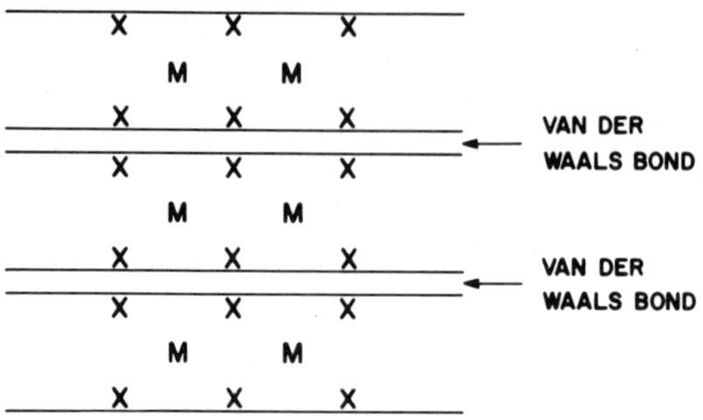

Figure 1. A schematic of the structure of MX_2 layered compounds shows the three atom thick sandwiches held together by relatively weak forces between adjacent sheets of X atoms.

one of a large number of metals from the periodic table. Here we will discuss primarily M = V, Nb, Ta (group Vb) and Ti (from group IVb). Since the anions are divalent, the electron configuration of the group Vb metals is d^1 and of IVb is d^0. The structure of the compounds, schematically illustrated in Fig. 1, is formed from 3 atom thick sandwiches. The top and bottom sheet of the sandwich is comprised of close packed chalcogenide (X) atoms, while the middle sheet is comprised of metal atoms. The bonding within a sandwich is strong (covalent or ionic), but between sandwiches (between adjacent X sheets) it is weak - usually labeled van der Waal's bonding. Consequently the physical properties of these compounds are anisotropic or "quasi-two dimensional". For example, these materials cleave easily, parallel to the sandwiches (or layers) much like graphite or mica. Many of the compounds are polymorphic, for two reasons: (a) The M atoms in a given sandwich are either all octahedrally coordinated (O) by X atoms or all trigonal prismatically coordinated (TP) (b) the layers can be stacked on top of one another in several different ways due to the weak interlayer forces. The unit cells, however, can all be described in the hexagonal system with the a-axis equal to the intralayer M-M distance and the c axis some multiple of the layer thickness. We will concern ourselves primarily with the two simplest polytypes: 1T - in which all the M atoms are O coordinated, and 2H - in which all M atoms are TP coordinated.

Other metals and many organic molecules (Lewis bases) can be inserted between the layers, i.e. intercalated.[(6-10)] Partly because of the weak interlayer bonding, intercalation usually proceeds rapidly at or near room temperature. The driving force for intercalation appears to be electron donation to the conduction band of the layers. Those layered compounds that form the most intercalation compounds have the largest electron affinity (i.e. work function), while the more active guest species are stronger electron donors. While these general principles of intercalation chemistry are empirically true, an adequate microscopic picture of how organic molecules bind to the layers is not available. Later we will talk about possible interactions between the CDW and the intercalated species.

Since the group Vb compounds are d^1, we expect them to be metallic because of the moderately close M-M distance (a $\simeq$ 3.3A) and the largely covalent nature of the bonds. More sophisticated theory, such as the APW band calculations of L. F. Mattheiss (11), leads to the same conclusion. The uppermost filled bands are primarily based on M d states, the density of states at the Fermi level for the group Vb compounds being 5 to 20 times that of Cu, for 1T and 2H polytypes respectively. Consequently, we expect these materials to be metallic conductors with conductivities that are approximately one order of magnitude smaller than that of Cu metal.

Some of the group Vb layered compounds are superconductors. 2H-$NbSe_2$ has the highest transition temperature (12) (7.2 K), while for most others T_c is below 1.0°K (2H-TaS_2 T_c = 0.8°K (13), 2H-$TaSe_2$ T_c = 0.2 K (13)). The transition temperature is changed by intercalation. For example, the T_c of 2H-TaS_2 can be increased to ∿ 5 K. (6,7) The occurrence of superconductivity leads one to expect that reasonably large electron phonon coupling occurs in these systems. The McMillan λ obtained from the T_c and other physical properties of 2H-$NbSe_2$ is 1.0. (14) The large oscillator strengths of the optically active lattice modes observed by infrared reflectivity in the group IVb compounds also lead one to expect a large electron-phonon coupling (15).

The resistivities (current parallel to the layers) of the 1T and 2H polymorphs of TaS_2 and $TaSe_2$ are shown in Fig. 2. The original investigators of the transport properties are given in the references: 1T-TaS_2 (16), 1T-$TaSe_2$ (17), 2H-TaS_2 (18), and 2H-$TaSe_2$ (19). While the resistivity (ρ) of the 2H-polytypes has a metallic like slope, there is a sharp decrease at low temperatures. The resistivity of the 1T polytypes, however, does not look like that of a simple metal, and there are sharp discontinuities at first order transitions.

The anomalous properties of these chalcogenide compounds were first noted in a magnetic susceptibility study by Quinn et al.[20] As might be expected, the anomalies seen in Fig. 2

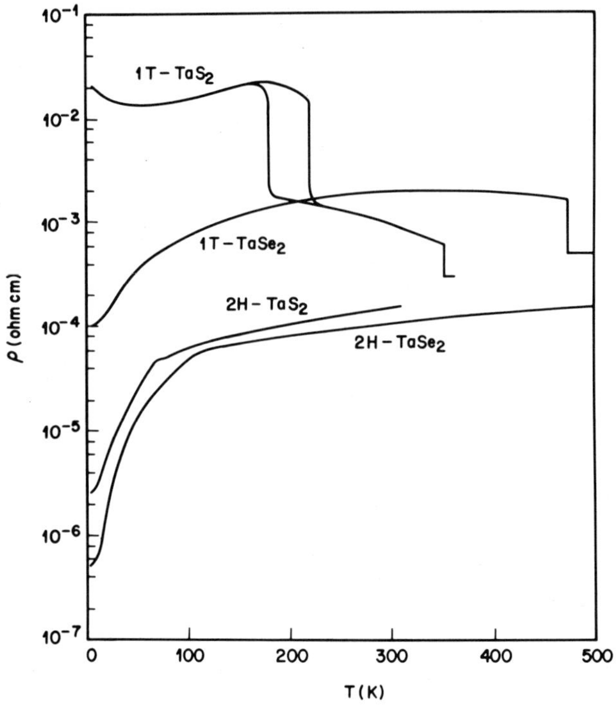

Figure 2. The electrical resistivity parallel to the layers of several layered compounds from 4.2K to 500K.

are observed in almost any other physical property; such as, magnetic susceptibility,(1) Young's modulus,(21) heat capacity,(22) and in microscopic measurements such as NMR(23,24,25) or in some cases XPS.(26,27)

At this point let me just assert that these anomalies in the physical properties (Fig. 2) are due to CDW formation and proceed to discuss what a CDW is and how it occurs, before attempting to explain these measurements.

Charge Density Wave instabilities were theoretically proposed by A. W. Overhauser in 1968(28), where he placed an emphasis on the correlation energy as the source of the instability. The source of the instability in these compounds appears to be somewhat different as outlined below.

A CDW is a static, coupled, periodic distortion of both the conduction electron density and the lattice. One is not the consequence of the other, but they are intimately tied together. We can see why, using a simple one dimensional model of a metal (Fig. 3). We consider a row of uniformly spaced positive ions and a uniform conduction electron density to preserve overall charge neutrality. If a sinusoidal perturbation of the conduction electrons occurs, the net charge, including the ions, oscillates from negative to positive at maxima and minima in the wave. The Coulomb energy of such a state is large. In fact, such an excitation is a plasmon, one quantum of which, in normal metals, costs on the order of 10 eV. However, if the positive ions move toward the maxima and away from the minima, the Coulomb energy can be greatly reduced and such a coupled distortion may become the stable ground state of the system. Consequently, a CDW is more likely to occur in systems with large electron-phonon (electron-lattice) coupling. Below the onset temperature of the CDW, T_0, the charge density and atomic displacements would be given by (for this simple model)

$$\rho(r) = \rho_0(r)[1+\phi \cos \vec{q} \cdot \vec{r}] \tag{1}$$

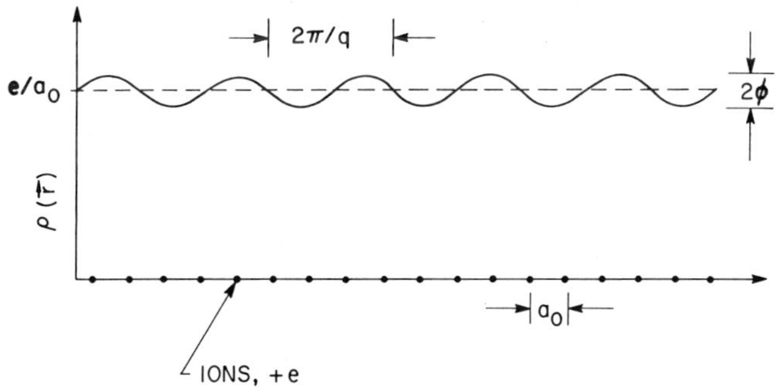

Figure 3. One dimensional model of a metal - used to show why electron phonon coupling is important and the CDW is a coupled distortion of the lattice and the conduction electron density.

$$\Delta\vec{x} = \vec{A} \sin \vec{q} \cdot \vec{r} \qquad (2)$$

where ϕ and $\vec{A}$ are the amplitude in charge density and atomic displacement respectively.

We further expect that lattice waves with this wave vector $\vec{q}$ and displacements parallel to $\vec{q}$ (longitudinal phonon) will be lowered in energy above T_o, becuase of the effective screening of the ionic charge by the electrons. Thus a dip, called a Kohn anomaly, will appear in the phonon dispersion curve (phonon energy vs. wave vector). Simple one dimensional models for this effect are published (29); some examples for the layered compounds will be given below.

Beside a large - electron phonon coupling, the shape of the Fermi surface (F.S.) is also important. In particular the F.S. determines the $\vec{q}$ of the distortions, and whether the CDW can be the ground state of the system. We can see what kind of F.S. is needed by considering a simple linear response model of the conduction electrons to a static perturbation of wave vector $\vec{q}$. First, we consider free electrons; later we will include interactions. In order to determine if an instability occurs, we use the following principle: If the response $\Delta\rho(\vec{q})$ to an infinitesimally small perturbation $V(\vec{q})$, becomes macroscopic, the system will spontaneously move to a distorted state. That is, if $\Delta\rho(\vec{q}) = \chi^o(\vec{q})\ V(\vec{q})$, a static distortion will spontaneously occur if $\chi^o(\vec{q}) \rightarrow \infty$.

For free electrons $\chi^o(\vec{q})$ is calculated in second order perturbation theory to be: (neglecting the matrix element $|\langle\vec{k}|e^{i\vec{q}\cdot\vec{r}}|\vec{k}+\vec{q}\rangle|^2$)

$$\chi^o(\vec{q}) \ \alpha \ \sum_{\vec{k}} \frac{f_{\vec{k}}\,(1-f_{\vec{k}+\vec{q}})}{\varepsilon_{\vec{k}+\vec{q}}-\varepsilon_{\vec{k}}} \qquad (3)$$

where

$\varepsilon_{\vec{k}}$ = energy of state with wave vector $\vec{k}$

$f_{\vec{k}}$ = Fermi occupation factor

We can see that $\chi^0(\vec{q})$ will become large for a given $\vec{q}$, which connects many filled states to empty states of the same energy. This can only occur if the two states are on the F.S. Figure 4 shows $\chi^0(\vec{q})$ for several situations.[30] F.S. with plane parallel sections, or equivalently, nesting electron and hole surfaces, will produce a $\chi^0(\vec{q})$ that diverges as $\ell n\ (E_F/kT)$ when $\vec{q}$ spans (or connects) the F.S. Other cases, such as saddle points in the band structure at the Fermi level, will also lead to a similar divergence in $\chi^0(\vec{q})$.[31] We see, then, that a free electron gas can be unstable only a T = 0.

If electron - phonon and electron - electron interactions are included, instabilities may occur at finite temperatures. When interactions are included, we consider the generalized susceptibility, $\chi(q)$. In the simplest approximation (the random phase approximation) we have

$$\chi(\vec{q}) = \chi^0(\vec{q})/(1 - X(\vec{q})\ \chi^0(\vec{q})), \qquad (4)$$

where $X(\vec{q})$ represents all the physics - i.e., the interactions.[32] Now we see that an instability will occur at the $\vec{q}$ where $X(\vec{q})\chi^0(\vec{q}) = 1$. (Note that the RPA expression is only valid where the denominator is close to 1, we take some license here in letting it decrease to zero in order to obtain some feel for the true physics). Consequently if the F.S. has parallel (or nearly so) sections, $\chi^0(q)$ will increase as T decreases, so that at some temperature, T_0, the denominator equals zero and the system becomes unstable. It is also true that $X(\vec{q})$ will have some structure, possibly <u>also</u> peaking when $\vec{q}$ spans the F.S. We will see later some calculations of $\chi^0(q)$ for some of the layered compounds.

Overhauser's original suggestion was that a CDW might occur when $\chi^0(q)$ had no peaks or was essentially free electron like (3 DIM-Fig. 4a). In that case, using Eq. 4 as a model, the divergence comes from $X(\vec{q})$. Indeed, Overhauser showed that the correlation energy for unscreened Coulomb potentials in a Hartree-Fock calculation diverges at $q = 2k_F$.[33] We return later to some discussion of the role of $X(\vec{q})$ and $\chi^0(\vec{q})$ in determining the instability. Note, however, that whichever term dominates in these models, the divergence in $\chi(\vec{q})$ will occur at a F.S. spanning vector.

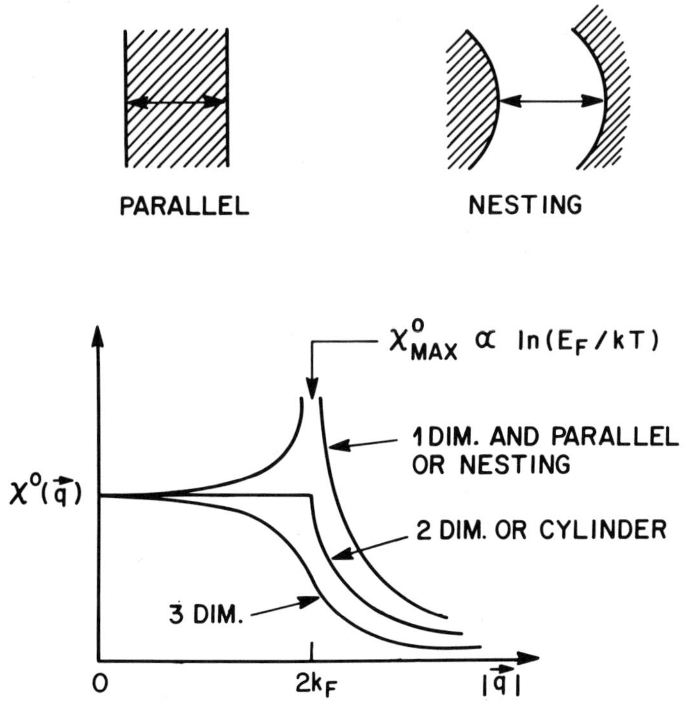

Figure 4. $\chi^{0}(\vec{q})$ for a number of free electron Fermi surfaces is shown vs. $|\vec{q}|$. 1, 2, and 3 DIM. refer to electrons with no periodic potential from the lattice in the given dimension. For real solids, some species of the band structure may have one dimensional dispersion. Two examples are shown: parallel and nesting pieces of Fermi surface, where the filled states are shaded.

Within the spirit of eq. 4, CDW instability occurs if the electron-phonon coupling dominates $X(\vec{q})$,(32) as suggested by the simple one dimensional model of Fig. 3. If $X(\vec{q})$ is dominated instead by the exhange, a spin density wave (SDW) will result(32), as in the metal Chromium(34).

Since $\vec{q}$ is determined by the F.S., the wavelength of the CDW, $2\pi/|\vec{q}|$, will usually be incommensurate with the lattice; that is, the wavelength will not equal a lattice translation. However, in many layered compounds, a first order transition to the commensurate state (CCDW) occurs at $T_d < T_o$. It is the incommensurate CDW (ICDW) that is the mark of a F.S. driven instability, at least in a metal where magnetic moments or permanent electric dipoles do not occur. In cases where only a commensurate distortion exists below T_o, it is more difficult to assess the role of the F.S. in "driving" the transition.

What happens below T_o? The instability produces a lattice distortion of wave vector $\vec{q}$, and a new lattice potential. This potential connects the states at the F.S. at $\vec{k}$ and $\vec{k}+\vec{q}$ in first order perturbation, splitting the states away from the Fermi level. That is, gap is produced at the F.S., just over those regions spanned by $\vec{q}$. Consequently these states contribute to χ^o with the gap energy Δ as a denominator, not zero as above T_o. Consequently, χ^o decreases in magnitude below T_o. Within simple models, the gap is expected to increase with decreasing temperature in exactly the same way as the BCS superconducting gap (29).

While this theoretical description of a CDW may be lacking, our purpose was mainly to introduce the parameters T_o, T_d, $\vec{q}$, ϕ, $\vec{A}$, and Δ; the last fcur of which are expected to be temperature dependent. We now return to the layered compounds and consider some of these parameters in more detail.

The CDW state below T_o can best be observed by diffraction techniques. Elastic scattering will occur at points in reciprocal space given by $\vec{k} = \vec{G} \pm n\vec{q}$ where $\vec{G}$ is any one of the reciprocal lattic vectors of the undistorted lattice that exists above T_o and n is any integer.[35,36] Thus, each main Bragg peak will have a series of satellite peaks about it with an intensity given from simple kinematic scattering theory by $I_n \sim (\vec{k}\cdot\vec{A})^{2n}/n!$ (Eq. (5)) for $\vec{k}\cdot\vec{A} << 1$. In fact, in the layered compounds, $\vec{A}$ is usually quite small, and the satellite peaks have only 10^{-3} or less of the intensity of the strongest main Bragg peak. This made the CDW's very difficult to find by simple power X-ray diffraction techniques. They were first discovered by electron diffraction, since the intensity of the satellite peaks can be greatly enhanced over that expected from the kinematic formula by dynamical scattering.[37] While the position of the satellite peaks is easy to obtain by electron diffraction (thus obtaining $\vec{q}$), the intensity cannot be easily used to obtain lattice displacements. The latter are obtained from X-ray or neutron diffraction measurements.

Table I lists a number of layered compounds with the CDW onset temperature, T_o, the lock-in temperature, T_d, and CDW wavelength in multiples of the a-axis. In all cases the CDW consists of a superposition of three CDW's with q vectors that are related by a 120° rotation about the c-axis (the normal to the layers). No ditellurides are listed in Table I. These compounds frequently show strong distortions at room temperature,(5) and there is some indication of phase transitions at high temperatures.(38) However, the ditellurides have a wide range of nonstoichiometry and are difficult to prepare at the exact MX_2 composition. This introduces another "parameter" in the problem, and we have not yet undertaken detailed study of these compounds.

A large variation of T_o is seen both with a change in polytype and between different compounds. A CDW has not been detected in 2H-NbS_2, but this particular compound is difficult to prepare, usually $Nb_{1+x}S_2$ is obtained. We will later see that this non-stoichiometry frequently lowers or eliminates T_o. Note also that the CDW wavelength is short - between 3 and 4 lattice parameters (or only 10 to 14 Å.)

Table I

Material	T_o(K)	T_d(K)	λ_{CDW}	Ref.
1T-$TaSe_2$	$\approx$600	473	∿3.5a	1, 17
2H-$TaSe_2$	122	∿95	∿3.0a	1, 39
1T-TaS_2	$\approx$600	$\approx$200	∿3.5a	1, 2, 40
2H-TaS_2	∿80	?	∿3.0a	1, 41
2H-$NbSe_2$	32	no	∿3.0a	39
1T-VSe_2	112	∿80	∿4.0a	2, 42

2H-$NbSe_2$ has the lowest T_o and the CDW remains incommensurate down to 4.2K. The CDW wavevector $\vec{q} = (1-\delta)\vec{a}^*/3$, where $\vec{a}^*$ is a reciprocal lattice vector in the plane and δ decreases smoothly from 0.02 at 32K to approximately 0.01 at 4.2K. The intensity of a first order satellite peak (n=1) in 2H-$NbSe_2$ as measured by neutron diffraction is shown in Fig. 5 vs. temperature (39). The transition at T_o is seen to be second order, since the intensity smoothly drops to zero. The atomic displacements at 4.2K calculated from this data are approximately 0.05 Å for Nb (parallel to the layers) and half that for the Se atoms (some component perpendicular to the layers.) The amplitude in charge density ϕ at 4.2K is estimated from NMR measurements to be 5% to 10% of the conduction electron density

(19, 20). Measurements of the specific heat[18] and Young's modulus[17] near T_o show that the density of states at the Fermi level is not significantly lowered by the presence of the CDW (i.e. the F.S. area is not significantly reduced by the presence of the CDW).

Next we consider the CDW transition in $2H-TaSe_2$. The results of neutron scattering measurements[32] are shown in Fig. 6. The inset in the upper left shows the scattering peaks in the (HK0) plane. The open circles are the main Bragg peaks and the dark circles are the satellite peaks. If we look in the main part of the figure about the position labeled 4 (position 4 is at $4\vec{a}^*/3$), we see just below T_o = 122K that $\vec{q} = (1-\delta)a^*/3$ and $\delta \sim 0.02$. As T decreases, δ decreases until it discontinuously goes to zero at $T_d \sim 95K$. Note also that a weak secondary peak appears on the other side of $4\vec{a}^*/3$ by 2δ. This peak occurs at the same position as the second order satellite coming from the Bragg peak at 6 (i.e. $2\vec{a}^* - 2\vec{q} = \frac{a^*}{3}(4+2\delta)$). However, the intensity calculated from Eq. (5) is much too small to be the source of this peak. Consequently, this is due to a second periodic distortion $\vec{q}_{2\delta} = (1+2\delta)\,\vec{a}^*/3$. A Landau free energy model has been developed by D. E. Moncton et al. (32) that shows how this secondary distortion occurs and its role in "pulling" $\vec{q}$ toward the commensurate value of $a^*/3$. W. L. McMillan has also developed more complicated Landau models to explain this behavior and other related phenomena,[43,44] which he will discuss at this meeting. The main point of these measurements and models is that the CDW lattice distortions are not purely sinusoidal, but include higher harmonics that are mixed in by anharmonic terms in the free energy. The displacement grows to $\vec{A} \sim 0.1$ Å for Ta and ~0.05 Å for Se at 4.2K. If we scale by the ratio of atomic displacements from the $2H-NbSe_2$ data, we estimate ϕ (4.2K) $\sim$ 0.2 (e/a).

At present, the most data concerning CDW properties has been obtained with $2H-TaSe_2$. Figure 7 shows the Kohn anomaly seen in the longtitudinal acoustic phonon at 300K.[39] The softening occurs over a wide region of this branch, and the minimum near q_o decreases with decreasing temperature as shown in Fig. 8.[45] The simpler models predict a soft mode with $\omega(q_o) = 0$ at T_o.[29] Instead a central peak develops as T approaches T_o. Below T_o new modes appear, one of which is the "stiffening" phonon seen in Fig. 8. This and other modes have been observed by Raman scattering below T_o[47,48], and will be discussed at this meeting by E. Steigmeyer. Infrared reflectivity measurements below T_o show weak gap-like features with Δ (4.2K) $\gtrsim 25kT_o$.[46] This number is much too large, since the BCS like theories give $\Delta \gtrsim 4kT_o$[29]. McMillan has shown that these difficulties can be resolved by including the phonon entropy (this conference).

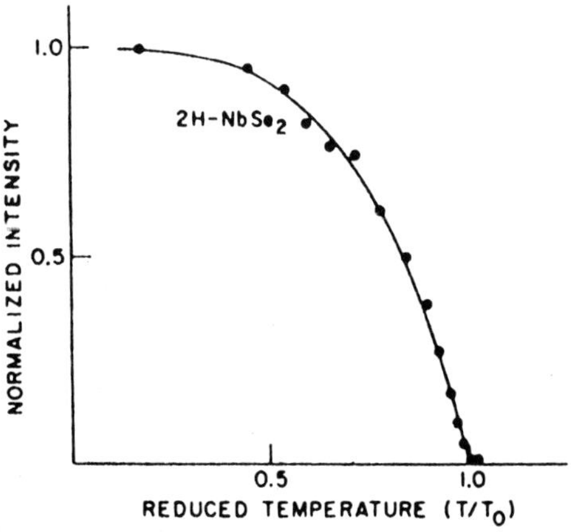

Figure 5. The satellite intensity vs. temperature for 2H-$NbSe_2$ below T_0 = 32K. The intensity is proportional to the lattice displacement squared. The lattice displacement may be used as an order parameter in a Landau theory.

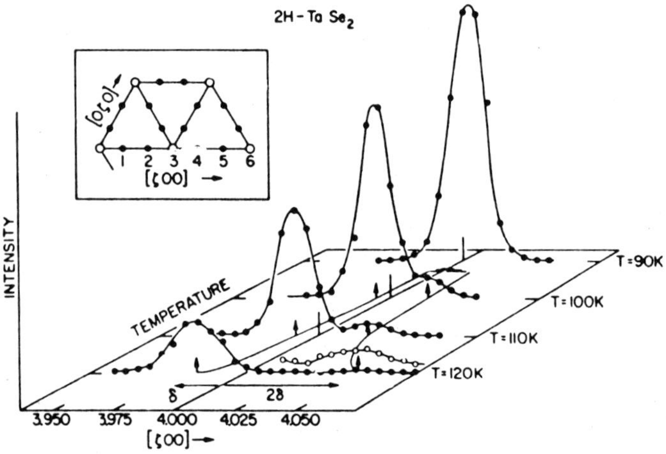

Figure 6. Elastic neutron scattering measurements of the satellite peak intensity and position in 2H-$TaSe_2$ vs. temperature.

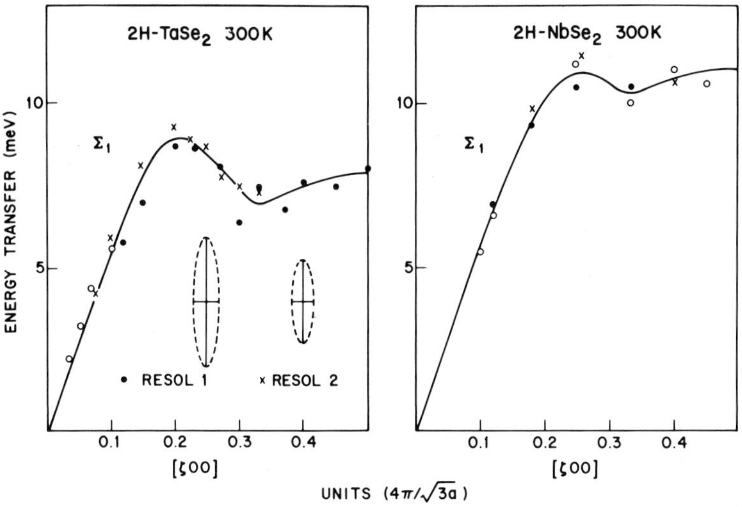

Figure 7. The longtitudinal acoustic phonon dispersion curves for 2H-$TaSe_2$ and 2H-$NbSe_2$ show pronounced but very broad dips near $q_o = 0.33\ a^*$. The curves are obtained by the constant Q method, the resolution functions shown are swept up parallel to the energy axis to obtain this data.

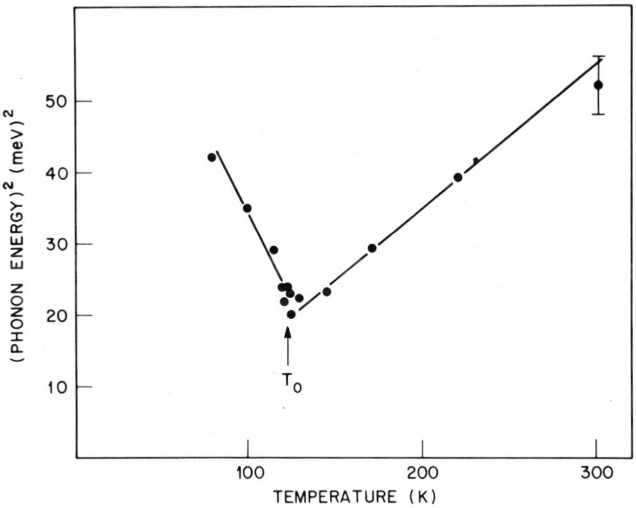

Figure 8. The energy of the longtitudinal acoustic phonon (squared) at $q_o = 0.33\ a^*$ in 2H-$TaSe_2$ decreases as T_o is approached from above or below. However, this energy does not decrease to zero and T_o, as simple models would suggest.

The CDW behavior of the 1T polytypes of the Ta dicholcogenides is quite different from that of the 2H. In 1T-$TaSe_2$ electron diffraction patterns of the (HK0) plane show a commensurate superlattice at room temperature (Fig. 9a) and an incommensurate CDW (Fig. 9b) above the first order transition apparent in the resistivity (Fig. 2) at T_d = 473K. In the incommensurate state $\vec{q}=0.285\vec{a}^*$ (i.e. $\vec{q}$ is parallel to the line joining main Bragg peaks). Also, strong diffuse scattering, in the form of circular rings is seen above T_d. This scattering is most likely due to CDW excitations involving the transverse displacement of $\vec{q}$. We expect that excitations of this sort will be soft, since at T_d $\vec{q}$ rotates by $13^o54'$ and shrinks slightly (∿2%) to produce the 3×1 superlattice apparent in Fig. 8a. This transition then involves primarily a rotation of $\vec{q}$. At room temperature and below, the atomic displacements are quite large, ≈0.25 Å for Ta. Further, the amplitude ϕ is about 1 e/a! This might be expected if we scale ϕ from 2H-$NbSe_2$ by the ratio of the onset temperatures. With such a large charge oscillation, the binding of the Ta core electrons shifts enough to be observable in ESCA (26,27,49). Splittings in the 4f binding energies of ∿0.5 eV are clearly observed at room temperature. This means that at room temperature and below, the CDW is not a weak perturbation of the F.S. Rather the commensurate phase may be thought of as a valence disproportionation of Ta^{4+} into Ta^{5+}, Ta^{4+}, Ta^{3+}. Summarizing the behavior of 1T-$TaSe_2$, we see that as the temperature is reduced from above T_o, the material passes through a number of states: a normal undistorted metal, then a second order transition to an incommensurate CDW state where, at least close to T_o, the amplitude ϕ or $\vec{A}$ is small, and finally, this state evolves through the transition at T_d to a valence disproportionation. As yet there are no adequate theories to explain the overall behavior of 1T-$TaSe_2$, although Landau models are able to qualitatively predict the sequence of transitions (43).

1T-TaS_2 is even more complicated than 1T-$TaSe_2$, as is apparent from the two first order transitions seen in the resistivity (Fig. 2). Below 200K the CDW shows the same commensurate state as 1T-$TaSe_2$ (2). Above the transition at T_d' = 350K, the CDW is incommensurate with $\vec{q} = 0.288\vec{a}^*$ and strong diffuse scattering is seen, very much as that shown for 1T-$TaSe_2$ in Fig. 9b (1,40). At 350K, $\vec{q}$ rotates by ∿12°, but stops 2° short of becoming commensurate! $\vec{q}$ continues to rotate with decreasing temperature, until at ∿200K it jumps the last fraction of a degree to become commensurate. This intermediate state between 200K and 352K (quasi commensurate state as we call it) appears to be stabilized by the presence of secondary distortions.[50] The physics of this situation is similar to that of 2H-$TaSe_2$ where secondary distortions appear immediately below T_o. In

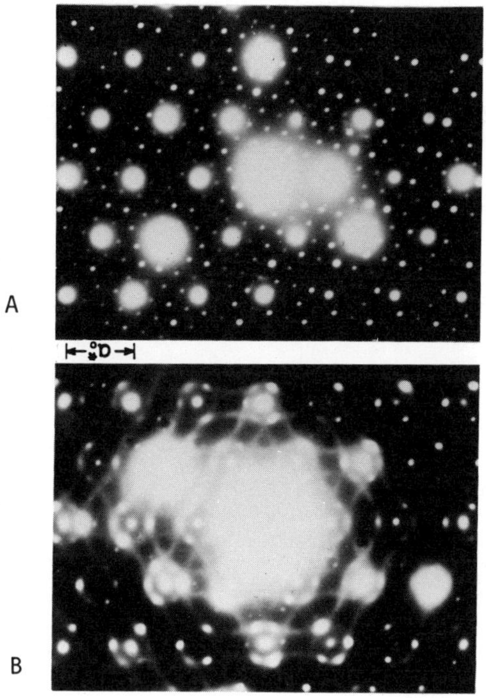

Figure 9. (a) The basal plane diffraction pattern of the CCDW in 1T-$TaSe_2$ shows the $\sqrt{13}$ a superlattice. Many of the main Bragg peaks appear very bright, or overexposed; the remainder of these peaks are easily found by "extending" the hexgonal pattern of these spots.

1T-$TaSe_2$, however, the amplitude of the CDW must be larger than some minimum value for the secondary distortions to occur. Again it is clear from ESCA and X-ray diffraction that the CDW amplitude at low temperatures is as large in 1T-TaS_2 as in 1T-$TaSe_2$.

We have talked about transitions to the CCDW, but have not tried to indicate why they occur. A hint to their origin can be obtained from a correlation between T_d and the crystallographic c/a ratio discovered by A. H. Thompson (51). Figure 10 shows the almost linear relation between T_d and c/a for 12 different compounds (or polytypes) that I will not bother to identify in detail.

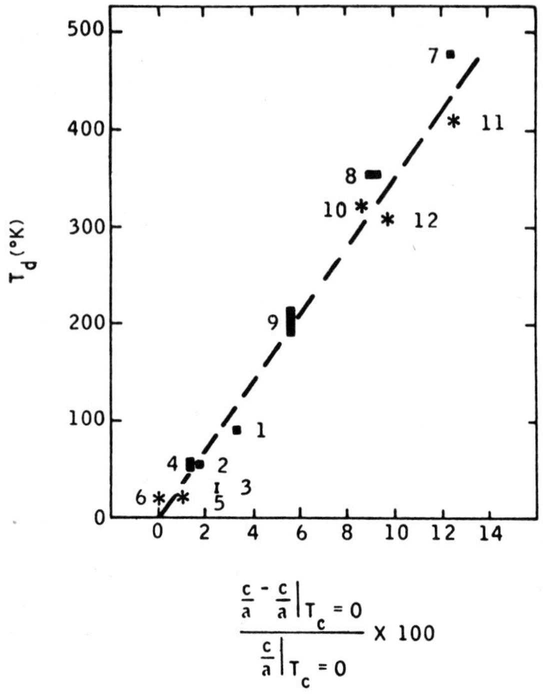

Figure 10. T_d is plotted vs. the crystallographic (normalized to a specific value) from Thompson (24). Each point represents a different layered compound or a different polymorph.

Previously, F. R. Gamble had shown that the c/a ratio in these compounds was related to the ionicity difference between the cation and anion (52). Consequently, we see that T_d is related to an ionicity difference. This result suggests that the driving force toward the commensurate state involves local ionic Coulomb or covalent bonding forces. One might see how these forces arise by considering a simple case. Suppose an incommensurate CDW exists in a two dimensional hexagonal packed sheet of metal atoms. Since the CDW is incommensurate, in general the charge will be increased at some nonsymmetrical position. Ionic or covalent energies will be maximized by placing this charge at a center of symmetry; such as, (a) on a metal atom, (b) half-way between two atoms to maximize the bonding charge, (c) at a geometrical center, like the center of a triangular set of three atoms, to make a bonded metal cluster, (d) etc. In the real compounds we must also consider the metal-nonmetal bonds, but this simple model gives the qualitative idea. Note that these bonding interactions can be expressed in terms of large electron phonon interactions. Consequently it is somewhat artificial to separate the driving forces for the ICDW and the CCDW, as pointed out by Thompson.[3] Rather in these materials a proper description of the conduction electrons is between the traditional covalent bond and metallic bond models, as pointed out by McMillan.[43] Apparently in these systems at small CDW amplitude the F.S. determines the nature of the distortions, while at larger amplitudes the bonding energies dominate.

So far we have talked about the CDW as a single layer phenomena. The CDW's interact from one layer to the next producing a given stacking sequence. This sequence is consistent with that obtained by minimizing interlayer Coulomb interactions (1,53). For example, if we consider the simplest CDW $\Delta\rho = \frac{\phi}{3} \times \sum_{i=1}^{3} \cos \vec{q}_i \cdot \vec{r}$, the contours of $\Delta\rho$ have hexagonal symmetry as shown in Fig. 11a. The maxima (or minima, depending upon the sign of ϕ) are at the cell edges, with minima at the center of each triangle. Using hexagonal notation, we label the maxima A and the two minima B and C as shown. In the incommensurate phase, the origin of the CDW can be placed arbitrarily at any point in the layer. If we choose point A in the first layer, then the Coulomb interaction is minimized with the next layer by placing its charge maxima over point B in the first layer. The third layer minimizes its Coulomb energy with <u>both</u> the first and second layer by putting its maxima over point C. This stacking sequence leads to a three layer repeat for the CDW as is found in the 1T polytypes. The interaction energy in the 2H polytypes is modified by the screw symmetry between adjacent layers and the CDW repeat appears to be two layers. In the commensurate phase, the origin of the CDW cannot be arbitrarily chosen. Rather, it appears that the CDW cell origin lies at a Ta site. Figure 11b shows that the CDW in the second layer can minimize its interaction energy with the first by translating the CDW origin by $2\vec{a}$. By continuing this sequence, the overall interaction energy is minimized. The origin of the CDW repeats itself every <u>13</u> layers. However, the true unit cell is triclinic with $a = \sqrt{13}a$, $b = \sqrt{13}a$ and $c = |\vec{c}+2\vec{a}|$ (3). In order to have long range stacking order both near neighbor and next-near-neighbor layer interactions must be included. In 1T-$Ta_{1-x}Zr_xSe_2$ the random impurity potential of Zr can dominate the next near neighbor interactions in the CCDW state,[53] producing a random stacking of near neighbor layers on the three sites near the B position shown in Fig. 11b.

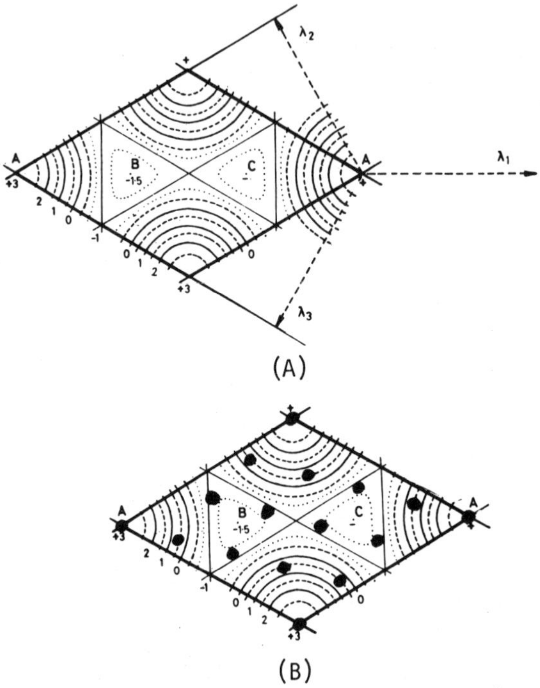

Figure 11. (a) The CDW pattern created by $\rho = \sum_{i=1}^{3} \cos \vec{q}_i \cdot \vec{r}$ with the 3 CDW wavelengths $\lambda_i = 2\pi/q_u$ shown. (b) The CDW pattern and metal atom positions expected for the CCDW in 1T-TaS_2/Se_2.

Having discussed in some detail the magnitude of the CDW parameters of some layered compounds, we briefly consider the resistivity. It is apparent from Fig. 2 that below the onset of the CDW the 1T polytypes become more resistive and the 2H polytypes less resistive. The simple picture introduced in discussing the driving force of the CDW leads one to expect the behavior observed for 1T-$TaSe_2$; that is, below T_0 the resistivity should increase (compared to the normal metal) as T decreases due to the formation of gaps at the F.S. At T_d the gaps increase discontinuously, further increasing the resistivity. Finally at low temperatures we expect metallic-like conductivity from those portions of the F.S. not destroyed by CDW gaps.

The resistivity of the 2H polytypes of TaX_2 rises slightly then decreases rapidly below T_o. A plausible model for this behavior is to assume that the sections of the F.S. eliminated by the CDW gaps are characterized by large effective masses. Above T_o, these carriers do not contribute much to the conductivity because of their large mass, but they do act as effective scattering sinks for lighter electrons on the F.S. Below T_o then, with the large mass states removed, the scattering rate can decrease leading to a decrease in resistivity even when some carriers are lost. This model is consistent with the nesting model of the CDW or the saddle point model (31) and the complex shape of the two sheeted F.S.,(1,11) but definite proof of its validity does not exist.

Finally we come to 1T-TaS_2. At each transition toward the commensurate state we expect the gaps to increase in size and the resistivity to increase. However, in the commensurate state the resistivity is very high and not at all metallic-like. Recent data show that the low temperature increase in resistivity (below 40K) is extrinsic.(54) The resistivity below 2K is described by variable range hopping and fits $\rho = \rho_o \exp (T_o/T)^{1/3}$. This form is due to Anderson localization by random impurities or defects.(55) Just why 1T-TaS_2 should be so sensitive to these impurities, and not 1T-$TaSe_2$, is not clear; but the high resistivity of 1T-TaS_2 just below 200K indicates that the remaining carriers have a very low mobility. Suggestions of Mott localization(56) and polaron formation(1) have been made, but these are untested hypotheses.

Also note from Fig. 1 that the resistivity of all of these compounds above T_o is on the order of 10^{-4} Ω/cm. This leads to an effective scattering time for the carriers of $\approx 2\times10^{-15}$ sec. We expect the average Fermi velocity to be less than that of (Cu), for example, so $V_F < 1.6\times10^8$ cm/sec. Consequently, the mean free path, is less than 30 Å. These small numbers indicate that the treatment of scattering by the usual Boltzman equation is questionable. A similar observation has been made for other transition metal compounds, in particular, the high superconducting transition temperature β-W compounds.(54) These difficulties are probably related to the large electron phonon interactions and the CDW instability of the layered compounds.

Next, we consider the effects of impurities. In particular, we consider the random substitution of the cation by other transition metals and the effect of this substitution on the transport properties and on CDW formation. This substitution causes randomness in the lattice potential and may also change the average conduction electron density, z. Each of these effects is related in different ways to changes in the CDW behavior.

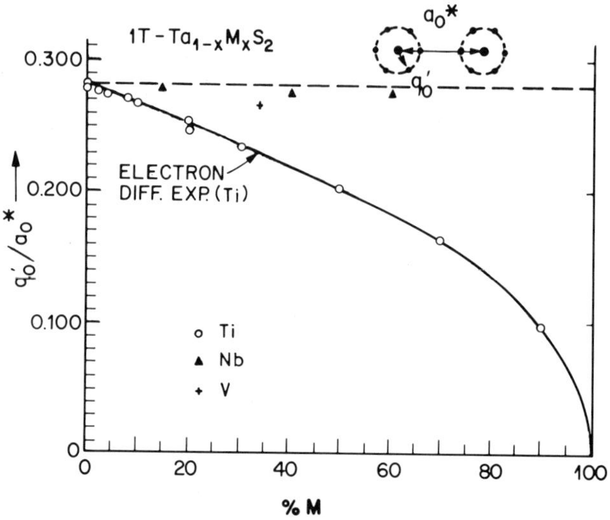

Figure 12. q/a^*, the wavelength of the CDW in the ICDW state divided by the reciprocal lattice vector, is shown vs. x for $1T\text{-}Ta_{1-x}M_xS_2$ where M = Ti, Nb, or V.

First, consider the effect of changing z: The calculated F.S. for the undistorted 1T polytypes has the shape of an ellipsoid in the plane of the layer with near perpendicular walls along c^*.(1,11) This F.S. then has sections that are near to parallel, leading to the CDW instability. If Ti is substituted for Ta, z decreases, and in the rigid band approximation the F.S. will shrink but remain an ellipsoidal cylinder. Consequently, in the ICDW phase we expect that q/a^* will decrease with increasing x in $1T\text{-}Ta_{1-x}Ti_xS_2$. The results of such measurements are shown in Fig. 12 (1,58). The solid line is a fit to the data which is close to that expected from the rigid band approximation. Also in Fig. 10 q/a^* is shown for $1T\text{-}Ta_{1-x}Nb_xS_2$ and $1T\text{-}Ta_{1-x}V_xS_2$. In these two cases q/a^* is close to constant, as expected, since Nb and V are isoelectric with Ta and z is constant. In the 1T polytypes, the effect of changing z is to smoothly change the F.S. and consequently q/a^*. The data shown in Fig. 12 are obtained at or above room temperature. Consequently T_o remains greater than 300K even for x $\gtrsim$ 0.7. It appears from magnetic susceptibility that T_o is reduced, but slowly, with increasing x. This is an expected effect of disorder (43) and occurs with Ti, Nb or V substitution (or indeed <u>any</u> cation disorder).

The disorder reduces T_o, but more rapidly suppresses the commensurate state. This can be seen in the resistivity of $1T\text{-}Ta_{1-x}Ti_xSe_2$ (Fig. 13). With increasing x, both the transition temperature from the incommensurate to the commensurate state, T_d, and the magnitude of the resistive anomaly at T_d decrease. This

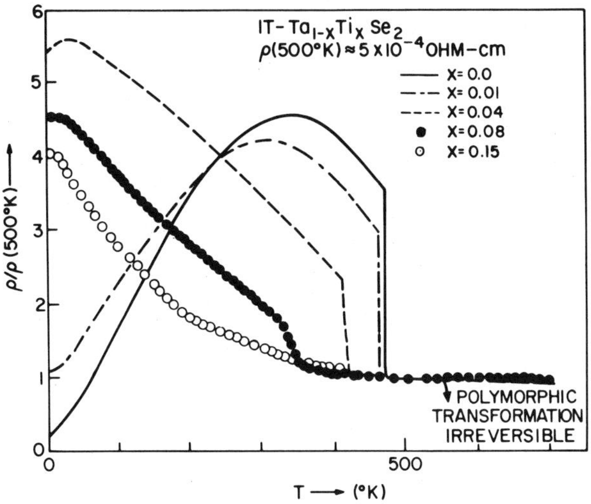

Figure 13. The electrical resistivity of $1T\text{-}Ta_{1-x}Ti_xSe_2$ shows the decrease in T_d with increasing x.

data, combined with measurements of the enthalpy of transition (58), show that for $x > x_c \sim 0.10$ the CCDW does not occur. We can see why this occurs with a simple chemical model. Consider a one dimensional string of atoms A and B that are randomly placed on lattice sites. We wish to compare the free energy of the CCDW and ICDW. Let us consider B to be the dilute species and assume that B is more electronegative than A. (We could assume it is more electropositive, but will obtain the same result). Because the B atoms are more electronegative, the free energy will be a minimum when the CDW charge maxima lie at B sites. If there is a CCDW and the alloy is random, many of the B atoms will not lie at maxima and we must pay some free energy proportional to the ionicity difference $(X_B-X_A)^2$. If there is an ICDW, the CDW can change its phase (or equivalently its local wavevector q) so that each B atom lies at charge maxima. In this case, we must pay some elastic energy to distort the CDW - but this turns out to be small. (The elastic energy to change the wave vector q from its commensurate to incommensurate value, must be relatively small for the CCDW to even exist in the pure material.) Consequently, we expect the increase in free energy with cation substitution to be larger in the CCDW than the ICDW, and thus the ICDW becomes stable (i.e.

T_d is suppressed and for $x \gtrsim x_c$ the CCDW does not exist.) McMillian's free energy model reaches the same conclusions (43).

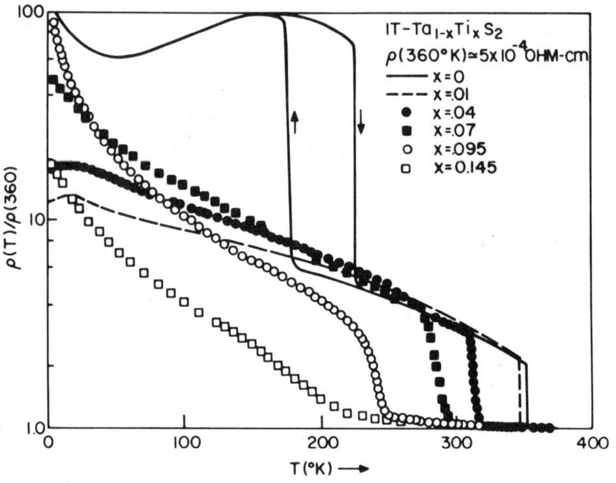

Figure 14. The electrical resistivity of $1T\text{-}Ta_{1-x}Ti_xS_2$ shows that both first order transitions toward the CCDW state are suppressed with increasing x, but the commensurate state is suppressed for $x \gtrsim 0.002$.

Similar effects are seen in $1T\text{-}Ta_{1-x}Ti_xS_2$ (Fig. 14). The transition at 200K to the commensurate state is suppressed for $x \gtrsim 0.002$, while the transition at T_d' (into the "quasi-commensurate" state) is slowly suppressed and is finally lost when $x \gtrsim 0.15$.

Fewer doping studies have been made for the 2H polytypes, because most cation substitution favors growth and retention of the 1T prototype. Rather, studies of the effects of anion mixing, as in $2H\text{-}TaSe_{2-x}S_x$, and of intercalation on the CDW are more common.(9) These studies show that disorder and intercalation (which at least in some cases causes both disorder and a change in the conduction electrodensity) reduce the CDW amplitude and/or T_o. Further, since the onset temperature is lower in these 2H polymorphs than in the 1T polymorphs, lower concentrations of "dopants" are necessary to suppress T_d or T_o. Such is also the case for the 1T polytypes, say in comparing $1T\text{-}VSe_2$ (T_o = 112K)(42) and $1T\text{-}TaSe_2$ ($T_o \approx$ 600K).(58)

We have discussed in some detail the magnitude and temperature dependence of the CDW in several layered compounds. Other studies of the properties of layered compounds will be mentioned shortly, but first we discuss the applicability of the simple model we will introduce to present the CDW parameters.

We focus on two aspects of the model: the bare susceptibility, and the spatial range of the interactions. Several calculations of $\chi^{o}(\vec{q})$ for $1T\text{-}TaS_2/Se_2$ and $2H\text{-}NbSe_2$ have been reported[59,60] and the results are reproduced in Fig. 15. These results are obtained from nonrelativistic, nonselfconsistent band calculations using muffin tin potentials, and therefore the $\chi^{o}(\vec{q})$ obtained should be taken as suggestive rather than definitive. First, consider $1T\text{-}TaSe_2$, where a peak is seen in $\chi^{o}(\vec{q})$ at the same wave vector as observed in the ICDW state. This peak is only 3 times the value that would be obtained from a three dimensional free electron gas of the same density at $q = 2k_F$. However, even in the three dimensional electron gas $\chi^{o}(q)$ is large compared to, Cu, for example because the electron density is lower than in an elemental metal. The calculated temperature dependence of this peak is small. This would suggest, in the language of Eq. 4, that the interaction term $X(q)$ has some temperature dependence in it and that T_o is not primarily determined by the temperature dependence of the bare susceptibility $\chi^{o}(q)$. Unfortunately no microscopic calculations of $X(q)$ of these compounds exists; we do not know, for example, if $X(q)$ also has a peak at the same place as $\chi^{o}(q)$. The extreme limit of this possibility, namely that $\chi^{o}(q)$ is free electron like, has been considered in explaining the phonon anomalies in Nb, NbC and NbN (ref. 61 & J. Hafner in this conference). These theories conclude that $X(q)$ is entirely responsible for the phonon anomalies, but these materials are not known to show CDW formation.

$\chi^{o}(q)$ for $1T\text{-}TaS_2$ shows two weak humps at wavevectors larger than the observed wavevector of the ICDW. It is not clear if this is an artifact of the calculation or represents the "true" $\chi^{o}(\vec{q})$. Again $\chi^{o}(q)$ does not have logarithmic like peaks and little temperature dependence is calculated.

The situation for the 2H polymorphs is even less clear. Here the obtained band structure is more sensitive to calculational details, since the conduction band is only about 1 volt wide and the F.S. is two sheeted. In any case, a calculated $\chi^{o}(q)$ shows only a broad maximum near $\vec{q} = 0.33\vec{a}^{*}$,[60] whose value is about 10 times that of a free electron gas of the same density. As in the 1T polymorphys the calculated $\chi^{o}(q)$ shows little temperature dependence. While Ricco et al.[60] argue that the electron

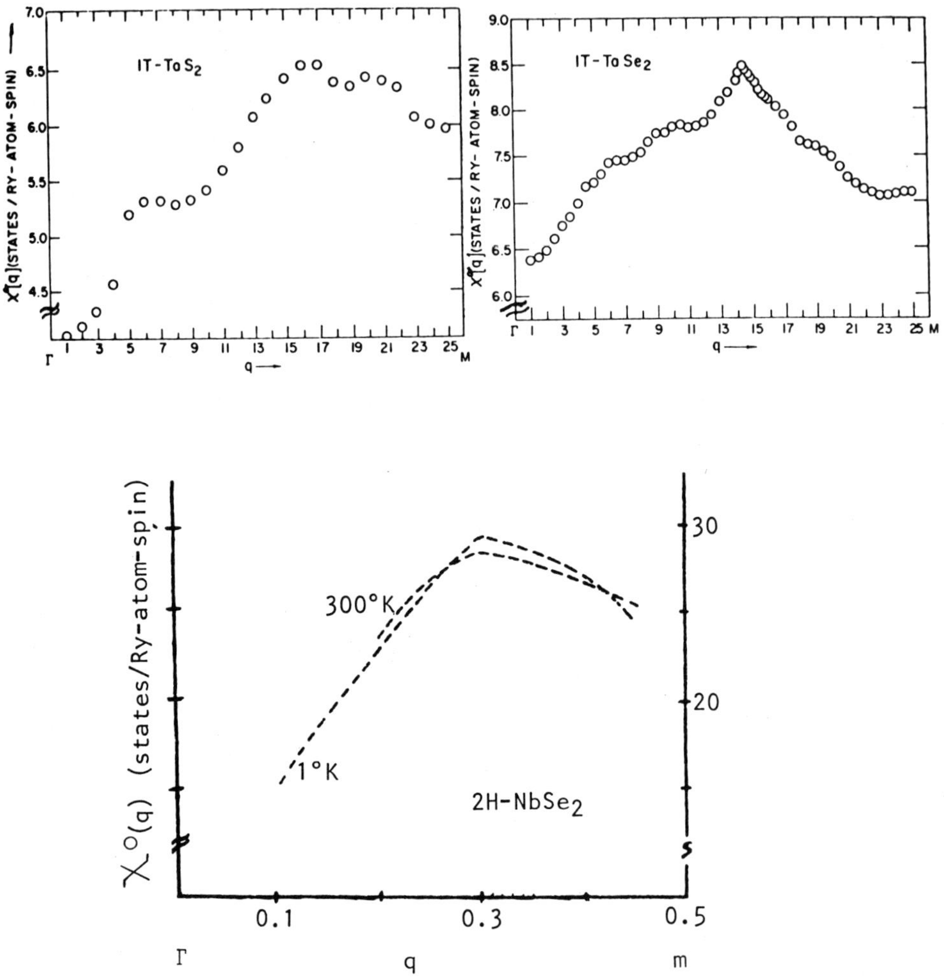

Figure 15. (top right) - The calculated $\chi^{0}(q)$ for 1T-$TaSe_2$ with q along the ΓM direction shows a weak peak at the observed ICDW wavevector.
(top left) - The calculated $\chi^{0}(q)$ for 1T-TaS_2 with q along the ΓM direction shows even weaker peaks, but these do not occur at the ICDW wavevector.
(bottom) - The calculated $\chi^{0}(q)$ for 2H-$NbSe_2$ along the ΓM direction shows a broad maximum near q_o. The calculated temperature dependence is very small.
These three figures suggest that while $\chi^{0}(q)$ is relatively large, prominent features expected from the simple model embodied in eq. 4 are not visible.

phonon interaction along with the large value of $\chi^{o}(2k_F)$ is sufficient to explain the instability, they do not explain what determines the CDW onset temperature T_o.

These calculations of $\chi^{o}(q)$ suggest that the interactions, $X(\vec{q})$, are more important than the simple model implies in determining the appearance of the CDW. At present our understanding of the magnitude and of the energy and wave vector dependence of these interactions is limited. A more serious difficulty with the simple theory has been pointed out by McMillan (Conference proceedings). McMillan has developed a model for the 2H-$TaSe_2$ CDW that takes into account the phonon entropy. Significant contributions to the entropy due to phonon softening is expected since such a large region of phonon states in the LA acoustic phonon branch are temperature dependent. (see Fig. 7). The inclusion of this term markedly modifies the instability condition (to determine T_o), and consequently a discussion of the instability in terms of eq. 4 may be misleading. This model, when compared with experimental data, leads to the prediction of a short coherence length for the CDW of about 10 Å, a value close to the mean free path above T_o and to the super lattice wavelength. This length is somewhat larger than found in antiferromagnetic transitions or in superfluid liquid He, but is much shorter than found in superconductors, where a mean field treatment of the thermodynamics is adequate.

These calculations of $\chi^{o}(q)$ and of the coherence length make the role of the F.S. less apparent than in the simple model. Yet is seems clear that the F.S. plays a role in determining the wave vector of the ICDW, but the microscopic details are still lacking. It is also clear, particularly for 1T-TaS_2 and $TaSe_2$, that weak coupling models of the CDW will be inadequate to describe the large amplitude of the CCDW. As has been emphasized by Thompson,[3] the general occurence of the CDW phenomena in these compounds makes it seem unlikely that the CDW is based on a subtle property of the band structure. It seems reasonable that the occurrance of the CDW is based on more general features, such as the large values of $\chi^{o}(q)$ and of the electron phonon interaction.[3,58]

Before closing this article, I would like to point out that I have only discussed a few of the layered compounds. I would like to briefly mention other phenomena that are connected with CDW formation or other phenomena that are of interest, at least to me. (1) 4Hb polymorphs, in which the layers alternate between octahedral and trigonal prismatic coordination, show separate uncoupled CDW transitions in the different symmetry layers[62] (2) the 2H polymorphs of group Vb are superconductors.

Several papers show that T_c increases as the CDW amplitude and/or T_0 is suppressed[9,24,63] (3) the semi metal $TiSe_2$ has a phase transition near 200K in which the hexagonal a and c axis double[54,65] (4) cation substitution in $1T-TaS_2$ and $1T-TaSe_2$ by Fe, Co, or Ni produces electron localization by the random impurity potential, even when the CCDW state is eliminated[66]. In the case of Fe a low spin to high spin conversion occurs as the temperature is increased[67] (5) TiS_2 shows an unusual T^2 resistivity from 10K to 400K[68]. At least two models have been proposed to explain this behavior[69,70]. (6) There may be technological use of layered and related compounds in rechargeable battery systems[71,72,73].

In conclusion, there now exists a large, but not exhaustive, body of emperical knowledge about CDW's in layered compounds. Largely through Landau models the data is at least qualitatively and some cases quantitatively connected. However, our understanding of this phenemona, particularly at the microscopic level is less well developed.

Finally, I wish to point out that this was not intended to be an exhaustive review of work on layered compounds. In particular, I have chosen to mainly emphasize points from my own work, and therefore the references used may not adequately give credit to many others who have contributed to this field.

References

1. J. A. Wilson, F. J. DiSalvo, and S. Mahajan, Adv. in Phys. 24, 117 (1975).

2. P. M. Williams, "Crystallography and Crystal Chemistry of Materials with Layered Structures" Vol. 2, ed. F. Levy Reidel Pub. (1976).

3. A. H. Thompson, Comments in Solid State Physics 7 125 (1976).

4. F. J. DiSalvo, 1976 NATO Conf. on Idim. conductors, Bolzano, Italy.

5. J. A. Wilson and A. D. Yoffe, Adv. in Phys. 18, 193 (1969).

6. F. R. Gamble, J. H. Osiecki and F. J. DiSalvo, J. Chem. Phys. 55, 3525 (1971).

7. F. R. Gamble, J. H. Osiecki, M. Cais, R. Pisharody, F. J. DiSalvo and T. H. Geballe, Science 174, 493 (1971).

8. F. J. DiSalvo, G. W. Hull, Jr., L. H. Schwartz, J. M. Voorhoeve and J. V. Waszczak, J. Chem. Phys. 59, 1922 (1973).

9. D. W. Murphy, F. J. DiSalvo, G. W. Hull, Jr., J. V. Waszczak, S. F. Meyer, G. R. Stewart, S. Early, J. V. Acrivos, and T. H. Geballe, J. Chem. Phys. 62, 967 (1975).

10. J. Rouxel, J. S. S. Chem. 17, 223 (1976).

11. L. F. Mattheiss, Phys. Rev. B8, 3719 (1973).

12. The first reports of super conductivity in these compounds are: M. H. Van Maaren and G. M. Schaeffer, Phys. Letts. 20, 131 (1966); 24A, 645 (1967).

13. F. R. Gamble, F. J. DiSalvo, R. A. Klemm, T. H. Geballe Science 168, 568 (1970).

14. R. E. Schwall, G. R. Stewart and T. H. Geballe, J. Low Temp. Phys. 22, 557 (1976); R. E. Schwall Ph.D. Thesis Stanford Univ. (1973).

15. G. Lucovsky, W. Y. Liang, R. M. White, K. R. Pischarody S. S. Comm. 19, 303 (1976).

16. A. H. Thompson, F. R. Gamble and J. F. Revelli S. S. Comm. 9, 981 (1971).

17. F. J. DiSalvo, R. G. Maines, J. V. Waszezak and R. E. Schwall S. S. Comm. 14, 497 (1974).

18. A. H. Thompson, F. R. Gamble and R. F. Koehler, Phys. Rev. B5, 2811 (1972).

19. H. N. S. Lee, M. Garcia, H. McKinzie and A. Wold, J.S.S. Chem. 1, 190 (1970).

20. R. K. Quinn, R. Summons, and J. J. Banewicz, J. Phys. Chem. 70 230 (1966).

21. M. Barmatz, L. R. Testardi and F. J. DiSalvo, Phys. Rev. B12, 4367 (1975).

22. J. M. E. Harper, T. H. Geballe and F. J. DiSalvo, Phys. Rev. B15, 2943 (1977).

23. E. Eherenfreund, A. C. Gossard and F. R. Gamble, Phys. Rev. B5, 1708 (1972).

24. C. Berthier, D. Jerome, P. Molinie, and J. Rouxel S. S. Comm. 19, 131 (1976).

25. J. A. R. Stiles and D. L. Williams, J. Phys. C. 9, 3941 (1976).

26. G. K. Wertheim, F. J. DiSalvo, and S. Chiang, Phys. Rcv. B13, 5476 (1976).

27. H. B. Hughes and R. A. Pollak, Comm. Phys. 1, 61 (1976).

28. A. W. Overhauser, Phys. Rev. 167 (1968).

29. M. J. Rice and S. Strassler, S.S. Comm. 13 125 (1973).

30. L. M. Roth, H. J. Zeigler and T. A. Kaplan, Phys. Rev. 149, 519 (1966).

31. T. M. Rice and G. K. Scott, Phys. Rev. Letts. 35, 120 (1975).

32. S. K. Chan and V. Heine, J. Phys. F3, 795 (1973).

33. A. W. Overhauser, Phys. Rev. 128, 1437 (1962).

34. W. M. Lomer, Proc. Phys. Soc. (London) 80, 489 (1962).

35. A. W. Overhauser, Phys. Rev. B3, 3173 (1971).

36. A. Guinier "X-ray diffraction in Crystals, Imperfect Crystals and Amorphous Bodies. Pub. W. H. Freeman 1963.

37. J. G. Allpress, 5th N.B.S. Symp. Mat. Res. Pub. 364, pg. 727 (1972).

38. J. Van Landuyt, J. Van Tendeloo, S. Amelinckx, Phys. Stat. Sol. A29, K11 (1975).

39. D. E. Moncton, J. D. Axe, and F. J. DiSalvo, Phys. Rev. Letts. 34, 734 (1975).

40. C. B. Scruby, P. M. Williams, and G. S. Parry, Phil. Mag. 31, 225 (1975).

41. J. P. Tidman, O. Singh, A. E. Curzon and R. F. Frindt, Phil. Mag. 30, 1191 (1974).

42. F. J. DiSalvo and J. V. Waszczak Journal de Physique Supplement C4, 37 157 (1976).

43. W. L. McMillan Phys. Rev. B12, 1187 (1975); Phys. Rev. B14 1496 (1976); and this conference.

44. S. A. Jackson, P. A. Lee and T. M. Rice Bull. Am. Phys. Soc. 22, 280 (1977).

45. D. E. Moncton, J. D. Axe and F. J. DiSalvo to be pub. in Phys. Rev. B15 (June 15, 1977).

46. A. S. Barker, Jr., J. A. Ditzenberger and F. J. DiSalvo Phys. Rev. B12, 2049 (1975).

47. E. F. Steigmeyer, G. Harbecke, H. Auderset, and F. J. DiSalvo, S.S. Comm. 20, 667 (1976).

48. J. A. Holy, M. V. Klein, W. L. McMillan, and S. F. Meyer Phys. Rev. Letts. 37, 1145 (1976).

49. G. K. Wertheim, F. J. DiSalvo, and S. Chiang, Phys. Letts. 54A 304 (1975).

50. Y. Yamada and H. Takatera, S.S. Comm. 21, 41 (1977).

51. A. H. Thompson, Phys. Rev. Letts. 34, 520 (1975).

52. F. R. Gamble, J.S.S. Chem. 9, 358 (1974).

53. D. E. Moncton, F. J. DiSalvo, J. D. Axe, L. J. Sham and B. R. Patton, Phys. Rev. B14, 3432 (1976).

54. F. J. DiSalvo and J. E. Graebner, to be published in S.S. Comm.

55. N. F. Mott, M. Pepper, S. Pollett, R. H. Wallis and C. J. Adkins, Proc. Roy. Soc. London A345, 169 (1975).

56. P. Fazekas and E. Tosatti, Proc. of XIII International Conf. on Physics of Semiconductors Rome 1976.

57. P. B. Allen, Phys. Rev. Letts. 37, 1638 (1976).

58. F. J. DiSalvo, J. A. Wilson, B. G. Bagley, and J. V. Waszczak, Phys Rev. B12, 2220 (1975).

59. H. W. Myron, J. Rath and A. J. Freeman, Phys. Rev. B15, 885 (1977).

60. B. Ricco, V. Heine, M. Schneiber, D. T. Henington and G. Wexler, to be published.

61. S. K. Sinha and B. N. Harmon, 2nd Rochester Conf. on Superconductivity in d and f band Metals (May 1976) ed D. H. Dougles (AlP, New York).

62. F. J. DiSalvo, D. E. Moncton, J. A. Wilson and S. Mahajan, Phys. Rev. B14 1543 (1976).

63. F. J. DiSalvo, Conf. on Low Lying Lattice Modes and their Relationship to Ferroelectricity and Superconductivity (Dec. 1975) Univ. of Puerto Rico.

64. J. A. Wilson and S. Mahajan, Comm. on Physics 2, 23 (1977).

65. F. J. DiSalvo, D. E. Moncton and J. V. Waszczak, Phys. Rev. B14, 4321 (1976).

66. F. J. DiSalvo, J. A. Wilson and J. V. Waszczak, Phys. Rev. Letts. 36, 885 (1976).

67. M. Eibschutz and F. J. DiSalvo, Phys. Rev. Letts. 36, 104 (1976).

68. A. H. Thompson, Phys. Rev. Letts. 35, 1786 (1975).

69. C. A. Kukhonen and P. F. Maldague, Phys. Rev. Letts. 37, 782 (1976).

70. J. A. Wilson, S.S. Comm. 22, 551 (1977).

71. M. S. Whittingham, J. Electrochem. Soc. 123, 315 (1976).

72. D. W. Murphy, F. A. Trumbore and J. N. Carides J. Electrochemical Society 124, 325 (1977).

73. J. N. Carides and D. W. Murphy, to be published in J. Electrochem. Soc.

LANDAU THEORY OF CHARGE DENSITY WAVES

W. L. McMillan

Dept. of Physics and Materials Research Lab

University of Illinois, Urbana, Ill. 61801 USA

A. THE LANDAU FREE ENERGY

The two theoretical methods which have proved so fruitful in the study of superconductivity have been applied to the charge density wave (CDW) phase transitions in the transition metal dichalcogenides. I will discuss the (Ginzburg-) Landau theory in this lecture and microscopic mean field theory in the next lecture. The Landau theory is due to Dave Moncton and myself.[2,3] The Landau theory approach is especially powerful in the discussion of position-dependent or time-dependent properties. We don't have an adequate dynamical theory yet and I will discuss only the static Landau theory.

In order to write down a Landau theory one must choose the proper order parameter and write down an expansion of the free energy in powers of the order parameter and the gradient of the order parameter. In the CDW phase one has a static periodic lattice distortion, a band gap in the electronic band structure, and an electronic charge density wave present simultaneously. Near the onset transition the three quantities are proportional to each other and we can choose any one as the order parameter. We choose the electronic charge density and since there are three CDW's present (thinking now of our canonical material, 2H-$TaSe_2$) we write

$$\rho(\vec{r}) = \rho_o(\vec{r})\ [1 + \alpha(\vec{r})] \qquad (1)$$

where

$$\alpha(\vec{r}) = \mathrm{Re}[\psi_1(\vec{r}) + \psi_2(\vec{r}) + \psi_3(\vec{r})] \tag{2}$$

and the $\psi_i(\vec{r})$ are three complex order parameters, one for each CDW. For the energy of one layer we write an expansion in powers of the order parameters, keeping all terms allowed by symmetry.

$$F_1 = \int d^2r[a(\vec{r})\alpha^2 - b(\vec{r})\alpha^3 + c(\vec{r})\alpha^4 + d(\vec{r})(|\psi_1\psi_2|^2 + |\psi_2\psi_3|^2 + |\psi_3\psi_1|^2)] \tag{3}$$

where $a(\vec{r})$, $b(\vec{r})$, $c(\vec{r})$ and $d(\vec{r})$ exhibit the periodicity of the crystal lattice. We will write, for example,

$$c(\vec{r}) = c_o + c_1 \sum_i e^{i\vec{K}_i\cdot\vec{r}} \tag{4}$$

where the six $\vec{K}_i$ are the six shortest reciprocal lattice vectors. We next include a random potential U(r), due to impurities

$$F_2 = \int d^2r\, U(\vec{r})\rho_o(\vec{r})\alpha(\vec{r}) . \tag{5}$$

The gradient terms are chosen so that the free energy of the three CDW's is minimum when they lie in the right directions and have the right wavelengths; any distortion from this optimal condition costs free energy.

$$F_3 = \int d^2r \left[e(\vec{r}) \sum_i |(\vec{q}_i \cdot \vec{\nabla} - iq_i^2)\psi_i|^2 + f(\vec{r}) \sum_i |\vec{q}_i \times \vec{\nabla}\psi_i|^2 \right] \tag{6}$$

where $|q_i| = 2\pi/\lambda$, λ is the wavelength of the incommensurate CDW, and the three $\vec{q}_i$ vectors lie in the ΓM directions 120° apart. The total free energy of one layer is the sum of these three terms

$$F = F_1 + F_2 + F_3 . \tag{7}$$

We could also include the coulomb interaction of CDW's in different layers. We assume that a_o changes sign at the onset transition

$$a_o = a'(T - T_{IN}) \tag{8}$$

and that the other parameters are constant. The Landau theory is a phenomenological theory and the parameters are to be determined from experiment.

B. PHASE TRANSITIONS

The simplest thing to discuss is the state of lowest free energy. One assumes that the order parameters are plane waves with definite amplitudes, phases and wave vectors, and one varies these quantities to minimize F. For $T > T_{IN}$ the amplitudes vanish and one has the normal metal phase. For T somewhat less than T_{IN} the quadratic term is negative and one finds a finite amplitude of all three CDW's and wave vectors determined by minimizing the distortion energy F_3. The wave vectors are incommensurate with the lattice and only the uniform terms in the free energy (i.e., a_o, b_o, c_o, and d_o) contribute. The lock-in energy

$$\sum_i \mathrm{Re}\left[\frac{C_1}{8}\int d^2r \quad \psi_i^3 e^{iK_i \cdot r}\right] \tag{9}$$

averages to zero. One can take advantage of the lockin energy to lower the free energy by setting the CDW wave vector equal to $\vec{K}_i/3$. This costs elastic energy and the transition temperature to the commensurate phase is governed by this energy balance. As we will see later on this treatment of the commensurate-incommensurate transition using uniform plane waves is oversimplified.

C. FLUCTUATION MODES

Small phase and amplitude distortions of the order parameters are the collective modes of the system. Since the periodic lattice distortion is proportional to the order parameter these modes appear in the lattice dynamic structure factor and can be measured by neutron scattering. For 2H-$TaSe_2$ there are three amplitude and three phase modes and these have been observed by Raman scattering.[4] The static Landau theory predicts the energy versus amplitude of these modes (but not the frequencies); one simply expands the free energy about the minimum for small distortions. Long wavelength phase distortions correspond locally to a translation of the CDW which costs very little energy in the incommensurate phase; the energy of the "phason" is proportional to k^2 where $\vec{K}$ is the phason wavenumber. In the commensurate phase the lock-in energy opposes translation of the CDW and the phason energy is proportional to a constant plus k^2. For the amplitude modes the energy is proportional to a constant plus k^2 with the constant proportional to $|T - T_{IN}|$ (within mean field theory). Thus one expects soft mode behavior near the onset phase transition with the amplitude modes stiffening in the incommensurate phase and the phase modes stiffening in the commensurate phase. Qualitatively this does happen; however the detailed behavior of the soft modes and the central peak near the phase transition are not well understood.

D. IMPURITY EFFECTS

Impurity effects may be calculated simply within Landau theory. In the normal phase a charged impurity attracts electronic charge and the system responds by dressing an impurity with a CDW cloud. The size of this cloud grows larger as the phase transition is approached and the correlation length of the CDW increases. A second impurity feels the potential of the CDW and there is an impurity-impurity interaction which tends to space impurities apart by one CDW wavelength. Since the impurities drive the CDW order parameter the incommensurate-normal metal phase transition is not sharp. In the incommensurate phase a phase distortion of the CDW is not costly and the CDW can lower its energy by distorting to place charge density peaks near impurities. This effectively pins the CDW to the impurities. In the commensurate phase the CDW is locked in to the lattice and cannot easily distort to take advantage of the impurity potential. Thus the incommensurate phase is favored energetically in the presence of impurities and the incommensurate-commensurate transition temperature is strongly depressed.

E. CDW DISLOCATIONS

In the incommensurate phase the cubic term in the free energy (proportional to c_0) fixes the relative phase of the three CDW's. The sign of this term is such that positively charged regions of the 3 CDW's add up on lattice sites of a hexagonal "lattice". The lattice spacing of this CDW "lattice" is approximately three times the crystal lattice spacing. The phason modes are just the phonon modes of the CDW "lattice". Vacancies or interstitials in the CDW "lattice" are not possible because the units making up the lattice are not discrete. However a CDW dislocation is possible and is closely analogous to the vortex line in superconductors or superfluid helium. We emphasize that the host crystal is assumed to be perfectly uniform and that only the CDW is distorted into a "dislocation". In order to introduce a dislocation we insert a vortex solution into two of the three CDW order parameters, with the phase of the order parameter changing by 2π as one encircles the dislocation. If one then examines the charge density map one row of charge peaks has been removed leaving a "lattice" dislocation.

F. DISCOMMENSURATIONS

The last concept I want to discuss is the discommensuration. For simplicity consider a single incommensurate CDW whose wavelength is 2% greater than three times the lattice spacing. If we take a uniform plane wave for our CDW then for 25 lattice spacings the CDW will be in phase with the lattice and the lock-in energy will be negative; then for the next 25 lattice spacings the CDW will be out of phase with the lattice and the lock-in energy will be repulsive. The spatial average of the lock-in energy will be zero. However, we can distort the CDW in two ways to take advantage of the lock-in energy. We can amplitude modulate the CDW making it larger in the in-phase region and smaller in the out-of-phase region. Or we can phase modulate the CDW making the in-phase region larger and the out-of-phase region smaller. Both distortions buy lock-in energy at the expense of distortion energy, with phase-modulation costing less distortion energy. If one considers phase modulation only (the weak coupling limit) the nonlinear problem is simple enough to solve exactly. Far from the lock-in phase transition one finds a weak phase modulation with the in-phase region slightly larger than the out-of-phase region. As one moves toward the phase transition the in-phase region expands and the average wavelength of the CDW shifts toward commensurability. Finally, very near the phase transition one has large in-phase regions which are locked in to the lattice and narrow out-of-phase regions which appear to be defects called discommensurations. The phase of the CDW relative to the crystal changes by $2\pi/3$ as one crosses the discommensuration. As one continues to lower the temperature the number of discommensurations decreases and vanishes in a continuous way at the commensurate-incommensurate phase transition. One interprets the transition as a defect melting transition in contrast to the more common order parameter type of continuous phase transition. Experimentally, the lock-in phase transition in 2H-$TaSe_2$ exhibits hysteresis and is first order; however the transition is very sensitive to impurities and super pure crystals are not available yet. Additional Bragg scattering from the distortion was predicted and observed by Moncton et. al.[1]

REFERENCES

1) D. E. Moncton, J. D. Axe, and F. J. DiSalvo, Phys. Rev. Lett. 34, 734 (1975).

2) W. L. McMillan, Phys. Rev. B12, 1187 (1975).

3) W. L. McMillan, Phys. Rev. B14, 1496 (1976).

4) J. A. Holy, M. V. Klein, W. L. McMillan, and S. F. Meyer, Phys. Rev. Lett. 37, 1145 (1976).

MICROSCOPIC MODEL OF CHARGE DENSITY WAVES IN 2H-$TaSe_2$

W. L. McMillan

Dept. of Physics and Materials Research Lab

University of Illinois, Urbana, Ill. 61801 USA

The ultimate goal of the theoretician in studying a particular phenomenon is to produce a quantitative microscopic theory from which one can calculate anything. That goal has been pretty well achieved in superconductivity. We are still struggling with the theory of charge density waves. The way that one progresses in solid state physics is to compute the properties of a theoretical model, compare the predictions with experiment, and modify the model when it fails. I want to discuss one iteration of this process today.

Our present understanding of CDW's in metals follows the early work of Peierls,[1] Fröhlich,[2] Overhauser,[3] and Chan and Heine[4] and is based on the following physical picture: one starts with a normal metal and introduces a periodic lattice distortion (a static phonon distortion of finite amplitude and wavevector $\vec{q}$). The new periodicity introduces a new Brillouin zone boundary and creates a band gap in the one-electron band structure near the zone boundary. If the zone boundary is near the Fermi surface (that is, if $\vec{q}$ spans nested portions of Fermi surface) the energy of many occupied electronic states will decrease and the total electronic energy will decrease. If the gain in electronic energy more than offsets the increase in elastic energy the distorted state will be the ground state. The occupied electronic states are nonuniform and there is a charge density wave with wavevector $\vec{q}$ which screens the lattice potential. Thus the three physical properties go together, the periodic lattice distortion, the energy gap in the band structure, and the charge density wave. I believe that this physical picture is correct for the layered compounds. However in order to calculate the properties of the CDW state at finite temperature the conventional approach is to include the entropy of electrons excited across the energy gap and to minimize the free energy. This is the right thing to do if the zero temperature correlation length ξ_0 (of the CDW) is long; it turns out to be incorrect for 2H-TaSe$_2$. What one must do at finite temperature is to include the lattice entropy; the electron energy gap is quite large and the electronic entropy is negligible.

I want to discuss three simple calculations and compare them to experiment.[5] The first is a model calculation of various energies at T = 0 and a prediction of lattice dynamics in the distorted state; the lattice frequencies have been measured by Raman scattering[6] and we can determine the magnitude of various contributions to the energy by fitting the Raman data. Next we examine the electronic energy model which, with some approximations, turns out to be identical to the BCS theory of superconductivity.[7] A comparison of this model with experiment yields some large discrepancies. Finally, we treat a lattice entropy model, which assumes a short coherence length, which is in semi-quantitative agreement with experiment.

At T = 0 we begin by writing for the displacement of the ith Ta atom in the ℓth layer from its lattice site $\vec{R}_i$

$$\vec{u}_\ell(\vec{R}_i) = \mathrm{Im} \sum_{j=1}^{3} \hat{q}_j \phi^j_\ell(\vec{R}_i) e^{i\vec{q}_j \cdot \vec{R}_i} \tag{1}$$

where $\vec{q}_j$ is the nesting vector of the jth CDW and $\phi^j_\ell(\vec{R}_i)$ is the local complex amplitude of the CDW. The three $\vec{q}_j$ form a star in the ΓM directions. We assume that the band gap $|2W|$ is proportional to the lattice distortion.

$$W^j_\ell(x) = \alpha \phi^j_\ell(x) \tag{2}$$

Only a portion of the Fermi surface is affected by the energy gap W^j_ℓ and we assume a simplified band structure for this nested portion: we assume perfect nesting with a Fermi velocity v_F and an electronic density of states (of one spin) of $N_\uparrow(0)$ (for each CDW). Then the electronic energy is

$$E_{e\ell} = \sum_{\ell j} \int d^2x [-N_\uparrow(0)|W^j_\ell(x)|^2 \log \frac{E_B^2}{|W^j_\ell(x)|^2}$$
$$+N_\uparrow(0)\xi_0^2 |\nabla W^j_\ell(x)|^2] \tag{3}$$

where E_B is the electronic band width and ξ_0 is of order $\hbar v_F/2W$. Electrons near the gap edge cannot respond to lattice vibrations with wave vectors further than $1/\xi_0$ from the nesting vector and we assume that $\phi_\ell^j(x)$ is a slowly varying function with a momentum space cutoff of $k_c = 1/\xi_0$. We rewrite (3) using phonon coordinates and include several other energies.

$$\begin{aligned} PE = \frac{1}{2}\sum_\ell \int \frac{d^2x}{\Omega} \{\sum_j [A|\phi_\ell^j|^2 - C|\phi_\ell^j|^2 \log|\phi_B/\phi_\ell^j|^2 \\ + C\xi_0^2|\nabla\phi_\ell^j|^2 - \mathrm{Re}(B_\ell \phi^{j^3} e^{i(3\vec{q}_j-\vec{G})\cdot\vec{x}}) \\ + F\ \mathrm{Re}(\phi_\ell^j\ \phi_{\ell+1}^j)] \\ + D(|\phi_\ell^1\phi_\ell^2|^2 + |\phi_\ell^1\phi_\ell^3|^2 + |\phi_\ell^2\phi_\ell^3|^2) \\ + \mathrm{Re}(E\phi_\ell^1\phi_\ell^2\phi_\ell^3)\} \end{aligned} \qquad (4)$$

where Ω is the area of the normal state unit cell in one plane. The first term is the unscreened elastic constant and the second and third terms are the electronic contributions from (3). The fourth term is the lock-in term and the fifth is the interlayer Coulomb interaction. The sixth term is a CDW interaction which arises when two CDW's compete to open an energy gap on the same portion of Fermi surface and the seventh is a weak CDW interaction permitted by symmetry.

The lattice kinetic energy is

$$KE = \frac{M^*}{4}\sum_{\ell j}\int \frac{d^2x}{\Omega} \left| \frac{\partial\phi_\ell^j(x)}{\partial t} \right|^2 \qquad (5)$$

where the Se atoms are assumed to adiabatically follow the Ta atoms with the same ratio of Se displacement to Ta displacement observed at low temperature; this yields $M^* = 206$ au. Equations (4) and (5) constitute a nonlinear dynamical model for the longitudinal acoustic phonons modes near $\vec{q}_j$.

To find the equilibrium distortion at T = 0 in the commensurate phase we substitute

$$\phi_\ell^j(x) = \phi_0 e^{i(\theta_\ell + \Delta\vec{q}\cdot\vec{x})} \qquad (6)$$

where $\Delta\vec{q} = \vec{G}_j/3 - \vec{q}_j$ and $\theta_\ell = -\frac{1}{3}$ phase (B_ℓ) and minimize PE with respect to ϕ_o to find ϕ_o.

In order to find the phonon frequencies we add a small phase or amplitude distortion to the static distortion.

$$\phi_\ell^j(\vec{x}) = [\phi_o + (\alpha_k + i\beta_k)\cos(k\cdot x + \frac{k_z \ell c}{2})]e^{i(\theta_\ell + i\Delta\vec{q}\cdot\vec{x})} \quad (7)$$

and expand PE to second order in α_k and β_k. The equations of motion are then harmonic and are simple to solve; for the amplitude modes we find

$$\frac{M^*\omega_k^2}{4} = 4C - 3|B|\phi_o + 8D\phi_o^2 + \bar{E}\phi_o + 2F - 2F\cos(\frac{k_z c}{2}) + C\xi_o^2 k^2 \quad (8a)$$

and

$$\frac{M^*\omega_k^2}{4} = 4C - 3|B|\phi_o - 4D\phi_o^2 - 2\bar{E}\phi_o + 2F - 2F\cos(\frac{k_z c}{2}) + C\xi_o^2 k^2 . \quad (8b)$$

where $\bar{E} = \mathrm{Re}[E\exp(i3\theta_o)]$. For the phase modes we find

$$\frac{M^*\omega_k^2}{4} = 9|B|\phi_o - 3\bar{E}\phi_o + 2F - 2F\cos(\frac{k_z c}{2}) + C\xi_o^2 k^2 \quad (8c)$$

and

$$\frac{M^*\omega_k^2}{4} = 9|B|\phi_o + 2F - 2F\cos(\frac{k_z c}{2}) + C\xi_o^2 k^2 . \quad (8d)$$

The (8b) and (8d) modes are doubly degenerate giving six modes. These six modes are observed in Raman scattering and from the Raman data we find

$$F = .03\ \mathrm{eV/\mathring{A}^2};\quad |B|\phi_o = .053\ \mathrm{eV/\mathring{A}^2};\quad |E\phi_o| = .036\ \mathrm{eV/\mathring{A}^2};$$

$$C = .29\ \mathrm{eV/\mathring{A}^2},\ D\phi_o^2 = .031\ \mathrm{eV/\mathring{A}^2};\ A/C + \xi_o^2\Delta q^2 - \log|\phi_B/\phi_o|^2 = -.90.$$

There are enough parameters in the theory to fit the data and there is no consistency check on the theory.

For the second calculation we include the entropy of electrons excited across the Peierls energy gap and calculate the transition temperature, heat capacity and magnetic susceptibility. If we neglect the CDW interactions the theory is equivalent to the BCS theory and we can borrow the following results from BCS. The energy gap at T = 0 is

$$2W(T = 0) = 3.52\ k_B T_{IN} \tag{9}$$

where T_{IN} is the incommensurate-normal metal transition temperature. The heat capacity jump at T_{IN} is

$$\Delta C_V = 3 \times 9.4\ N_\uparrow(0) k_B^2 T_{IN} \tag{10}$$

and the change in susceptibility due to the band gap is

$$\Delta\chi = 1.3 * 4\mu^2 N_\uparrow(0) \tag{11}$$

including an electron-electron enhancement factor of 1.3. From DiSalvo's[8] value $\Delta\chi = 55 \cdot 10^{-6}$ emu/mole we find $N_\uparrow(0)$ = .33 states/eV/Ta atom. Using (10) the model predicts a specific heat jump of 0.8 joules/mole K. Craven[9] finds a specific heat jump of 4 joules/mole K. Including the CDW interaction increases this discrepancy. The theory predicts a Peierls gap of 2W(0) = .037 eV whereas Barker _et al_.[10] find an energy gap of .25 eV. We conclude that the electronic entropy model doesn't work for $2H\text{-}TaSe_2$.

For the third calculation let's do the theory in the opposite limit, that of small correlation length. For small ξ_0 the phonon frequencies are modified over a large region of momentum space and the phonon entropy is large. The phonon entropy depresses the transition temperature so that $2W(0) \gg 3.52\ k_B T_{IN}$; then, the electronic entropy is negligible and one can forget about the electrons altogether. The electronic ground state energy provides the energy surface for the motion of the atoms; the Born-Oppenheimer or adiabatic approximation is valid. Therefore all we need to know is the nonlinear Hamiltonian (4) and the theory is completely specified. Unfortunately, in spite of the recent progress in understanding critical behavior, there are no theoretical tools which allow us to compute either the static or dynamic behavior of (4) assuming either classical or quantum statistics for the lattice. For the moment we will have to be satisfied with a rather primitive mean field theory. We first transform to a lattice model with the number of lattice sites equal to the number of modes of the continuum model with its cutoff $k_c = 1/\xi_0$. We take a square lattice in one layer with a lattice spacing of $\pi\xi_0$. We define the lattice order parameter by the value of the continuum order parameter at the lattice site

$$\phi^j_{\ell mn} = \phi^j_\ell\,(\vec{x}_{nm}) \tag{12}$$

Replacing the gradient term in (4) by the finite difference we find

$$PE = \frac{1}{2}\frac{(\pi\xi_o)^2}{\Omega} \sum_{\ell mn} \{ \sum_j [A|\phi^j_{\ell mn}|^2 - C|\phi^j_{\ell mn}|^2 \log|\phi_B/\phi^j_{\ell mn}|^2$$

$$+ \frac{C}{\pi^2}(|\phi^j_{\ell mn} - \phi^j_{\ell m+1n}|^2 + |\phi^j_{\ell mn} - \phi^j_{\ell mn+1}|^2)]$$

$$+ D(|\phi^1_{\ell mn}\phi^2_{\ell mn}|^2 + |\phi^2_{\ell mn}\phi^3_{\ell mn}|^2 + |\phi^3_{\ell mn}\phi^1_{\ell mn}|^2)\} \tag{13}$$

where we have neglected small terms in the energy.

We now make the mean field approximation in which each local mode moves in a potential due to its neighbors.

$$V_1(\phi) = \frac{1}{2}\frac{(\pi\xi_o)^2}{\Omega}[A|\phi|^2 - C|\phi|^2 \ln|\phi_B/\phi|^2$$

$$+ \frac{4C}{\pi^2}(|\phi|^2 - 2\phi^*\langle\phi\rangle) + 2D|\phi|^2\langle|\phi|^2\rangle] \tag{14}$$

Near the phase transition k_BT is greater than the phonon frequencies and we can use classical statistics. The self consistency conditions for the order parameters are then

$$\langle\phi\rangle = \int d^2\phi\, \phi e^{-V_1(\phi)/T} / \int d^2\phi e^{-V_1(\phi)/T} \tag{15}$$

and

$$\langle|\phi|^2\rangle = \int d^2\phi\, |\phi|^2 e^{-V_1(\phi)/T} / \int d^2\phi e^{-V_1(\phi)/T} \tag{16}$$

which we can solve numerically and predict the temperature dependence of the order parameters, the thermodynamic properties and the lattice dynamics. The predicted heat capacity jump is 1.67 k_B per mode and the transition temperature is

$$T_{IN} = .296\, C(\pi\xi_o\phi_o)^2/2\Omega \tag{17}$$

From the observed heat capacity jump of .48 k_B per Ta atom and T_{IN} = 122°K we find $\pi\xi_o \approx 10$ Å and ϕ_o = .16 Å. The length $\pi\xi_o$ = 10 A is equal to the superlattice unit cell size and is the shortest correlation length which is physically reasonable. This correlation length is consistent with the extent in k-space of the Kohn anomaly in the longitudinal acoustic branch[11] and with the fluctuation heat capacity observed by Craven so that it is safe to conclude that 2H-TaSe$_2$ is the short ξ_o limit. Moncton has measured

ϕ_0 directly by neutron crystallography and finds ϕ_0 between .05 Å and .09 Å.

With such a short correlation length the energy gap is the same order of magnitude as the Fermi energy. We can estimate the magnitude of the energy gap very crudely as follows. From Matthiess' 12/ band structure the Fermi energy is .35 eV and the (isotropically averaged) basal plane band mass is about 5 electron masses. From $\xi_0 \approx \hbar v_F/2W$ we estimate $2W \approx .3$ eV which is in order of magnitude agreement with the weak absorption edge (at .25 eV) observed by Barker et al.

The phonon entropy model works well in semi-quantitative comparisons with experiment. However it does not quantitatively reproduce Monctons measurement of $\langle\phi\rangle$ versus temperature and the phonon dynamics appear to be more complicated than that predicted by the mean field calculation. Both of these discrepancies may be due to the mean field approximation and we need to develop more powerful theoretical techniques to treat the nonlinear Hamiltonian.

The physical picture of the normal state is more complicated in the phonon entropy model because the order parameter fluctuations are quite large. The mean displacement $\langle\phi\rangle$ and the mean energy gap $\langle W\rangle$ vanish at the phase transition. However, the mean square local displacement $\langle|\phi|^2\rangle$ and the mean square local energy gap $\langle|W|^2\rangle$ are almost as large at the phase transition as at $T = 0$. The strongly fluctuating energy gap modifies one's physical picture of the normal state and the implications for resistivity, susceptibility, etc., have not been explored. Another important theoretical problem which has not been attacked yet is the derivation of the electronic energies in (4) from a realistic band structure in the distorted state.

REFERENCES

1) R. E. Peierls, Quantum Theory of Solids (Clarendon, Oxford, 1955).

2) H. Fröhlich, Proc. Roy. Soc. (London) A233, 296 (1954).

3) A. W. Overhauser, Phys. Rev. 167, 691 (1968); 3B, 3173 (1971).

4) S.-K. Chan and V. Heine, J. Phys. F: Metal Phys. 3, 795 (1973).

5) W. L. McMillan, Phys. Rev., in press.

6) J. A. Holy, M. V. Klein, W. L. McMillan and S. F. Meyer, Phys. Rev. Lett. 37, 1145 (1976).

7) J. Bardeen, L. N. Cooper and J. R. Schrieffer, Phys. Rev. 108, 1175 (1957).

8) J. A. Wilson, F. J. DiSalvo and F. Mahajan, Adv. Phys. (GB) 24, 117 (1975).

9) R. Craven, to be published.

10) A. S. Barker, Jr., J. A. Ditzenberger and F. J. DiSalvo, to be published.

11) D. E. Moncton, J. D. Axe and F. J. DiSalvo, Phys. Rev. Lett. 34, 734 (1975); D. E. Moncton, Ph.D. thesis (MIT, 1975).

12) L. F. Mattheiss, Phys. Rev. B8, 3719 (1973).

LIGHT SCATTERING BY CHARGE DENSITY WAVE MODES IN KCP AND 2H-$TaSe_2$

E.F. Steigmeier, G. Harbeke and H. Auderset

Laboratories RCA Ltd., Zürich, Switzerland

Charge density waves (CDW's) are known to occur in both quasi one- and two-dimensional materials the best studied examples of which are $K_2Pt(CN)_4Br_{0.3}\cdot 3.2H_2O$ (KCP) and 2H-$TaSe_2$, respectively (1). For studying the dynamic excitations of the CDW's, called CDW modes, light scattering turns out to be a most useful technique. Its wave-vector resolution is of the order of 5×10^{-5}Å^{-1}, which is 10^3 times better than inelastic neutron scattering; such a high resolution is needed for an unambiguous analysis of the eigenfrequencies because of the considerable dispersion near the distortional wavevector.

Raman measurements on KCP (2) and KCP* (deuterated KCP) (3) show a sharp line at 44 cm^{-1} and 38 cm^{-1} (4K), respectively, of A_1 symmetry (xx=yy = 0, zz$\neq$ 0). It is attributed to the amplitude mode of the CDW. This mode is observed at all temperatures and its frequency never tends to zero (it rather increases slightly with temperature) suggesting that KCP is distorted at all accessible temperatures. The linewidth of the mode increases strongly with increasing temperature. The light scattering results are in good agreement with inelastic neutron scattering measurements (4) if the limited wavevector resolution of the latter are taken into account.

One of the most important findings of Raman scattering in KCP is the strong isotope effect of the amplitude mode frequency (3). This evidences that the water, situated in between the Pt chains, participates considerably in the eigenvector of the CDW mode, which therefore, by no means is of purely one-dimensional character. From the behaviour of the water stretching mode with temperature, it appears that the water may be crucially involved also in the temperature dependence of the interchain correlation.

In the hexagonal layer structure material 2H-$TaSe_2$ the light scattering measurements (5) show several modes clearly associated with the CDW transitions which occur at T_{inc} = 122K, from the normal to the incommensurate, and at T_{com} = 90K, from the incommensurate to the commensurate phase. Three of these modes are observed only below T_{com}, namely the one of A_{1g} symmetry at 43 cm^{-1}, of E_{2g} at 63 cm^{-1} and of A_{1g} at 82 cm^{-1} (4K values) (Figure 1). One mode of E_{2g} symmetry at 49 cm^{-1} (4K) persists in the incommensurate phase and disappears only at T_{inc}. Based on the fact that in the incommensurate phase the phase mode can be of zero frequency while the amplitude mode cannot, we have reason to believe this E_{2g} 49 cm^{-1}

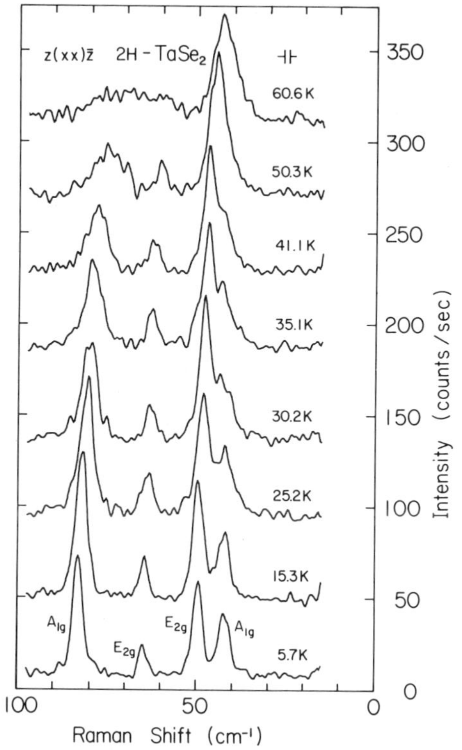

Figure 1: Raman spectrum of 2H-$TaSe_2$ at various temperatures. (Note that in ref. 5 overheating was less than 0.2 deg, while in the present figure the increased resolution was obtained at the cost of an overheating of about 5-10 deg above the stated nominal temperature.)

mode to be an amplitude mode (6). All four modes soften and broaden with increasing temperature. These results represent the first observations of a softening of CDW modes in a distorted phase.

The damping constant of the E_{2g} (49 cm^{-1}) mode increases quite drastically in approaching T_{inc}, approximately as $(T_{inc}-T)^{-1}$, while the damping constants of the other three modes behave the same way in approaching T_{com}, approximately as $(T_{com}-T)^{-1}$. Increases of this extent are quite unusual.

Of particular interest is the detailed temperature dependence of the E_{2g} (49 cm^{-1}) mode frequency which varies as $(T_{inc}-T)^{1/3}$. Further, it is proportional to the order parameter, for which the coupling coefficient of the light to the CDW mode is a direct measure (7). This non-classical value of the exponent suggests that there exists quite a large region near T_{inc} where the simple meanfield theory (8), despite its qualitative success, is of limited use for describing quantitatively the thermodynamics of the CDW's.

References

(1) For introduction to this topic the reader is referred to the lectures by F.J. Di Salvo and by W.L. McMillan.

(2) E.F. Steigmeier, R. Loudon, G. Harbeke, H. Auderset and G. Scheiber, Solid State Comm. 17, 1447 (1975).

(3) E.F. Steigmeier, D. Baeriswyl, G. Harbeke, H. Auderset and G. Scheiber, Solid State Comm. 20, 661 (1976).

(4) K. Carneiro, G. Shirane, S.A. Werner and S. Kaiser, Phys. Rev. B 13, 4258 (1976).

(5) E.F. Steigmeier, G. Harbeke, H. Auderset and F.J. Di Salvo, Solid State Comm. 20, 667 (1976).

(6) This assignment differs from the one made in the lecture of W.L. McMillan. See also J.A. Holy, M.V. Klein, W.L. McMillan and S.F. Meyer, Phys. Rev. Letters 37, 1145 (1976).

(7) Previously such a proportionality between soft mode frequency and order parameter had been reported for the antiferrodistortive transition in $SrTiO_3$ by E.F. Steigmeier, H. Auderset and G. Harbeke in Anharmonic Lattices, Structural Transitions and Melting, ed. by T. Riste, NATO-ASI (Noordhoff Publishing Leiden 1974) p. 153.

(8) Lecture of W.L. McMillan.

SYMMETRY CLASSIFICATION OF MODULATED STRUCTURES

P.M. de Wolff

Technische Hogeschool Delft, Lab. v. Techn.Nat.

Lorentzweg 1, Delft (the Netherlands)

1. DEFINITION OF MODULATED STRUCTURES

In a normal crystal, symmetry is defined by a group of operations, each of these belonging to the much larger group of all proper and improper movements. The symmetry translations

$$\vec{r}\,' = \vec{r} + \vec{n} \qquad (1)$$

where $\vec{n} = n_1\vec{a} + n_2\vec{b} + n_3\vec{c}$ (n_i = integer, i = 1,2,3) constitute an invariant subgroup of every crystal symmetry group.

We shall now try to find similar operations for modulated crystals. In an earlier paper (De Wolff, 1974, to be referred to as (I) from now), the possible point groups for such crystals have been derived from a description in four-dimensional space; the actual crystal is a section of this model, lying in the hyperplane which is constituted by the physical three-dimensional space. The same approach can very well be followed for the present purpose, viz. the derivation of symmetry operations and lattice types. We shall not do so (apart form an occasional reference) but rather emphasize the description both of the crystal and of the operations in direct three-dimensional space. The reason for choosing this alternative approach is firstly, that it is more convincing and easier to follow, the more so since the basis for the point group enumeration will be derived in passing so that (I) need not necessarily be consulted. Secondly, it is better adapted to a precise definition of the symmetry operations and of the modulated crystals themselves.

The latter are characterized by two properties: _a_) To each modulated structure corresponds a basic structure, that is, a "normal" crystal structure from which the modulated one can be derived by finite modifications, which are periodically ordered. We take displacive modulation as an example - other types of modulation yield the same results, and the displacive case is the more complicated one since here the modifications have vector character. In fact they consist of a displacement $\vec{u}$ of each atom, which depends upon:

- the kind of atom, each atom in the unit cell of the basis structure being considered as a separate kind so that the relevant index i runs from 1 to N, if N is the number of those atoms.
- the position of the atom in the basic structure, defined by a vector $\vec{r}^{\,o}_{ip}$ where the index p symbolizes the integers p_1, p_2 and p_3 numbering the unit cell from an origin in a fixed cell. Thus if the fractional coordinates of the i-th atom in the basic structure are x^o_{ij} (j = 1,2,3), the components of $\vec{r}^{\,o}_{ip}$ are $x^o_{ij} + p_j$.
- a fixed vector $\vec{k}$ in reciprocal space, with components k_1, k_2 and k_3 with respect to the reciprocal basis of the basic structure. It is conceivable that structures exist modulated with more than one k-vector simultaneously. Here we shall restrict ourselves to the case of a single k-vector. The more general case is treated e.g. by Janner & Janssen (1977).
- N periodic and continuous vector functions $\vec{u}_i(\alpha)$, each with unit period, that is, invariant for an integer shift in α. These functions define the displacement of the atom at position $\vec{r}^{\,o}_{ip}$ in the basic structure by

$$\vec{u}(i,p) = \vec{u}_i(\vec{k}\cdot\vec{r}^{\,o}_{ip}) = \vec{u}_i\{\sum_{1}^{3} k_j(x^o_{ij} + p_j)\} \qquad (2)$$

For convenience's sake we shall normalize the functions $\vec{u}_i$ by the condition that the average displacement vanishes:

$$\int_0^1 \vec{u}_i(\alpha)\, d\alpha = \vec{0}, \qquad i = 1...N. \qquad (2a)$$

so that the x^o_{ij} define the average position of the i-th atom.

b) At least one of the components k_1, k_2 and k_3 depends in a continuous manner upon external parameters such as the temperature. It can therefore be considered as an irrational number, in contrast with the rational $\vec{k}$-components of superstructures.

Property b was introduced as such in (I). Property a corresponds to the assumed existence of a lattice of "main reflections" in (I). The latter, however, are the diffraction image of the average structure (smeared atoms), not of the basic structure. Another competing structure is that of the solid phase without modulation, usually prevailing in a higher temperature range. It may be isostructural with the basic structure but again is not identical to it, since the temperature difference changes both unit cell and position parameters. Relations between these several structures will be given in section 3.

2. SYMMETRY OPERATIONS AND -TRANSLATIONS

A symmetry operation is usually defined as a movement of the crystal, which brings it into coincidence with itself. An alternative definition leaves the crystal where it is and defines as a symmetry operation each mapping $\vec{r} \to \vec{r}'$ for which

$$\phi(\vec{r}') = \phi(\vec{r}) \quad \text{for all } \vec{r}, \tag{3}$$

ϕ being any local function, such as the electron density. For symmetry translations, the mapping is given by (1). This "mapping definition" is so obviously equivalent to the "movement definition" that the distinction is hardly ever made.

For modulated structures, however, the "movement definition" leads nowhere, but the definition (3) can readily be extended. To begin with, (2) is replaced by

$$\vec{u}(i,p) = \vec{u}_i(\vec{k}\cdot\vec{r}^{\,o}_{ip} + t). \tag{4}$$

The new parameter t is an overall phase parameter for all $\vec{u}_i$-functions simultaneously. For an infinite crystal, the introduction of t - or, more generally, a change Δt in t - is no more than an infinitesimal structural change. This fact follows from property b : Indeed the change Δt is equivalent to a mere change of the origin cell by a vector $n_1\vec{a} + n_2\vec{b} + n_3\vec{c}$, provided that $\vec{k}\cdot\vec{n} = \Delta t$ (mod 1). Because of the irrationality of at least one k_j, this equation can be fulfilled to any desired degree of accuracy by a judicious choice of the integers n_j. Stated in less mathematical terms: the environment of a given atom is of course modified if t changes, but the new environment could have been found before the change for an atom elsewhere - not exactly, but with any desired degree of precision. Therefore it is plausible to extend (3) as follows: An "MS-(modulated structure-)symmetry operation" is a mapping $\vec{r} \to \vec{r}'$, $t \to t'$ such that

$$\phi(\vec{r}',t') = \phi(\vec{r},t) \quad \text{for all } \vec{r} \text{ and } t. \tag{5}$$

A mathematically exact symmetry translation can now be derived for modulated crystals, starting from (1) and adding a "compensating" change in t:

$$\vec{r}' = \vec{r} + \vec{n} \tag{6a}$$

$$t' = t - \vec{k}\cdot\vec{n}. \tag{6b}$$

$\vec{n}$ = basic structure lattice vector.

In the sense of symmetry definition (5), this operation can be interpretated most easily by looking first at (6b). Writing $\vec{u}(i,p,\tau)$ for the displacement of the atom at $\vec{r}^{\,0}_{ip}$ for $t = \tau$, we find $\vec{u}(i,p,t') = \vec{u}_i(\vec{k}(\vec{r}^{\,0}_{ip}-\vec{n})+t) = \vec{u}(i,p-n,t)$. Hence (6b) signifies that all displacements are carried over to atoms removed by a shift $\vec{n}$ from the atom to which they referred originally, the shift of course being performed on the corresponding atoms in the basic structure. Then it is obvious that (6a) indeed relates a site $\vec{r}$ before the execution of (6b), to a site at $\vec{r}\,'$ with identical ϕ after that operation, cf. fig.1. Because of the periodicity of the $\vec{u}_i$-functions, adding an integer to t is also a symmetry operation. Therefore we can now write down the complete set of <u>MS-symmetry translations</u>:

$$\vec{r}\,' = \vec{r} + \vec{n} \qquad t' = t - \vec{k}\vec{n} + s \tag{7}$$

$\vec{n}$ = basic structure lattice vector; s = integer.

Though the set (7) clearly defines a group of translations in four-dimensional ($\vec{r}$, t)-space, we shall continue to use words like "space" and "vector" in the three-dimensional sense. Moreover, we shall drop the term "pseudosymmetry" used in (I), since the approximative nature suggested by that term applies neither to translations (7) nor to the operations which will be discussed in the next sections. Instead, the term "MS-symmetry translations, MS-symmetry operations" etc. will be used for operations which include a specified mapping $t \to t'$ (even if t' should be equal to t) in order to distinguish these from normal symmetry operations.

3. PROPERTIES OF MS-SPACE GROUP OPERATIONS

Besides translations (7), a modulated crystal may have other symmetry operations. In our extended concept of symmetry, defined by (5), we shall have to specify the general nature of these operations $\vec{r} \to \vec{r}\,'$ and $t \to t'$. Beginning with t', we observe that this parameter can depend only upon the former value t - not on $\vec{r}$, because both t and t' are defined by (4) as overall phase parameters constant in space. The actual form of this dependence has to preserve the symmetry translations consisting of integer shifts in t, hence they must be of the form $t' = \varepsilon t + \delta$, where $\varepsilon = \pm 1$ and δ is a constant.

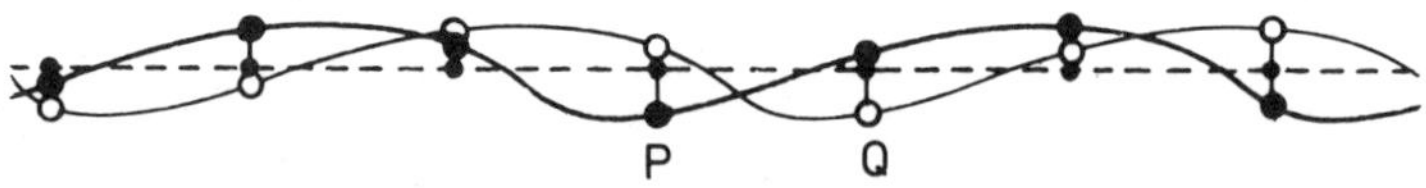

fig.1. Operation t' = t - 1 moves each displacement of the modulated chain (thick line) to the next atom, e.g. from P to Q.

The other part $\vec{r} \to \vec{r}'$ of a symmetry operation has to be a normal space group operation. As a matter of fact it belongs to the space group of the average structure. This follows if we take the average of both sides of (5) over one period of t (which according to the above is also a period of t'):

$$\phi_{av}(\vec{r}') = \phi_{av}(\vec{r}).$$

Accordingly a MS-symmetry operation has the general form

$$\vec{r}' = S\vec{r} + \vec{\rho} \tag{8}$$

$$t' = \varepsilon t - \vec{k}\vec{\rho} + \eta \tag{9}$$

where (8) is an element of the space group of the average structure. In (9) we have replaced the above constant δ by $-\vec{k}\vec{\rho} + \eta$, in analogy with (7), and in order to simplify further relations.

Finally it should be noted that (8) is also an element of the space group of the basic structure. This follows from the fact that, just as with symmetry translations (7), equation (8) connects points with identical electron density before and after the application of (9). Hence, it also connects corresponding atomic centra, which means that an atom at $\vec{r}_{ip}(t)$ is imaged by (8) in some atom at $\vec{r}_{jq}(t')$ see fig.1b.

$$\vec{r}_{jq}(t') = S\vec{r}_{ip}(t) + \vec{\rho}$$

or

$$\vec{r}^{\,0}_{jq} + \vec{u}(j,q,t') = S\vec{r}^{\,0}_{ip} + S\vec{u}(i,p,t) + \vec{\rho}. \tag{10}$$

Since by (2a) the average of $\vec{u}$, and therefore of $S\vec{u}$ as well, is zero, we obtain by averaging (10) over t and t'

$$\vec{r}^{\,0}_{jq} = S\vec{r}^{\,0}_{ip} + \vec{\rho} \tag{11}$$

which establishes the operation (8) as a normal symmetry operation of the basic structure.

Moreover, by substituting (11) in (10) we obtain

$$\vec{u}(j,q,t') = S\vec{u}(i,p,t), \tag{12}$$

an important starting point for the derivation of the relation between symmetry-equivalent $\vec{u}_i$'s which will be given in section 5.

Now that both the average and the basic structure have been shown to possess the symmetry elements (8), the question may arise what the relation is between the space group G evidently formed by these elements, and the space groups G_{av} and G_b of the average and basic structure, respectively. Since the centre of an atom in the basic structure is the centre of gravity of the corresponding smeared atom in the average structure, each symmetry operation of the latter applies to the former as well. Combining this result with those derived above we obtain

$$G \subseteq G_{av} \subseteq G_b \tag{13}$$

In most of the structures which are known the three groups are identical.

4. REDUCED FORM OF POINT GROUP OPERATIONS

Obviously the homogeneous parts of (8) and (9) taken together form an element of a finite group K_4, which plays the same rôle as the point group of normal crystals. We shall number these elements with an index m=1...g, where g is the group order:

$$\vec{r}\,' = S_m \vec{r} \qquad t' = \varepsilon_m t \tag{14}$$

On the other hand, the operators S_m clearly constitute a normal point group K. The question then arises to what extent K is different from K_4. It was shown in (I) that the one-dimensional representation ε_m of K_4 is contained in S_m. Hence K and K_4 are isomorphous. Moreover the corresponding reduction of S_m to

$$S_m = \begin{pmatrix} a_m & b_m & 0 \\ c_m & d_m & 0 \\ 0 & 0 & \varepsilon_m \end{pmatrix} \tag{15}$$

occurs for a suitably chosen basis of basic structure lattice vectors. The proof of (15) will be repeated here in terms of a new approach, using the properties of the vector $\vec{k}$.

5. EQUIVALENCE AND INVARIANCE OF $\vec{k}$-VECTORS

So far, the vector $\vec{k}$ has been considered as a constant not changed by the action of a symmetry operation. However, it is clearly not a unique vector. We may guess that more or less equivalent $\vec{k}$-vectors can be derived in three ways from a given $\vec{k}$:
a) by adding a basic structure reciprocal lattice vector $\vec{n}^*$
b) by inversion, yielding $-\vec{k}$
c) by letting the point group operations S act on it.

a) The first procedure will now be shown to lead to what we shall call "t-equivalent" $\vec{k}$-vectors. Two vectors $\vec{k}$ and $\vec{k}\,'$ are <u>t-equivalent</u> if there exist functions $\vec{u}_i'$ such that the description of the structure by (4), based on $\vec{k}$, can be replaced by a corresponding one based on $\vec{k}'$:

$$\vec{u}_i(\vec{k}\cdot\vec{r}^{\,o}_{ip} + t) = \vec{u}'_i(\vec{k}'\vec{r}^{\,o}_{ip} + t). \tag{16}$$

Hence the new functions $\vec{u}'_i(\alpha)$ have to obey

$$\vec{u}'_i(\alpha) = \vec{u}_i\{\alpha - (\vec{k}' - \vec{k})\vec{r}^{\,o}_{ip}\} \tag{17}$$

for all unit cells p. Since $\vec{u}_i$ is periodic with period 1, two different unit cells, lying a B.S. lattice vector $\vec{n}$ apart, can satisfy (17) only if $(\vec{k}' - \vec{k})\cdot\vec{n}$ = integer; and this has to be true for all lattice vectors $\vec{n}$. Such a Laue-type condition has the well-known consequence

$$\vec{k}' - \vec{k} = \vec{n}^* \tag{18}$$

where $\vec{n}^*$ is a reciprocal lattice vector of the basic structure. Hence, $\vec{k}'$ and $\vec{k}$ are t-equivalent only if they obey (18). Conversely, substitution of (18) in (17) yields the phase correction $-\vec{n}^*\cdot\vec{r}^{\,o}_{ip}$ to be applied to $\vec{u}_i$ in order to accomodate a shift (18), so (18) is both a sufficient and a necessary condition for t-equivalence of $\vec{k}'$ and $\vec{k}$.

b) The second of the above three procedures, inversion, never yields a t-equivalent vector, since for inversion (18) requires $2\vec{k} = \vec{n}^*$ which is impossible for the incommensurate $\vec{k}$'s considered here. This result may seem strange because $\vec{k}$ and $-\vec{k}$ are obviously equivalent vectors e.g. in the description of the diffraction image. However, t-equivalence is a mathematical concept, not a physical one. (The inversion $\vec{k}' = -\vec{k}$ can be accomodated by a condition like (16) but only with reversed sign of t on the right-hand side of that equation. Indeed t-equivalence would be physically significant if modulation were a wave phenomenon, with t proportional to the time, which would make a distinction between $\vec{k}$ and $-\vec{k}$ meaningful.)

c) Finally we investigate the kind of equivalence which arises from the action of S on $\vec{k}$. We shall show that there is t-equivalence between $\vec{k}$ and $\varepsilon_m S_m \vec{k}$ (not $S\vec{k}$!), where the index m numbers the point group operations as in (14), for m = 1...g. Starting from (12) and letting S_m^{-1} act on both sides we obtain

$$\vec{u}_i(\vec{k}\cdot\vec{r}^{\,o}_{ip} + t) = S_m^{-1}\vec{u}_j(\vec{k}\cdot\vec{r}^{\,o}_{jq} + t'). \tag{19}$$

The right-hand side of $\vec{u}_j$ in (19) is now brought in the form required by (16) first by substitution of (11) and (9)

$$\vec{k}\cdot\vec{r}^{\,o}_{jq} + t' = \vec{k}\cdot(S_m\vec{r}^{\,o}_{ip} + \vec{\rho}) + \varepsilon_m t - \vec{k}\vec{\rho} + \eta = \vec{k}\cdot S_m\vec{r}^{\,o}_{ip} + \varepsilon_m t + \eta.$$

Since $\vec{k}(S\vec{r}) = (S^{-1}\vec{k})\cdot\vec{r}$ we obtain

$$\vec{k}\cdot\vec{r}^{\,o}_{jq} + t' = \varepsilon_m\{(\varepsilon_m S_m^{-1}\vec{k})\cdot\vec{r}^{\,o}_{ip} + t\} + \eta.$$

Substitution in (19) leads to

$$\vec{u}_i(\vec{k}\cdot\vec{r}^{\,o}_{ip} + t\ ) = S^{-1}\vec{u}_j\varepsilon_m\{(\varepsilon_m S_m^{-1}\vec{k})\cdot\vec{r}^{\,o}_{ip} + t\} + \eta$$

in which equation we recognize (16) with

$$\vec{k}' = (\varepsilon_m S_m)^{-1}\vec{k}, \quad \text{or} \quad \vec{k} = \varepsilon_m S_m \vec{k}' \tag{20}$$

$$\vec{u}_i'(\alpha) = S^{-1}\vec{u}_j(\varepsilon_m\alpha + \eta), \quad \text{or} \quad \vec{u}_j(\alpha) = S\vec{u}_i'\{\varepsilon_m(\alpha - \eta)\} \tag{21}$$

so any two $\vec{k}$-vectors related by (20) are indeed t-equivalent. According to (18) this means that

$$\varepsilon_m S_m \vec{k} - \vec{k} = \vec{n}^*. \tag{22}$$

We now consider the "projection operator" $\frac{1}{g}\sum_1^g \varepsilon_m S_m = P$. Applied to any vector $\vec{v}$, the result is a vector $P\vec{v}$ invariant under all operations $\varepsilon_m S_m$ (m = 1...g). Such vectors $P\vec{v}$ constitute a linear vector space L_t (line, plane or whole space) and P actually performs the geometric projection of $\vec{v}$ onto L_t. The one thing we do not yet know is whether L_t is not of zero dimension, that is, whether there exists a vector $\vec{v}$ so that $P\vec{v}$ is not zero. The answer is that $\vec{k}$ is such a vector. This is readily found by writing down the identity:

$$\vec{k} = P\vec{k} - (1/g)\sum_1^g(\varepsilon_m S_m \vec{k} - \vec{k}). \tag{23}$$

From (22) it follows that the second term is a rational vector. Hence, the first term, $P\vec{k}$, cannot be zero since that would make the left-hand side $\vec{k}$ rational as well. So $\vec{k}$ is a vector for which $P\vec{k} \neq 0$.

<u>If $\vec{k}$ is in L_t</u>, (20) reduces to $\vec{k}' = \vec{k}$ and (16) to $\vec{u}_i(\alpha) = \vec{u}_i'(\alpha)$. Then (21) immediately yields the important relation (de Wolff, 1977)

$$\vec{u}_j(\alpha) = S_m\vec{u}_i\{\varepsilon_m(\alpha - \eta)\} \tag{24}$$

between displacements of symmetry-related atoms.

<u>If $\vec{k}$ is not in L_t</u>, it can be written

$$\vec{k} = P\vec{k} + \vec{k}_0.$$

From (23) and (22) we conclude that $\vec{k}_0$ is a rational vector. It will be shown in section 8 that $\vec{k}_0$ can be disregarded - so that $\vec{k}$ becomes a vector in L_t - by introducing a new type of symmetry translations as well as the corresponding new Bravais lattice types.

The proof that every S_m of the point group K can be written in the form (15) is obvious now: we only have to choose the $\vec{c}$ basis vector among (multiples of) the rational vectors $P\vec{n}$ left invariant by each $\varepsilon_m S_m$, that is, in L_t. If L_t is three-dimensional, (15) reduces to $\pm \mathbb{1}_3$; if not, the other basis vectors can always be chosen to obtain a form of (15) as well, because if $\vec{n}$ is a lattice vector clearly both the mutually perpendicular vectors $P\vec{n}$ and $\vec{n} - P\vec{n}$ are rational vectors.

6. POINT GROUPS

It follows from (15) that there is just one ε_m-value possible for each S_m within a given group K_4 of operators (14), that is, K_4 is isomorphous with K. The notation for groups K_4 can therefore be made unambiguous by using the Hermann-Mauguin symbols for the corresponding point group K, adding a prime (') to those symmetry elements which are generated by an operation for which $\varepsilon = -1$. The possible elements are enumerated below, with indices // and $\perp$ denoting their position with regard to L_t.

Generating operation has:	$\varepsilon = 1$	$\varepsilon = -1$
L_t = whole space	1	$\bar{1}'$
L_t = plane	$m_{//}$ or 1	$2'_{\perp}$ or $\bar{1}'$
L_t = line	$m_{//}, 1, 2_{//}$,	$2'_{\perp}, \bar{1}', m'_{\perp}$
	$3_{//}, 4_{//}$ or $6_{//}$	$\bar{3}'_{//}, \bar{4}'_{//}$ or $\bar{6}'_{//}$

In order to enumerate the point groups K_4, one can start by looking at the possible groups K and use the isomorphism between K and K_4 to arrive at groups K_4 afterwards. Regarding the groups K we observe that any normal point group which has a reducible vector representation (that is one, which leaves at least one line invariant) may occur as a group K. As a matter of fact, if this line is inverted by an element S of such a group, then the corresponding element of K_4 will have $\varepsilon = -1$, and otherwise $\varepsilon = +1$, so that εS leaves the vectors along the line invariant and the line will serve as (part of) L_t. In this way for each of these groups, a corresponding group K_4 can be constructed.

Since all point groups except the 5 cubic groups fulfill the above condition, the group K can be any of the 27 non-cubic point groups. However a given group among these 27 can lead to more than one group K_4 if there are several inequivalent choices possible for the invariant line(s). There is no problem in the groups with 3-, 4- or 6-fold axes, which have just this axis as the invariant line, nor with the triclinic system. The orthorhombic groups have three

Table 1. Point Groups

System	Normal	Magnetic	Modulated (K_4)	Group of εS
Triclinic	1	1	1	1
	$\bar{1}$	$\bar{1}$	$\bar{1}'$	1
Monoclinic II	2	2'	2'	m
	m	m'	m	m
	2/m	2'/m'	2'/m	m
Monoclinic III	2	2	2	2
	m	m	m'	2
	2/m	2/m	2/m'	2
Orthorhombic (L_t//c)	222	2'2'2	2'2'2	mm2
	mm2	m'm'2	mm2	mm2
	mmm	m'm'm	mmm'	mm2
	m2m	m'2'm	m2'm'	m2m
Trigonal	3	3	3	3
	$\bar{3}$	$\bar{3}$	$\bar{3}'$	3
	32	32'	32'	3m
	3m	3m'	3m	3m
	$\bar{3}$m	$\bar{3}$m'	$\bar{3}'$m	3m
Tetragonal	4	4	4	4
	$\bar{4}$	$\bar{4}$	$\bar{4}'$	4
	4/m	4/m	4/m'	4
	422	42'2'	42'2'	4mm
	4mm	4m'm'	4mm	4mm
	$\bar{4}$2m	$\bar{4}$2'm'	$\bar{4}'$2'm	4mm
	4/mmm	4/mm'm'	4/m'mm	4mm
Hexagonal	6	6	6	6
	$\bar{6}$	$\bar{6}$	$\bar{6}'$	6
	6/m	6/m	6/m	6
	622	62'2'	62'2'	6mm
	6mm	6m'm'	6mm	6mm
	$\bar{6}$2m	$\bar{6}$2'm'	$\bar{6}'$2'm	6mm
	6/mmm	6/mm'm'	6/m'mm	6mm

invariant lines, but only for mm2 the axes are inequivalent. This leads to K_4-groups mm2 and m2'm' for L_t parallel to $\vec{c}$.

A much more fundamental distinction can be made for monoclinic groups K, depending on the choice of invariant line(s) perpendicular to the binary axis, or parallel to that axis. There is good reason to distinguish two systems of monoclinic groups K_4 correspondingly, viz.

system "monoclinic II" (L_t = plane): point groups 2', m and 2'/m

system "monoclinic III" (L_t = line): point groups 2, m' and 2/m'.

The roman numerals stem from the numbers of the corresponding four-dimensional groups, cf. (I). The reason for having two different monoclinic systems is that different types of lattice obtain for each, cf. table 2. There is no such complication for the other systems so that K_4-groups can be classified conventionally in those. In total we find 27+1+3 = 31 groups K_4. They have been listed in (I). These groups are in a 1-1 correspondence with the 31 "admissible magnetic groups" (Opechowski and Guccione, 1965), consisting of combined time- and space operations which leave a magnetic moment vector invariant. Indeed the time inversion is formally equivalent to our $\varepsilon = -1$, and the difference between the positions of primes in the two lists is caused by the axial character of the magnetic vector as compared with our polar $\vec{k}$-vector.

7. RATIONAL AND IRRATIONAL NON-ZERO COMPONENTS OF $\vec{k}$

The vector Pk, lying in L_t, has essentially more than one non-zero coordinate in two systems, viz. two in monoclinic II, and three in the triclinic system. In these cases it should be noted that our definition of a moculated structure requires at least one irrational coordinate. The remaining one(s) can be rational, though that is not plausible from a physical point of view except for e.g. structures with a strong layer-like character.

The component $\vec{k}_0$ of $\vec{k}$ perpendicular to L_t, if not zero, has to be a rational vector in order to satisfy (22). It is easily determined in most systems merely by substituting for εS the generating rotation about the unique axis, if any. The result often depends on the kind of Bravais lattice of the structure. Firstly, since (22) is valid for a description on a primitive base, the vector $\vec{n}^*$ in it has to be a reciprocal lattice vector not extinguished by centring conditions for a non-primitive base. Secondly, a solution $\vec{k}_0$ of (22) is not significant if there exists a reciprocal lattice vector $\vec{n}_1^*$ with the same projection along L_t. (As a matter of fact, the equations

$$\vec{n}_1^* = P\vec{n}_1^* + \vec{k}_0 \quad \text{and} \quad \vec{k} = P\vec{k} + \vec{k}_0$$

yield $\vec{k} - \vec{n}_1^* = P(\vec{k} - \vec{n}_1^*)$ as a vector which can replace $\vec{k}$ since according to (18) it is a t-equivalent vector, and which lies in L_t.)

This restriction invalidates many solutions for centred lattices. For instance, in system monoclinic II the solution $(0\frac{1}{2}0)$ for $\vec{k}_0$ is easily found, but when applied to a C-centred lattice, the corresponding vector (now written conventionally as (010)) coincides with (110) in the projection along L_t which in this case is the projection on $\vec{b}$. The various possibilities for all systems will now be enumerated in full, they are shown in fig.2.

For the hexagonal system one finds that (22) with $\varepsilon S = R_{\pi/3}$ already excludes any fractional vector $\vec{k}_0$. In the tetragonal system $\vec{k}_0 = (\frac{1}{2}\frac{1}{2}0)$, and in the trigonal system $(\frac{1}{3}\frac{1}{3}0)$ fulfill the condition, but they are both significant only for P-type Bravais lattices. In monoclinic III, with $\varepsilon S = R_\pi$ about the unique b-axis, $(\frac{1}{2}00)$, $(\frac{1}{2}0\frac{1}{2})$ and $(00\frac{1}{2})$ are equivalent possibilities; the latter two are valid for both P- and C-lattices. Monoclinic II, with εS = mirror with respect to the a, c-plane, yields $(0\frac{1}{2}0)$ for P-, but no significant solution for C-lattices as shown above. In the triclinic system $\vec{k}_0 = \vec{0}$ since $P\vec{k} = \vec{k}$ for any vector. Finally, with L_t along c the orthorhombic system yields $(\frac{1}{2}00)$ for P, A, C- and F-lattices, $(\frac{1}{2}\frac{1}{2}0)$ for P-lattices only and no solution for I-lattices, where it must be stressed that the coordinates refer to an orthogonal base (100), (010) of the p- or c-net of non-extinct points of the reciprocal net l=0 (for the centred lattices this differs from the convention which e.g. for a C-lattice would call this base: (200), (020)).

Both table 1 and the drawings of fig.2 are easily interpreted if one remembers that the operations εS leave a vector invariant. For each group K_4 therefore, they form one of the "pyro-electric groups", as indicated in the last column in table 1.

8. NECESSITY OF INTRODUCTION OF BRAVAIS LATTICE TYPES WITH IMPROPER TRANSLATIONS

In this section it will be shown that it is possible to assign a $\vec{k}$-vector lying in L_t to any given modulated structure notwithstanding the just-mentioned possibility of non-vanishing perpendicular components, by accounting for such components through the introduction of a new type of symmetry translations. We shall begin by showing that such a seemingly complicated procedure is necessary for the enumeration of inequivalent MS space groups.

The MS space groups are to a large extent similar to those of normal crystals. Equations (8) and (9) indicate clearly the recipe for enumeration one should follow : each generator of the basic structure's space group*) (given by S and $\vec{\rho}$) must be completed with

*) We use this term in a loose sense. Actually the group G is meant, which in theory may be of lower symmetry than G_b, cf.(13).

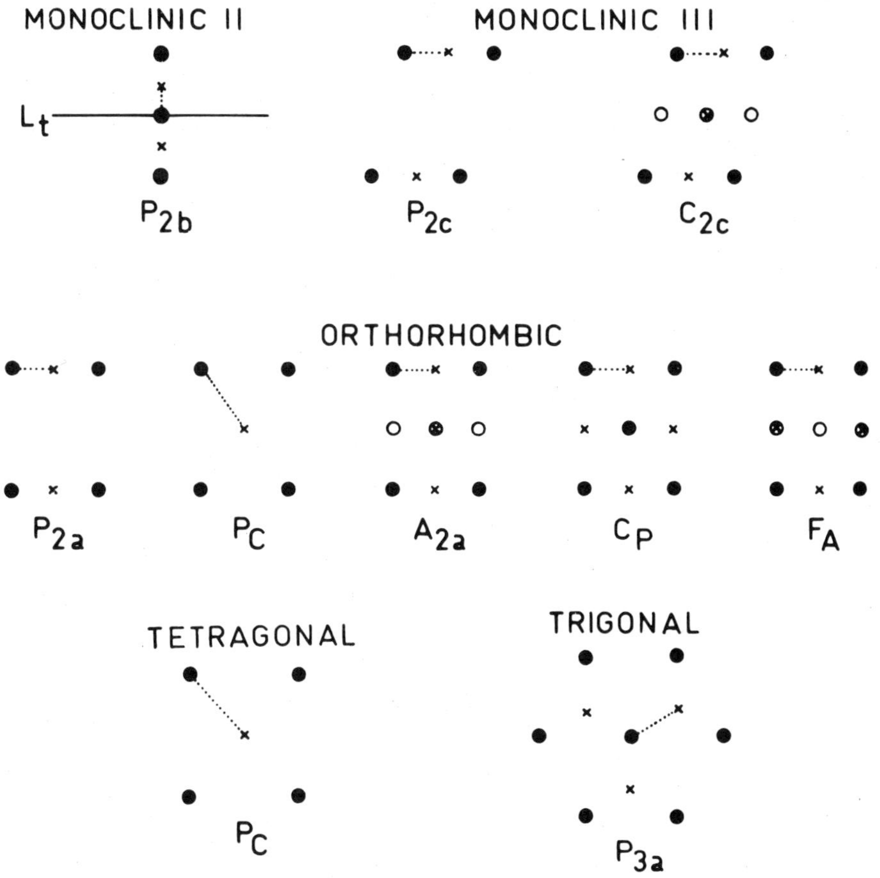

fig.2. The $\vec{k}$-vectors not in L_t, corresponding to the 10 new types of Bravais lattices in table 2. In each case L_t is normal to the paper; it is a plane for the system monoclinic II and a line otherwise. The dotted line shows the component $\vec{k}_0 \perp L_t$ of the k-vector.
●,× basic structure reciprocal lattice points at levels 0 and $\frac{1}{2}$.
o,⊗ satellites, at levels Pk and Pk+$\frac{1}{2}$.

some indication of the corresponding ε and η, and the lattice symbol of the basic structure must be added. For ε , the prime convention has been borrowed from magnetic groups already in (I), as an indication of $\varepsilon = -1$.

However, the results of the foregoing section show that one more distinction has to be made, namely between the possible values of $\vec{k}_0$. This rational component of $\vec{k}$ may well influence the number of inequivalent choices for the shift η in t in the rotational generators. For instance if in the system monoclinic III the space group of the basic structure*) is P2, the rotation over π ($\varepsilon = +1$) may or may not be accompanied by a shift $\eta = \frac{1}{2}$ in t, so that we obtain two MS-space groups when $\vec{k}_0 = \vec{0}$.

On the other hand, if $\vec{k}_0 = \frac{1}{2}\vec{a}^*$ the product of the same rotation with the translation $\vec{a}$ already yields such a combined operation of R_π and a shift of $\frac{1}{2}$ in t. Hence for $\vec{k}_0 = \frac{1}{2}\vec{a}^*$ there is only one MS-space group, while there are two for $\vec{k}_0 = \vec{0}$. The situation is entirely comparable to normal space groups for a monoclinic structure: we have P2, $P2_1$ and C2, but $C2_1$ is equivalent with C2. Indeed it can be shown that $\vec{k}_0 = \frac{1}{2}\vec{a}^*$ corresponds to a centring in four-dimensional space.

The effect of $\vec{k}_0$ can be accounted for entirely and unambiguously by an extension of the lattice types. This is particularly simple for the non-trigonal systems. The case $\vec{k}_0 = \frac{1}{2}\vec{a}^*$ for instance, signifies according to (7) that the translation $\vec{a}$ increases t by $\frac{1}{2}$. The alternative way of looking at such translations which we now propose is a) to ignore $\vec{k}_0$, so that $\vec{k} - \vec{k}_0$ replaces $\vec{k}$; b) to account for the t-shift by allowing s in (7) to equal $\frac{1}{2}$ when n_1 is odd**). The t-shift is thereby incorporated in the lattice type. In itself it is an operation of order 2, and it commutes with all symmetry translations of the basic structure. Therefore it offers the same extension of lattice types as that other external binary element: time inversion, well known in magnetic symmetry.

The difference with the magnetic case is, that t-shifts of $\frac{1}{2}$ have to be considered only for the axes perpendicular to L_t. Nonetheless we can make full use of the existing enumeration of magnetic lattice types and their nomenclature, as given by Opechowski and Guccione (1965), cf. section 10. In the trigonal system, the case $k_0 = (\frac{1}{3}\frac{1}{3}0)$ obviously calls for a different approach. It leads to only one new type of lattice, shown in fig.7, which has a primitive hexagonal lattice of the basic structure, and a shift of $\frac{1}{3}$ in t along $\vec{a}$. In analogy with the above-mentioned nomenclature, we propose the symbol P_{3a} for this lattice. The ensuing list of MS-lattice types is given in table 2, cf. section 10.

9. TWO-DIMENSIONAL EXAMPLE OF IMPROPER TRANSLATIONS

In order to elucidate the foregoing sections, we discuss the two-dimensional MS-plane groups p1m and $p_{2b}1m$ illustrated in fig.3a and 3b. Both are derived from a basic structure with plane group pm.

*) We use this term in a loose sense. Actually the group G is meant, which in theory may be of lower symmetry than G_b, cf.(13).

**) Such translations are termed "improper" by de Wolff (1977).

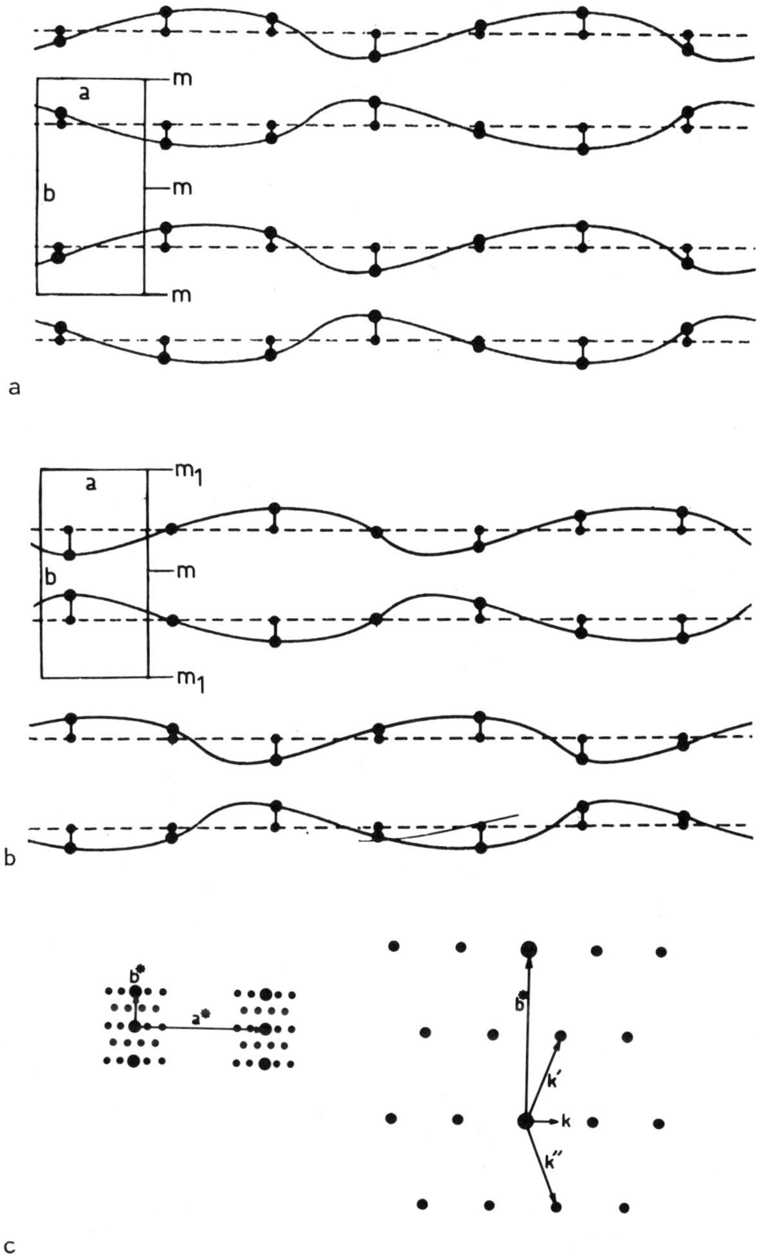

fig.3. Plane groups derived from the normal group pm. a) p1m b) $p_{2b}1m$ c) reciprocal lattice of $p_{2b}1m$; left: overall picture, right: enlarged portion. Large dots are main reflections.

The more elaborate notation (p1m means m perpendicular to b) is not absolutely necessary but it helps to prevent confusion. The fact that there is no prime in either symbol means that m has $\varepsilon = +1$, hence the vectors in L_t are those parallel to m. The simplest case, therefore, is the one with $\vec{k}$ parallel to m, that is, in the direction of $\vec{a}$: MS-plane group p1m.

Just as in the three-dimensional case, the fact that L_t is the site of a twofold symmetry element allows a second possibility for the k-vector, called k', with a component $\vec{k}_0 = \frac{1}{2}\vec{b}^*$ normal to L_t. This is rendered by the lattice type p_{2b}. The diffraction image in fig.3c clearly shows that the remaining k parallel to L_t is unique and is in agreement with the orthogonal symmetry, whereas the original $\vec{k}'$-vector is ambiguous (its image $\vec{k}''$ with respect to m is an equivalent choice) and clashes with the point group symmetry. The lattice type p_{2b} eliminates these drawbacks. It does so at the cost of an unusual extinction rule, viz. presence of m-th order satellites for

$$2k + m = \text{even}$$

where one has to admit half-integer values of the k-index.

As explained in section 6, there is no choice if one wants to express the influence of $\vec{k}_0$ on MS-space groups available for a given basic structure. In this respect the situation for basic structure space group P2 sketched there is exactly as in the present case: apart from p1m and p_{2b}1m there existst just one further extension, viz. p1m_1 (cf. fig.4; m_1 is a mirror coupled with a shift 1/2 in t), and no separate group "p_{2b}1m_1" (the corresponding symbols in (I), table 2, are Pm for p1m, Am for p_{2b}1m and Pc for p1m_1).

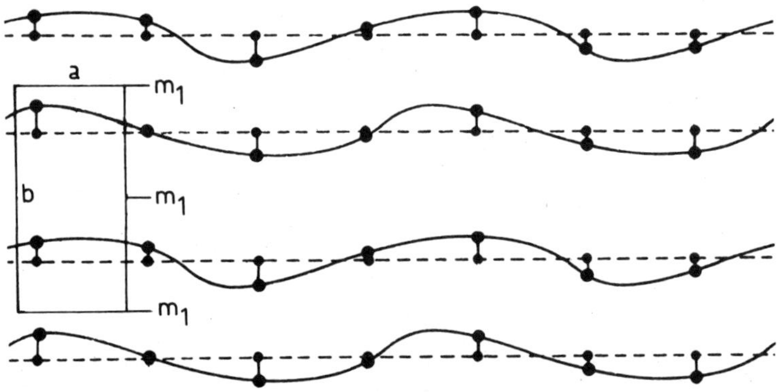

fig.4. The plane group p1m_1

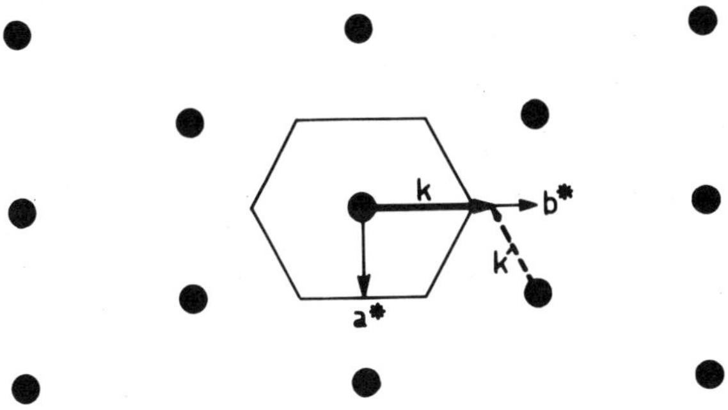

fig.5. Reciprocal centred net, L_t is parallel to $\vec{b}$. The vector $\vec{k}$, when chosen in L_t, extends beyond the first Brillouin zone shown in the figure.

10. ENUMERATION OF LATTICE TYPES

The last three sections have made it clear that among the possible criteria for normalizing $\vec{k}$, the choice of a vector belonging to L_t is by far the most prompting in order to avoid complications. It should be stressed that such a choice is by no means equivalent to a preference for the first Brillouin zône. To illustrate this, fig.5 shows a two-dimensional reciprocal net of the centred type. With orthogonal axes a > b and L_t along $\vec{b}$, one finds that $\vec{k}$ vectors lying within L_t for which

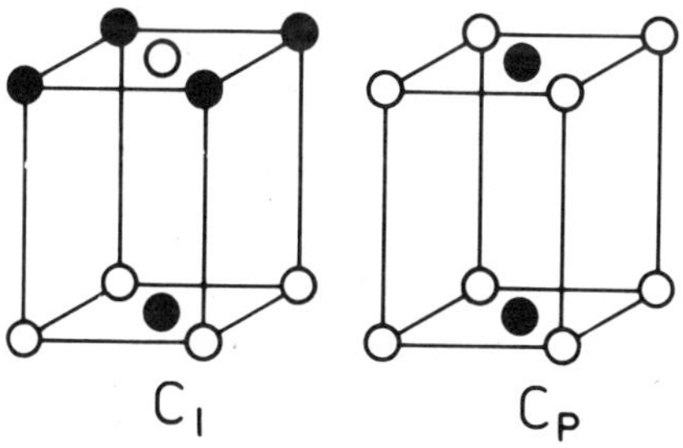

fig.6. Two magnetic Bravais lattice types which are equivalent when applied to modulated structures.

Table 2

MS-system	vectors in L_t	Lattice types		$\vec{k}$ for lattice of basic structure (irr=irrational)		
		basic structure	modulated structure	k_1	k_2	k_3
triclinic	all	P	P	irr	irr	irr
monoclinic II	b	P	P	irr	0	irr
		P	P_{2b}	irr	½	irr
		C	C	irr	0	irr
monoclinic III	//b	P	P	0	irr	0
		P	P_{2c}	0	irr	½
		C	C	0	irr	0
		C	C_{2c}	0	irr	½
orthorhombic	//c	P	P	0	0	irr
		P	P_{2a}	½	0	irr
		P	P_C	½	½	irr
		A	A	0	0	irr
		A	A_{2a}	½	0	irr
		C	C	0	0	irr
		C	C_P	1	0	irr
		I	I	0	0	irr
		F	F	0	0	irr
		F	F_A	1	0	irr
hexagonal	//c	P	P	0	0	irr
trigonal	//c	P	P	0	0	irr
		P	P_{3a}	1/3	1/3	irr
		R	R	0	0	irr
tetragonal	//c	P	P	0	0	irr
		P	P_C	½	½	irr
		I	I	0	0	irr

$$\tfrac{1}{2}\{1 + (b/a)^2\} < k_2 < 1,\ k_1 = 0$$

fall outside the first Brillouin zône. Preference for the latter would lead to an oblique vector k' not in L_t.

Cases in which $\vec{k}_0$ is essentially non-zero (section 7) have recently been found to occur among actual modulated structures. An example is the monoclinic III-structure of TTF-TCNQ between 54K and 47K which has $\vec{k} = (\tfrac{1}{2}, 0.295, 0)$ so that it corresponds - after interchanging a and c - with the case P_{2c} in fig.2 (Bak, 1977). Hence,

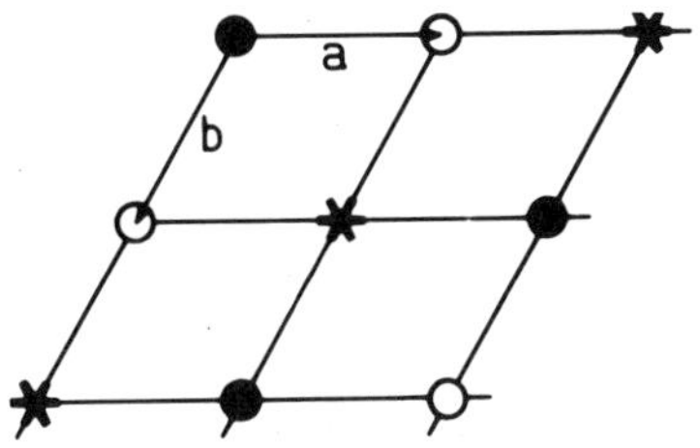

fig.7. The lattice type P_{3a}: symmetry translations perpendicular to $\vec{c}$. The shift in t is $\frac{1}{3}$ or $\frac{2}{3}$ in going from a black dot to a circle or an asterisk, respectively.

enumeration of lattice types including all such cases is of more than academic interest.

The types listed in table 2 have been derived in three ways: a) As in section 7, from the possible vectors $\vec{k}_0$, including zero. b) for the non-trigonal systems: by studying the type or types corresponding to each magnetic lattice. Some of these lead to more than one MS-type. For instance, if L_t in an orthorhombic lattice is along $\vec{c}$, then clearly the lattices C and A are not equivalent. On the other hand, many magnetic lattices are superfluous as an MS-type. An example is C_I, illustrated in fig.6. It has C-centring with a shift $\frac{1}{2}$ in t accompanying the centring translation as well as the c-translation. With L_t again in the $\vec{c}$-direction, however, the latter shift merely changes k_3 into $1 - k_3$. Though this may in some cases bring $\vec{k}$ within the first Brillouin zône, we prefer to disregard that criterion as we have done above, and to replace C_I by the simpler type C_P which has a t-shift for the centring translation only.

REFERENCES

BAK, P. (1977) These proceedings, p. 66-87.

JANNER, A. & JANSSEN, T. (1977), Phys.Rev., 15, 643-658.

OPECHOWSKY, W. & GUCCIONE, R. (1965), "Magnetic symmetry" in Magnetism IIa, ed. Rado & Suhl. New York, Academic Press.

WOLFF, P.M. DE (1974), Acta Cryst. A30, 777-785.

WOLFF, P.M. DE (1977), Acta Cryst. A33, 493-497.

SUPERSPACE GROUPS FOR THE CLASSIFICATION OF MODULATED CRYSTALS

T. Janssen

Institute for Theoretical Physics, University of Nijmegen

Toernooiveld, Nijmegen, the Netherlands

I. Introduction

During this meeting one has heard a lot about crystals with an incommensurate phase. Characteristic for these is the absence of space group symmetry. However, as de Wolff explained already (ref.1), it is possible to extend the notion of symmetry. Our approach differs somehow from his and originates from a study of the space-time symmetry of vibrating crystals. The symmetry of a vibration mode is, in general, irrelevant, unless this mode plays a predominant role, e.g. if it is a softening mode. However, we will see that the symmetry considerations can also be applied to modulated crystals, both static and dynamic.

Consider a crystal vibrating in a single mode. For simplicity we take a one-dimensional Bravais crystal (fig. 1). The displacement of the n-th atom in the chain is given by $u_n = u \sin(qna-\omega t)$. The pattern of world lines in the x-t-plane shows invariance under a lattice of translations denoted by Σ. A basis of Σ is formed by $a_1 = (a,qa/\omega)$ and $a_2 = (0,2\pi/\omega)$. For fixed t the positions of the atoms do not have translation symmetry, but form a crystal with a displacive modulation. The difference between the structures at two different times is just an overall phase shift.
Hence, if we identify ωt with the phase ϕ, one can see the modulated crystal as a section of a periodic pattern in the $x\phi$-plane. In this way we have imbedded the crystal in position space (denoted by V_E) into a larger space (called superspace) which is the sum of V_E and an internal space V_I.

In the superspace V_S the reciprocal lattice Σ^* has a basis with $a_1^* = (2\pi/a,0)$ and $a_2^* = (-q,\omega)$. The projection of Σ on V_E consists of the vectors $k = n(2\pi/a) + mq$ (n,m integers). This means that the diffraction peaks of the modulated crystal belong to this projection: the points $n(2\pi/a)$ are the main reflections, the other

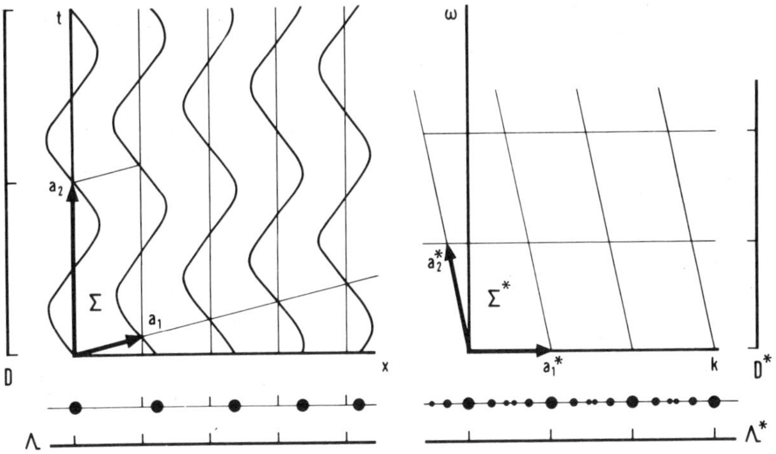

Fig. 1

ones (with $m \neq 0$) the satellites. If we denote the lattice of the undistorted crystal by Λ, the lattice in V_I spanned by $(0,2\pi/\omega)$ by D, their reciprocal lattices by Λ^* and D^*, the projections on V_E and V_I by π_E and π_I, resp., then one has the following important properties:

$$\Sigma \cap V_I = D,\ \Sigma^* \cap V_E = \Lambda^*,\ \pi_E\Sigma = \Lambda,\ \pi_I\Sigma^* = D^* \qquad (1.1)$$

Apart from the translations there is another symmetry element of the pattern in V_S: the 180° rotation which is a combination of the reflection $x \to -x$ (which is a symmetry element of the undistorted crystal) and the operation $t \to -t$ (which is a transformation in V_I). The symmetry group of the pattern is the space group p2. The elements of this group are combinations of space group elements in V_E with transformations of the internal space. This extension of the class of considered transformations is not uncommon. In the theory of non-rigid molecules, e.g., the symmetry elements are also combinations of orthogonal transformations with internal transformations. As an example, the symmetry group of C_2H_6 has 36 elements (fig.2) and is generated by i) a 120° rotation of the whole molecule, ii) a 120° rotation of the top part with respect to the bottom part (an internal transformation) and iii) a reflection followed by an internal rotation. The only Euclidean transformations are the rigid 120° rotations.

One can generalize the concepts introduced above. The displacement is not necessarily sinusoidal, but is described by a periodic function: $u_n = u(qna-\phi)$ with $u(x+2\pi) = u(x)$.
A modulated crystal with such a modulation in n dimensions (usually n=3) can be imbedded into a (n+1)-dimensional superspace. For a superposition of modulation waves the modulated crystal can be im-

bedded into (n+d)-dimensional superspace, when d is determined as follows. The diffraction pattern of such a crystal consists of k-vectors with

$$\underline{k} = \sum_{i=1}^{n} n_i \underline{a}_i^* + \sum_{j=1}^{d} m_j \underline{b}_j^* \qquad (1.2)$$

where $\underline{a}_i^*$ are main reflections and $\underline{b}_j^*$ basic satellites. Finally, not only displacive modulation can be described in this way, but also a continuous density distribution for which the spectrum is given by eq. (1.2). The different more general cases are discussed in ref. 2.

II. Superspace groups

All the cases mentioned at the end of the first section can be described by superspace groups, generalisations of the two-dimensional space groups, found for the simple n=1, d=1 example. The mathematical definition is as follows. A superspace group G is

a) a subgroup of E(n) x E(d) such that
b) the translations in G form an (n+d)-dimensional lattice Σ, and
c) the intersection of the reciprocal lattice Σ^* with V_E is an n-dimensional lattice Λ^*.

Condition a) means that the elements g of G are pairs (g_E, g_I) of Euclidean transformations in, resp., n and d dimensions.
Condition b) implies that G is a space group. Condition c) gives the space group additional structure. It implies that one can choose standard bases for Σ and Σ^*. A standard basis for Σ is one for which the last d basis vectors $a_{n+1}, \ldots, a_{n+d}$ belong to V_I. A standard basis for Σ^* is one where $a_1^*, \ldots, a_n^*$ belong to V_E.

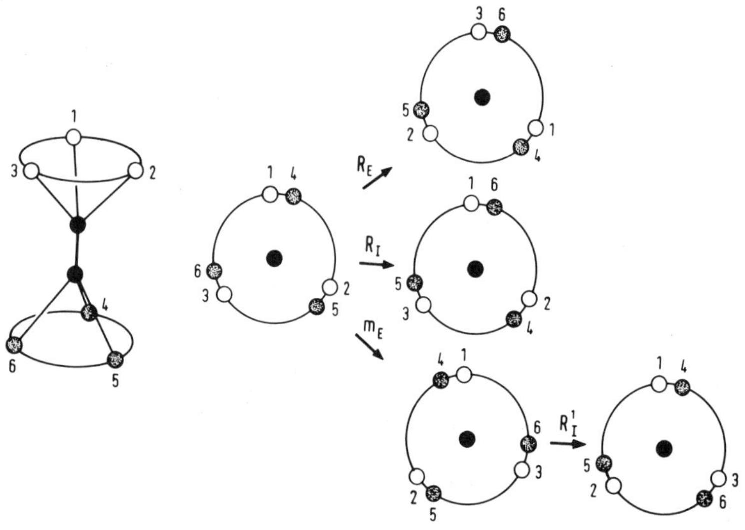

Fig.2

Because the satellites of the zero vector belong to the projection of the lattice spanned by $a^*_{n+1},\ldots,a^*_{n+d}$ on V_E, the property that the modulation is incommensurate can be formulated by the condition:

d) the intersection of Σ with V_I is the zerovector.

Condition d) is not essential for the consequences and is not included in the definition. For convenience, however, we shall assume only the incommensurate case in the following.

A superspace group is given by its lattice translations, its point group and the nonprimitive translations v(R), where $R=(R_E,R_I)$ belongs to the point group K. A standard basis for Σ is given by

$$\begin{aligned} a_i &= (\underline{a}_i, -\underline{\Delta a}_i) \;, \quad (i = 1,\ldots,n), \underline{a}_i \in \Lambda \;, \; \underline{\Delta a}_i \in V_I \\ a_{n+j} &= (0, \underline{b}_j) \;, \quad (j = 1,\ldots,d), \underline{b}_j \in D \end{aligned} \tag{2.1}$$

If $\underline{a}^*_i$ (i=1,..,n) form the reciprocal basis of Λ^* and $\underline{b}^*_j$ that of D^*, the reciprocal basis of Σ^* is

$$\begin{aligned} a^*_i &= (\underline{a}^*_i, 0) \;, \quad (i = 1,\ldots,n) \\ a^*_{n+j} &= (\underline{\Delta^* b}^*_j, \underline{b}^*_j) \;, \quad (j = 1,\ldots,d), \underline{\Delta^* b}^*_j \in V_E \end{aligned} \tag{2.2}$$

Because D and Λ^* span V_I and V_E, resp., one can express $\underline{\Delta a}_i$ and $\underline{\Delta^* b}^*_j$ as

$$\begin{aligned} \underline{\Delta a}_i &= \sum_{j=1}^{d} \sigma_{ji} \underline{b}_j \\ \underline{\Delta^* b}^*_j &= \sum_{i=1}^{n} \sigma_{ji} \underline{a}^*_i \;. \end{aligned} \tag{2.3}$$

An arbitrary $k = (\underline{k}_E, \underline{k}_I)$ of Σ^* with $k = \Sigma n_i a_i$ has a projection on V_E given by

$$\underline{k}_E = \sum_{i=1}^{n} n_i \underline{a}^*_i + \sum_{j=1}^{d} n_{n+j} \underline{\Delta^* b}^*_j \tag{2.4}$$

Comparing this with eq. (1.2), one finds that the $\underline{\Delta^* b}^*_j$ are the basic satellites. Hence the entries of the dxn matrix σ in eq. (2.3) are the coordinates of these basic satellites with respect to the basis $\underline{a}^*_1,\ldots,\underline{a}^*_n$ of Λ^*.

For an incommensurate modulation it is easy to show that the point group K is isomorphic to the group K_E of elements R_E. With respect to a standard basis the point group elements R are represented by matrices

$$\Gamma(R) = \begin{pmatrix} \Gamma_E(R) & 0 \\ \Gamma_M(R) & \Gamma_I(R) \end{pmatrix} , \quad R \in K \tag{2.5}$$

The matrices $\Gamma_E(R)$ form the group K_E, the matrices $\Gamma_I(R)$ the group K_I. They are n-, resp. d-, dimensional point groups. The dxn matrix $\Gamma_M(R)$ is given by

$$\Gamma_M(R) = \sigma\Gamma_E(R) - \Gamma_I(R)\sigma . \qquad (2.6)$$

Because $\Gamma(R)$ (hence also $\Gamma_E(R)$, $\Gamma_I(R)$ and $\Gamma_M(R)$) are integral matrices, eq. (2.6) puts strong restrictions on the possible σ's (i.e. possible modulation vectors) for given K_E and K_I. Two matrices σ can give rise to the same $\Gamma_M(R)$, i.e. to the same point group. There is a unique decomposition $\sigma = \sigma^i + \sigma^r$ such that $\sigma^i\Gamma_E(R) - \Gamma_I(R)\sigma^i = 0$, (all $R \in K$). The matrices σ^i for given $\Gamma_E(R)$ and $\Gamma_I(R)$ form a real vector space, corresponding to L_t in de Wolffs talk (ref.1). The matrices σ^r have rational coefficients.

Starting from the knowledge of the point groups in n and d dimensions, one can determine all possible σ's, i.e. all possible point groups of superspace groups. Then one can determine all superspace groups using a method discussed in ref.3.

III. Equivalence classes

In the usual crystallography one identifies isomorphic space groups. In this way one obtains 219 nonequivalent space groups. Superspace groups have an additional structure. Therefore, one defines: two superspace groups G and G' are equivalent if and only if a) they are isomorphic with an isomorphism that b) maps a standard basis for the lattice Σ of G on a standard basis for G'. Just as for usual crystallography this implies an equivalence relation for lattices: the lattices Σ and Σ' belong to the same Bravais class if and only if there are standard bases for both Σ and Σ' such that the holohedries (the symmetry point groups of the lattices) have the same matrices. This means that these point groups are related by $\Gamma'(R) = S^{-1}\Gamma(R)S$ (all R in K) with

$$S = \begin{pmatrix} S_E & 0 \\ S_M & S_I \end{pmatrix},$$

where S_E, S_M and S_I are integral matrices. With this equivalence relation the number of Bravais classes is finite. To give an idea, in table I is given the number of Bravais classes for n=3,d=0,1,2,3. Since the matrices $\Gamma_E(R)$ of the holohedry form an n-dimensional point group which belongs to an n-dimensional (usual) Bravais class, one can assign each Bravais class in superspace to a Bravais class in n dimensions. To this correspond the different columns in table I: Tr = triclinic, M = monoclinic, O = orthorhombic, T = tetragonal, Tg = trigonal, H = hexagonal, C = cubic.

A Bravais class is characterized by its holohedry, i.e. by the elements $\Gamma_E(R)$, $\Gamma_I(R)$ and $\Gamma_M(R)$ or equivalently by the corresponding arithmetic point groups K_E, K_I and the matrix σ^r. Since σ^r is rational a lattice can be considered as a centering (denoted by C) of a lattice with $\sigma^r = 0$ (a P lattice). For n=3, d=1 the Bravais classes with K_E in the primitive monoclinic Bravais are given in table II.

Table I: Number of Bravais classes; n=3								
d	Tr	M	O	Tg	H	T	C	total
0	1	2	34	1	1	2	3	14
1	1	7	10	2	1	3	-	24
2	1	16	39	8	4	15	-	83
3	1	26	122	18	4	30	14	215

Table II: P-monoclinic Bravais classes; n=3;d=1					
System	Bravais class	σ-matrix	centering	generators	holohedry
$\frac{2/m}{\bar{1}\ 1}$	$P^{P2/m}_{\bar{1}\ 1}$	αβ0	-	$\bar{1}$ 0 0 0 0 $\bar{1}$ 0 0 0 0 1 0 0 0 0 $\bar{1}$	1 0 0 0 0 1 0 0 0 0 $\bar{1}$ 0 0 0 0 1
	$C^{P2/m}_{\bar{1}\ 1}$	αβ½	00½½	$\bar{1}$ 0 0 0 0 $\bar{1}$ 0 0 0 0 1 0 0 0 1 $\bar{1}$	1 0 0 0 0 1 0 0 0 0 $\bar{1}$ 0 0 0 $\bar{1}$ 1
$\frac{2/m}{1\ \bar{1}}$	$P^{P2/m}_{1\ \bar{1}}$	00γ	-	$\bar{1}$ 0 0 0 0 $\bar{1}$ 0 0 0 0 1 0 0 0 0 1	1 0 0 0 0 1 0 0 0 0 $\bar{1}$ 0 0 0 0 $\bar{1}$
	$C^{P2/m}_{1\ \bar{1}}$	½0γ	½00½	$\bar{1}$ 0 0 0 0 $\bar{1}$ 0 0 0 0 1 0 $\bar{1}$ 0 0 1	1 0 0 0 0 1 0 0 0 0 $\bar{1}$ 0 1 0 0 $\bar{1}$

The symbol for an arithmetic point group (in this case a holohedry) consists of 3 parts. The topline gives the point group K_E. To each element of K_E corresponds an element of K_I. The bottom line gives K_I (for d=1 consisting of elements ± 1). In front is a symbol characterizing the centering, i.e. σ^r.

The elements g_E form, if $g = (g_E, g_I)$ belongs to a superspace group, an n-dimensional space group G_E. The symbol for a superspace group consists also of 3 parts. The top line gives G_E. In the bottom line are the corresponding elements g_I. These do not form a space group. For d=1 the possible elements are 1, -1 and s (denoting a nonprimitive translation in V_I). The superspace groups for n=3, d=1 with G_E belonging to the primitive monoclinic Bravais class are given in table III.

The superspace groups are space groups in n+d dimensions. However, because of the additional structure, the equivalence classes are different from those of ordinary space groups.

Table III: Superspace groups for the primitive monoclinic Bravais class, n=3, d=1

G_E	Bravais class			
	$P^{P2/m}_{\bar{1}\,1}$	$C^{P2/m}_{\bar{1}\,1}$	$P^{P2/m}_{1\,\bar{1}}$	$C^{P2/m}_{1\,\bar{1}}$
Pm	P^{Pm}_{1}, P^{Pm}_{s}	C^{Pm}_{1}	$P^{Pm}_{\bar{1}}$	$C^{Pm}_{\bar{1}}$
Pb	P^{Pb}_{1}	C^{Pb}_{1}	$P^{Pb}_{\bar{1}}$	$C^{Pb}_{\bar{1}}$
P2	$P^{P2}_{\bar{1}}$	$C^{P2}_{\bar{1}}$	P^{P2}_{1}, P^{P2}_{s}	C^{P2}_{1}
$P2_1$	$P^{P2_1}_{\bar{1}}$		$P^{P2_1}_{1}$	$C^{P2_1}_{1}$
P2/m	$P^{P2/m}_{\bar{1}\,1}$, $P^{P2/m}_{\bar{1}\,s}$	$C^{P2/m}_{\bar{1}\,1}$	$P^{P2/m}_{1\,\bar{1}}$, $P^{P2/m}_{s\,\bar{1}}$	$C^{P2/m}_{1\,\bar{1}}$
$P2_1/m$	$P^{P2_1/m}_{\bar{1}\,1}$, $P^{P2_1/m}_{\bar{1}\,s}$		$P^{P2_1/m}_{1\,\bar{1}}$	$C^{P2_1/m}_{1\,\bar{1}}$
P2/b	$P^{P2/b}_{\bar{1}\,1}$	$C^{P2/b}_{\bar{1}\,1}$	$P^{P2/b}_{1\,\bar{1}}$, $P^{P2/b}_{s\,\bar{1}}$	$C^{P2/b}_{1\,\bar{1}}$
$P2_1/b$	$P^{P2_1/b}_{\bar{1}\,1}$		$P^{P2_1/b}_{1\,\bar{1}}$	$C^{P2_1/b}_{1\,\bar{1}}$

This means that one can not use the knowledge of space groups in dimension 2,3 (Tables of X-ray crystallography) and 4 (as determined by Fast and myself and by Wondratschek, Neubüser and Brown). Up to now superspace groups have been determined for n=2,3 and d=1.

IV. Examples

As examples we consider two compounds discussed earlier during this conference. The structure of γ-Na_2CO_3 has been determined by de Wolff and co-workers (ref.4). One has a case n=3, d=1.

space group of basic structure: C2/m;
wave vector of modulation: $\underline{q} = \alpha(\underline{a}_1^* + \underline{a}_2^*) + \beta\underline{a}_3^*$, or $\sigma = (\alpha\alpha\beta)$,
where $\alpha \approx 0.091$ and $\beta \simeq 0.318$;

superspace group $P^{C2/m}_{\bar{1}\,s}$

The 4-dimensional pattern is left invariant by

i) the translations $(\underline{a}_1, -2\pi\alpha)$, $(\underline{a}_2, -2\pi\alpha)$, $(\underline{a}_3, -2\pi\beta)$, $(0, 2\pi)$, where $2\pi\alpha = \Delta\underline{a}_1$, etc.;
ii) a two-fold rotation along the unique axis combined with inversion of the phase;
iii) the mirror in the perpendicular plane combined with a phase shift π: nonprimitive translation $\frac{1}{2}a_4$.

The data for 1T-TaS_2 can be found in ref.5:

space group of basic structure: $P\bar{3}m1$;
there are 3 modulation waves: $(\alpha 0 1/3), (0\alpha 1/3), (\bar{\alpha}\bar{\alpha} 1/3)$ with $\alpha \approx 0.285$; since the third is a linear combination of the first two and a_3^*, one has d=2 and

$$\sigma = \begin{pmatrix} 0 & \alpha & 1/3 \\ \alpha & 0 & 1/3 \end{pmatrix},$$

the lattice in superspace belongs to the Bravais class $C^{P\bar{3}m}_{p6m}$;
this is a centering 001/3 1/3 1/3 of $P^{P6/mmm}_{p61mm}$;
the only superspace group with $G_E = P\bar{3}m1$ belonging to this Bravais class is $\boxed{C^{P\bar{3}1m}_{p61m}}$

This group has a 5-dimensional lattice: the translation $\underline{a}_1$ in position space is combined with a shift $-\alpha\underline{b}_1$ in internal space, $\underline{a}_2$ with $-\alpha\underline{b}_2$, $\underline{a}_3$ with $-(\underline{b}_1+\underline{b}_2)/3$; the modulation function as a whole can be shifted over $\underline{b}_1$ and over $\underline{b}_2$. The roto-inversion $\bar{3}$ is combined with a 6-fold rotation in internal space, the mirror which interchanges $\underline{a}_1$ and $\underline{a}_2$ is to be combined with a mirror which interchanges $\underline{b}_1$ and $\underline{b}_2$. Of course, one cannot determine the superspace group from only the $\underline{q}$-vectors. A precise analysis must give an answer to the question if $C^{P\bar{3}/m}_{p6/m}$ or only a subgroup is the superspace group.

V. Conclusions

Like the ordinary space groups, superspace groups can be used for the classification of structures, for selection rules and for the characterisation of excitations. The description of structures has been discussed in section IV. Selection rules follow from the properties of the Fourier components of a distribution invariant under a superspace group. If the function $\rho(r)$ in superspace is invariant under $g = \{R|v(R)\}$, then $\hat{\rho}(Rk) = \hat{\rho}(k) \exp\{i(Rk)v(R)\}$. This has consequences for the intensities of diffraction spots: they have point group symmetry and $\hat{\rho}(k) = 0$ if $Rk = k$ and $kv(R) \neq 2\pi n$.
In this way one can explain systematic extinctions in structures not having space group symmetry.

The excitations of modulated crystals can be characterized with irreducible representations of the superspace group. As an example, phasons transform indeed according to such an irreducible representation. Phonons in modulated crystals should be characterized by irreducible representations of the superspace group G, not by the space group of the basic structure, which is no longer a symmetry group. However, one can show that the lattice D is an invariant subgroup of G and that G/D is isomorphic to G_E. Hence representations of the (ordinary) space group G_E are also representations of G. Moreover since K and K_E are isomorphic, often labels corresponding to G_E can be used. However, one has to use another k-vector labelling, because in position space there is no Brillouin zone left.

The results presented above have been obtained in a research together with prof. A. Janner. We have profited very much from stimulating discussions with prof. P.M. de Wolff.

References

1. P.M. de Wolff: these proceedings and Acta Cryst. A30(1974)777.
2. A. Janner and T. Janssen: Phys.Rev. B15 (1977) 643.
3. T. Janssen, A. Janner, E. Ascher: Physica 42 (1969) 41.
4. W.v. Aalst, J. den Hollander, W.J.A.M. Peterse, P.M. de Wolff: Acta Cryst. B32 (1976) 47.
5. F.J. di Salvo: these proceedings; J.A. Wilson, F.J. di Salvo, S. Mahajan: Adv. Phys. 24 (1975) 117.

STRUCTURAL PHASE TRANSITIONS AND SUPERCONDUCTIVITY IN A-15 COMPOUNDS

L. R. Testardi

Bell Laboratories

Murray Hill, New Jersey 07974

I. INTRODUCTION

Ten years after the discovery by Hardy and Hulm[1] of high superconducting transition temperatures in A-15 structure materials evidence of their structural instability emerged. Shull[2] in neutron diffraction work, and Batterman and Barrett in more extensive x-ray studies found that V_3Si underwent a structural transformation at temperatures not far above the superconducting T_c ($T_c \approx 17K$). The transition from cubic to tetragonal structure shown by the x-ray data of Fig. 1 begins at $T_m \sim 20.5K$ and progresses rapidly (though apparently continuously) on cooling down to $\sim$17K where the onset of superconductivity arrests the progress of the transformation. The tetragonal distortions are relatively small, $(c/a-1) \approx 2.2 \times 10^{-3}$ and with $\Delta c/c \sim -2\Delta a/a$ so that there is little change in volume from the cubic state. Structural domains (of differing c axes orientations) occur below T_m.

Mailfert et al.[3] and Vieland et al.[4] later reported a cubic to tetragonal transformation in Nb_3Sn ($T_c \approx 18K$) similar to that in V_3Si but with the important differences of i) $(a/c-1) \approx 5.2 \times 10^{-3}$ (opposite tetragonality though still with approximately no volume change), ii) a (first order type) discontinuity in tetragonality at T_m (but no observable latent heat), and iii) $T_m \approx 45K$.

The (apparent) thermodynamic second order nature of the transformation in V_3Si was noted by Anderson and Blount[5] who showed that a cubic to tetragonal transformation should be first order in the absence of a new internal order parameter. A sublattice distortion has been observed in Nb_3Sn from neutron diffraction studies

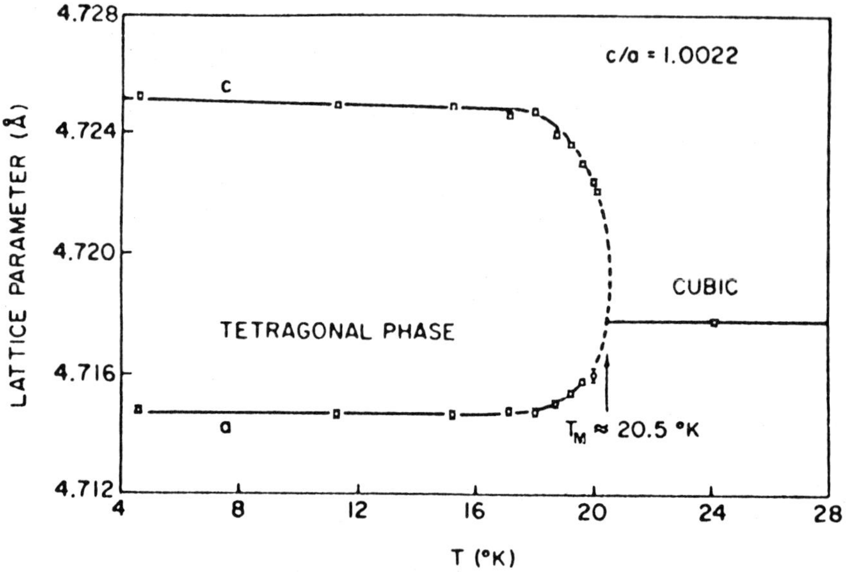

Fig. 1 The lattice parameters of V_3Si vs T showing the cubic to tetragonal transformation (after Batterman and Barrett[2]).

by Shirane and Axe.[5] (Problem steming from V make a comparable determination in V_3Si too difficult.) In the undistorted A-15 structure (compound formula A_3B) the transition metal atoms, A, from the orthogonal linear chains (see Fig. 2). The sublattice distortion observed by Shirane and Axe in Nb_3Sn involves (along two of the chains) a pairing of Nb atoms in a manner similar to that expected for a Peierls distortion in a one dimensional system (see Fig. 3).

The structural transformation has now been observed in at least some samples of almost all of the high T_c ($\gtrsim$ 15K) A-15 superconductors but has never been seen in the isostructural compounds having relatively low ($\lesssim$ 10K) T_c's. (For further references and data see the review articles of references 6-8, herein.) These experimental findings constitute part of the correlation and the conjectured causal relation between structural instability and high temperature superconductivity. We present other evidence below.

II. Instabilities and Transformation Effects on the Physical Behavior

There are numerous "anomalous" temperature dependences for the

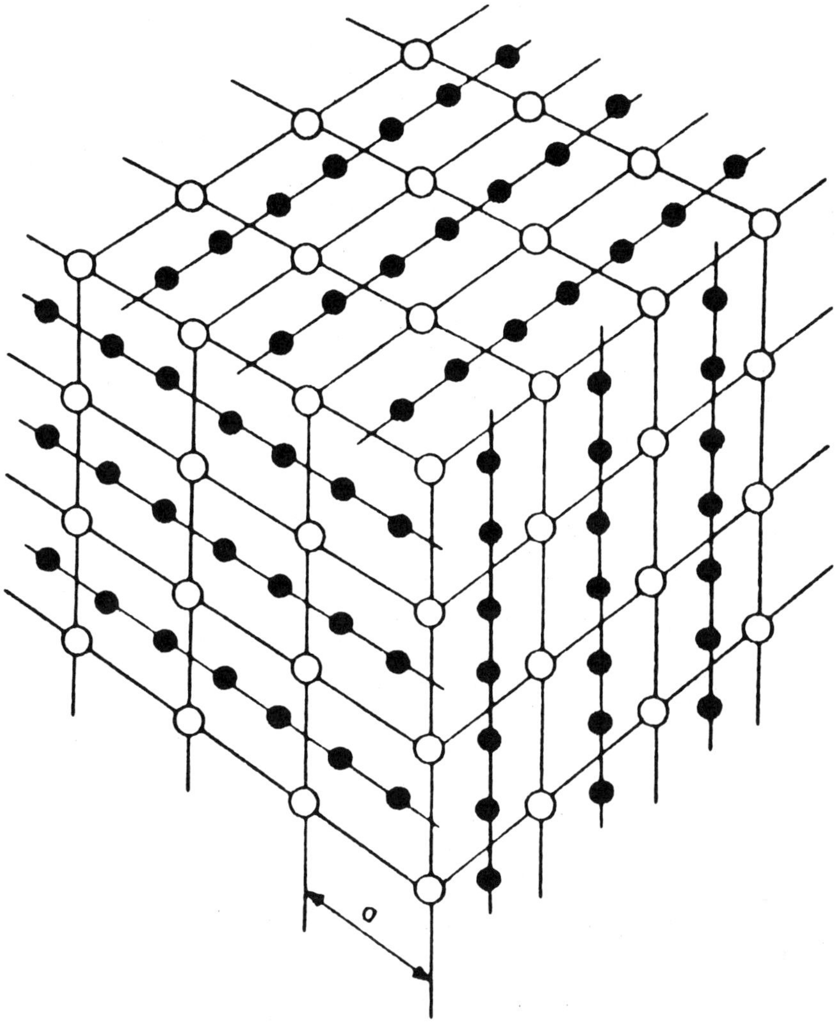

Fig. 2 The A-15 structure for compound formula A_3B. A atoms are transition metals and form 3 linear orthogonal chains. B atoms are usually nontransition metals (in high T_c compounds) and occur at the bcc sites (center position not shown).

behavior of A-15 compounds, many of which are now reviewed as manifestations of the instability and precursors of the transformation (see ref. 6 to 8 for futher details).

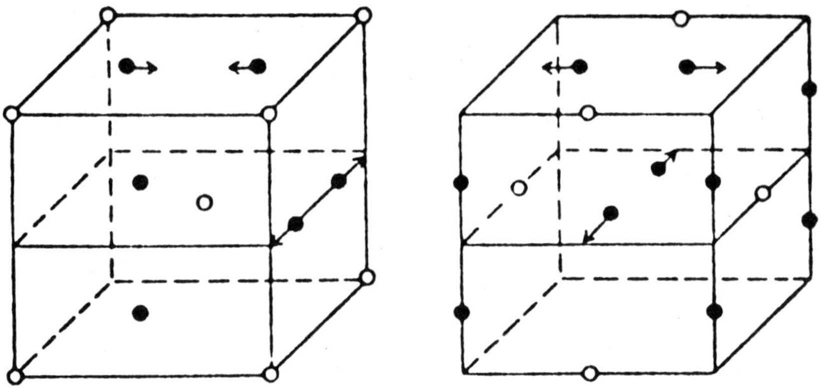

Fig. 3 The sublattice distortion due to the structural transformation in Nb_3Sn. (After Shirane and Axe[5].)

In V_3Si the elastic modulus $(c_{11}-c_{12})/2$, which defines the shear restoring force for [110] transverse waves with [$1\bar{1}0$] polarization, shows a positive temperature coefficient. This modulus softens so greatly on cooling that it would appear (by extrapolation) ready to vanish between 10K and 20K (see Fig. 4). For samples which exhibit the Batterman-Barrett transformation the softening is arrested at T_m but the occurrence of domains complicates the ultrasonic experiments in which these data were obtained.

Not all samples exhibit the transformation (The metallurgical factors are complicated but experiments show that transforming samples have higher resistance ratios and some second phase inclusions compared to nontransforming ones.). For V_3Si samples not exhibiting the transformation it is superconductivity which arrests the softening and, presumably, the need for the structural transformation. A theoretical discussion of this observation has recently been made by Ting and Birman.[10]

The deformation associated with soft shear modulus $(c_{11}-c_{12})/2$ is consistent with the (tetragonal) symmetry of the transformed phase and with the lack of volume change associated with the transformation. In this sense it is a specific precursor as well as a driving force for the transformation.

The observations that superconductivity arrests the growing structural transformation in a transforming sample, and arrests the softening in a nontransforming sample, shows the similarity of the interactions responsible for the structural instability and the

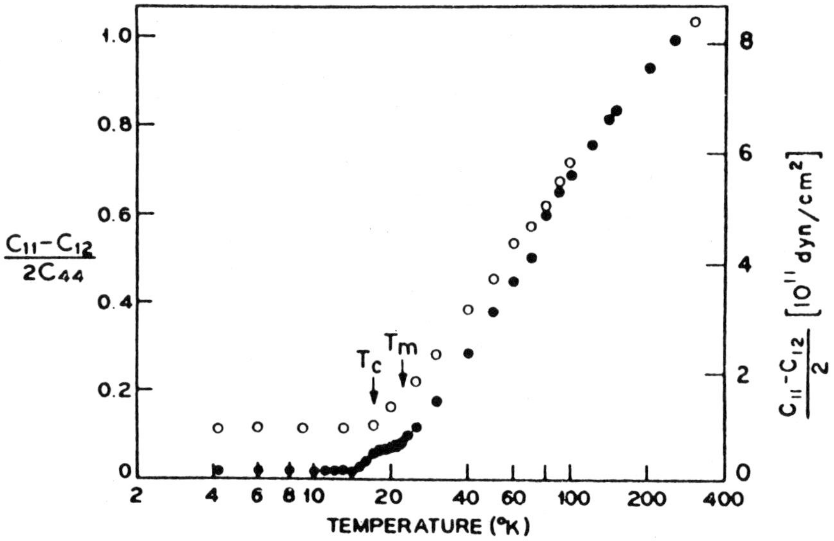

Fig. 4 Elastic modulus $(c_{11}-c_{12})/2$ vs T for transforming and non-transforming V_3Si (after Testardi et al.[9]).

high temperature superconductivity.

Keller and Hanak,[11] and Rehwald[12,13] observed a similar softening of $(c_{11}-c_{12})/2$ in Nb_3Sn. Again, several distinctive differences occurred. The modulus $(c_{11}-c_{12})/2$ was found to recover its stiffness below T_m (and above T_c) in Nb_3Sn (it does not in V_3Si where T_m and T_c differ by only $\sim 4°K$), and the modulus c_{44} shows considerably greater softening on cooling to 4°K in Nb_3Sn ($\Delta c/c \sim -50\%$) than in V_3Si ($\Delta c/c \sim -6\%$). The latter is not a trivial observation. Many theoretical treatments of the A-15 compounds assume noninteracting chains of transition metal atoms. Such a model will produce no anomalous temperature dependence for the c_{44} cubic face shear modulus.

Correlations of mode softening with superconductivity are observed. Shear mode softening is observed at least qualitatively (i.e. in polycrystalline samples) in all high T_c A-15 compounds (where investigated) but in none of the low T_c compounds.[9]

The occurrence of a near vanishing modulus in the ultrasonic experiments indicates that the basic instability for these compounds is macroscopic (q=0) rather than microscopic (q>0). Nevertheless our present ideas on the microscopic source of superconductivity require knowledge of the behavior of high frequency phonons to indicate some relationships of the structural instability and the superconductivity. Shirane and Axe[5] obtained the phonon dispersion relation for the soft (q=0) mode in Nb_3Sn shown in Fig. 5. They find significant softening for $q > 0$ but considerably less than that observed in the ultrasonic measurements. A "central

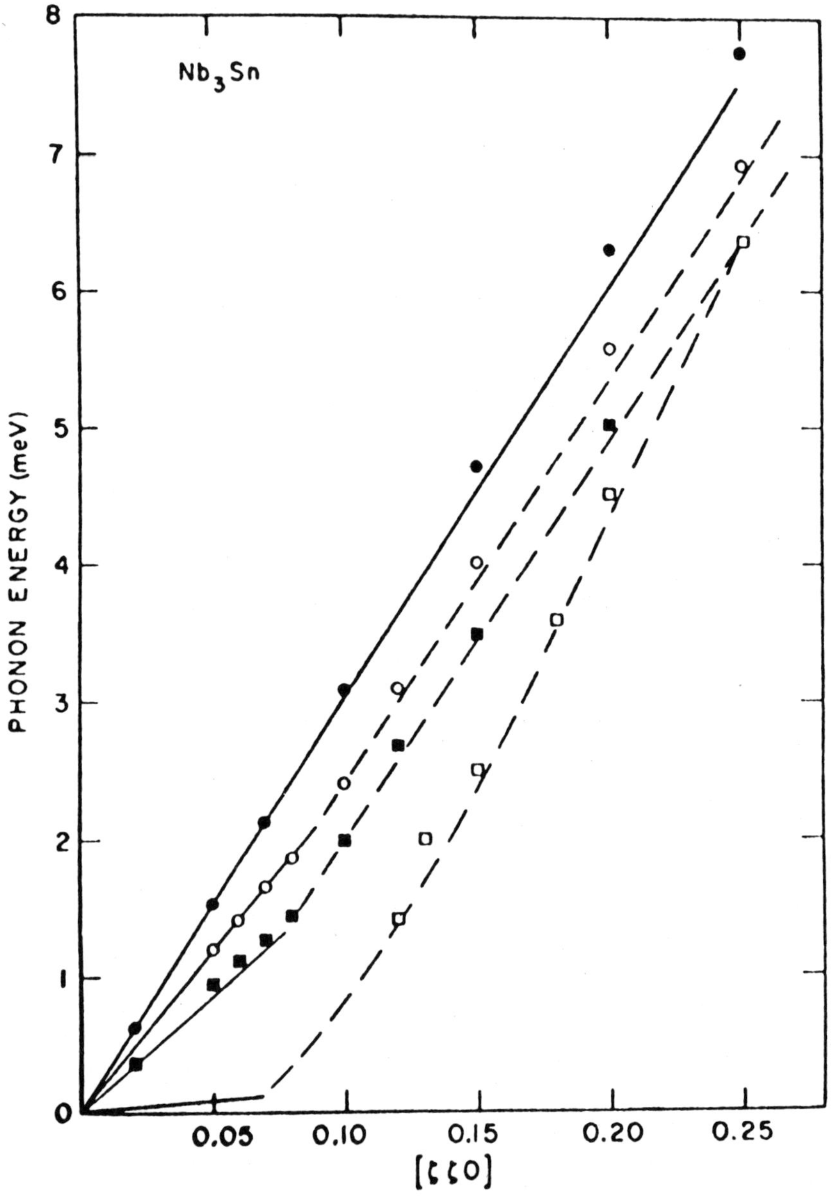

Fig. 5 Acoustic phonon dispersion curves for [110] waves with [$1\bar{1}0$] polarization in Nb_3Sn (after Shirane and Axe[5]).

peak" is also observed on approaching the structural transformation from above.

The Karlsruhe group[14] (N. Nücker, W. Reichardt, H. Rietschel, E. Schneider, P. Schweiss, and V. Tripadus) have obtained the total phonon density of states F for V_3Si, V_3Ge, V_3Ga, Nb_3Al, and Nb_3Sn at room and low temperatures. Some (but not considerable) mode softening is found in the total density of states (7/8ths of which is from optic modes) with evidence of optic as well as acoustic mode softening. The results for Nb_3Sn as well as α^2F from tunneling and α^2 (derived) are shown in Fig. 6. Note that the electron-phonon interaction α^2 shows considerable variation with energy and is strongest for the low frequency acoustic modes.

III. More on the Relation of Structural Instability and High Temperature Superconductivity

It has been suggested[7,8,9] that structural instability - those microscopic conditions which make a change in phase imminent - promotes high temperature superconductivity. The structural transformation, however, since it relieves these conditions, causes a reduction in the T_c otherwise achievable. While a microscopic theoretical justification in terms of soft modes and enhanced electron-phonon interactions (α^2) is lacking, several experimental tests support the empirical relation.

One such test is the relative variations of T_c and T_m with chemical changes or stress. The former is more complicated and limited results are available. Vieland and Wicklund[15] found that $\sim$4% Al added to transforming Nb_3Sn prevented the transformation and caused T_c to increase by $\sim$.5K. Chu and Testardi[16] find that for V_3Si hydrostatic pressure decreases T_m (while increasing T_c). In Nb_3Sn pressure increases T_m while decreasing T_c. If instability favors superconductivity one expects that whatever causes T_m (always $> T_c$) to decrease/increase such that the instability is greater/smaller at T_c will consequently cause T_c to increase/decrease. The experimental observations are at least consistent with this conjecture. Other data, including the pressure dependence of the soft shear modulus and the strong anharmonic behavior of these solids, is discussed in refs. 6 and 8.

IV. Instabilities, Unstable Phases, and Superconductivity

It is possible, then, that the structural instabilities which occur at solid state phase transformations may be attended by conditions favorable to high temperature superconductivity. If these

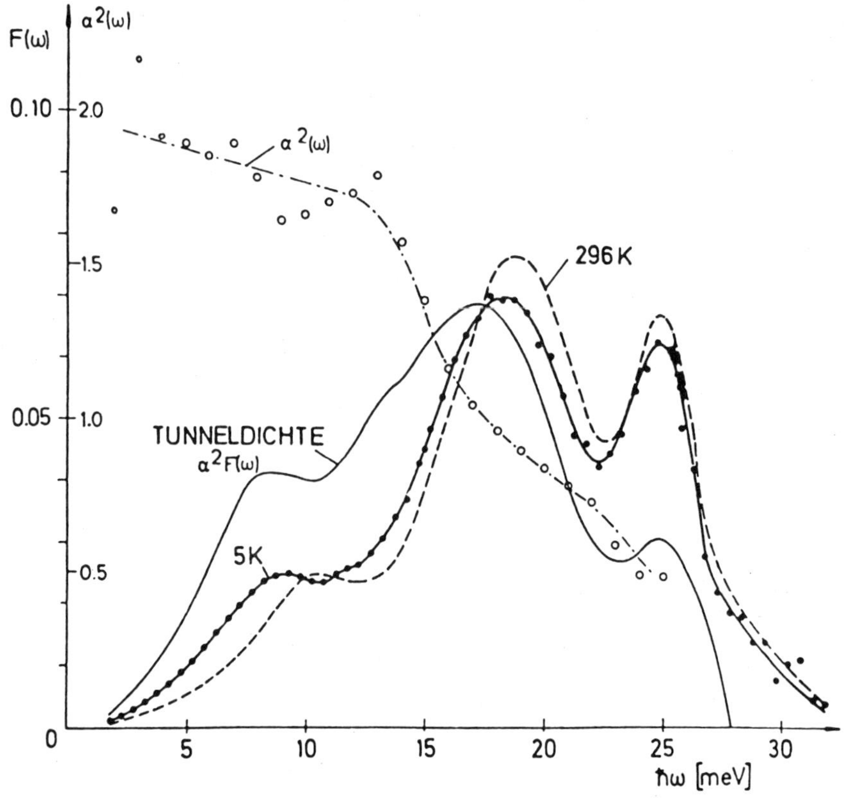

Fig. 6 Total phonon density of states, α^2F (from tunneling) and α^2 (derived) for Nb_3Sn (after Karlsruhe group[14]).

conditions can be frozen in, rather than the transformation products which relieve the instability, higher T_c's may result. The first deliberate test of this was achieved[18] by sputtering (as a function of temperature) through the eutectoidal transformation temperature ($\sim$1100°C) in a portion of the Mo-Re phase diagram (see Fig. 7). Sputtering at the eutectoidal boundary has frozen in a metastable structure with enhanced T_c. Similar though less dramatic effects have been seen in a large number of alloys.[17] Gavaler[18] was able to form metastable high T_c Nb_3Ge by hot substrate sputtering in high argon atmosphere.

V. Defects, Instabilities, and Superconductivity

The complexity of the A-15 compounds and the difficulty of

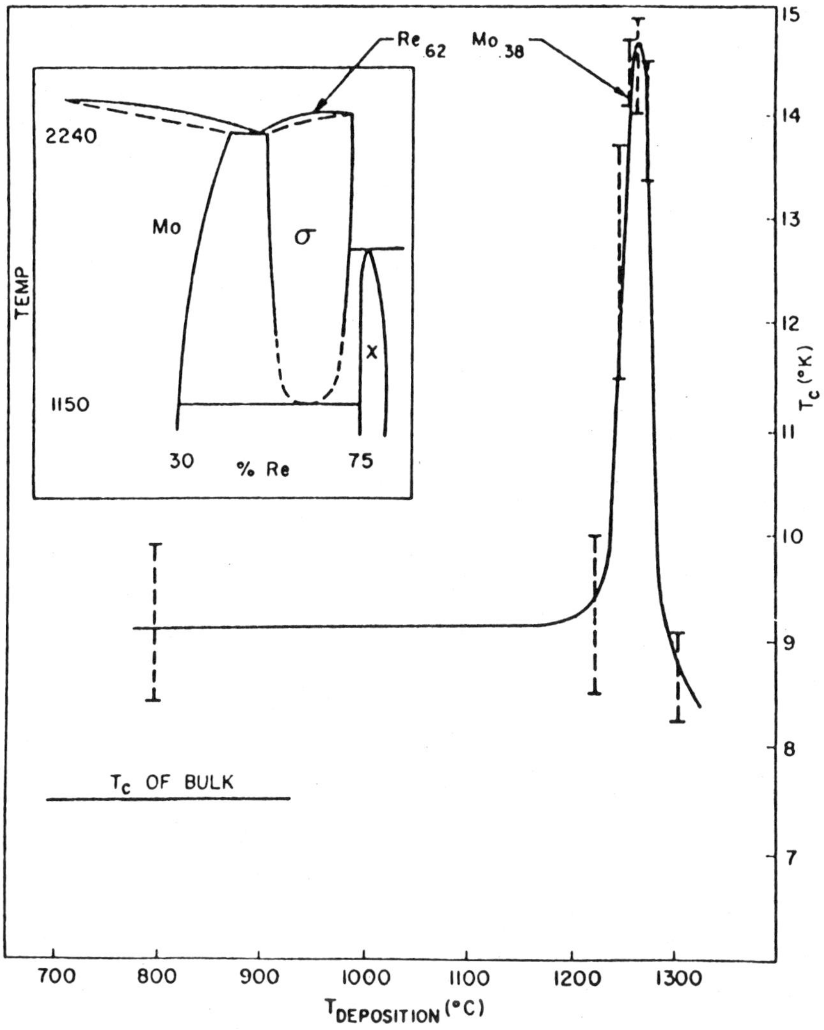

Fig. 7 T_c vs film deposition temperature for $Mo_{.38}Re_{.62}$ (after Testardi et al.[17]).

achieving near exact reproducibility in the physical properties of these materials has caused a number of investigators to consider the role of defects (see, for example, Hein,[19] and Testardi[20] for further references). "Defects" are generally described as i) non-stoichiometry, ii) antisite defects (A atoms on B sites and vice versa in the A_3B structure), iii) vacancies, interstitials, and impurities, iv) second phase inclusions, and v) strains.

Nonstoichiometry is often considered a crucial "defect" although recent work[21] indicates that, while important, these defects may not be as extremely detrimental to T_c as expected.

Blaugher et al.[22] showed from x-ray measurements that the unit cell compressibility of V_3Si at room temperature was considerably greater than that calculated from sound velocity data for pressures 0-10 kbar but was in agreement with expectations at higher pressures. The failure to agree with the ultrasonic predictions, which should apply near zero pressure, indicates that the physical process responsible for the additional mechanical compliance must require times too long to be observed at ultrasonic frequencies (20 MHz). Varma et al.[23] suggested that the result was due to a pressure dependent vacancy concentration.

A correlation between T_c (as-grown) and the electrical resistance ratio $\rho(300K)/\rho$ (25K) has been found in a number of A-15 compounds[21] (see Fig. 8 for Nb_3Ge data similar results obtain for other A-15's). This correlation, more general than that between T_c and composition, suggests that a key factor responsible for the wide range in T_c for these materials is the occurrence of a defect with universal character in A-15 compounds. Good evidence for this comes from the behavior of an initially high T_c film irradiated by 2 MeV ^4He particles where we find we can reproduce the as-grown correlation by varying the defect concentration only at constant chemical composition.[24] (The ^4He particles do not stop in the film.)

Sweedler et al.[25] have found that neutron radiation damage causes a large reduction in T_c to occur in a universally similar manner for all A-15 superconductors. They deduce from Bragg peak intensities that the effect is due to antisite defects. Similar reductions of T_c have been found with ^4He damage by Poate et al.[24] but it is concluded that the crucial defect lies, in part, in small bond distortions.

The defects have significant effect on the electrical resistivity and the lattice parameter as well. Increasing defect concentration causes not only a reduction in T_c and an increase in residual resistance but a reduction in the thermal part of the electrical resistivity as well.[24] Thus the defects strongly influence not only the superconducting properties but also the normal state properties which reflect the electron-phonon interaction responsible for the superconductivity.

The exact nature of the defect has not yet been established. Channelling measurements on V_3Si before and after defect formation indicate a strong tendency for bond distortions and quasi-amorphous structure. This behavior is more reminiscent of the covalently bonded group IV semiconductors rather than metallic bonding.

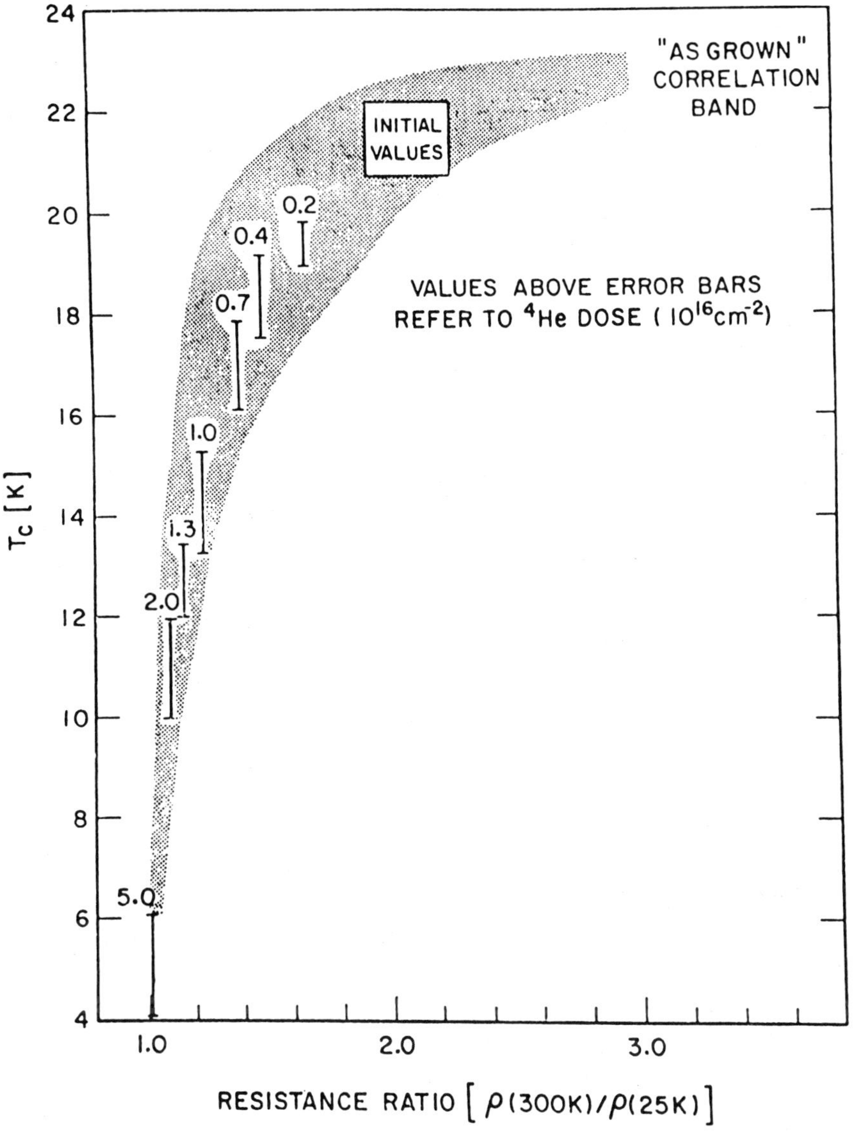

Fig. 8 T_c vs electrical resistance ratio for about 200 samples (shaded band) of V_3Si. Also shown is an initially high T_c film after various doses of 2 MeV ^4He damage (after Poate et al.[24]).

The ready tendency of the A-15 materials to form defects is another manifestation of their structural instability. The defect problem, however, appears to be a deterrent to achieving the more unstable A-15 compounds having, perhaps, T_c's higher than any presently available.

References

1. C. G. Shull, MIT Annual Report-Research in Materials Science (1963-1964).
2. B. W. Batterman and C. S. Barrett, Phys. Rev. Lett. 13, 390 (1964); Phys. Rev. 149, 296 (1966).
3. R. Mailfert, B. W. Batterman, and J. J. Hanak, Phys. Lett. A24, 315 (1967); Phys. Status Solidi 32, K67 (1969).
4. L. J. Vieland, R. W. Cohen, and W. Rehwald, Phys. Rev. Lett. 26, 373 (1971).
5. G. Shirane, and J. D. Axe, Phys. Rev. B4, 2957 (1971); J. D. Axe and G. Shirane, Phys. Rev. B8, 1965 (1973).
6. L. R. Testardi, in "Physical Acoustics", ed. by W. P. Mason and R. N. Thurston (Academic Press, N.Y.) Vol. X, p.193 (1973) and Vol. XIII (an undate, to be published 1978).
7. M. Weger and I. B. Goldberg, in "Solid State Physics", ed. by F. Seitz and D. Turnbull (Academic Press, N.Y.) Vol. 28, p.1 (1973).
8. L. R. Testardi, Rev. Mod. Phys. 47, 637 (1975).
9. L. R. Testardi, W. A. Reed, R. B. Bateman, and V. G. Chirba, Phys. Rev. Lett. 15, 250 (1965); Phys. Rev. 154, 399 (1967); L. R. Testardi and T. B. Bateman, Phys. Rev. 154, 402 (1967).
10. C. S. Ting and J. L. Birman, Phys. Rev. B12, 1093 (1975).
11. K. R. Keller and J. J. Hanak, Phys. Lett. 31, 263 (1966); Phys. Rev. 154, 628 (1967).
12. W. Rehwald, Phys. Lett. A27, 287 (1968).
13. W. Rehwald, M. Rayl, R. W. Cohen, and G. D. Cody, Phys. Rev. B6, 363 (1972).
14. Karlsruhe group, in Progress Report KFK 2054 (1974) and KFK 2183 (1975), Gesellschaft für Kernforschung, M. B. H., Karlsruhe, Germany.
15. L. J. Vieland and A. W. Wicklund, Phys. Lett. A34, 43 (1971).
16. C. W. Chu and L. R. Testardi, Phys. Rev. Lett. 14, 766 (1974); C.W. Chu, Phys. Rev. Lett. 33, 1283 (1974).
17. L. R. Testardi, J. J. Hauser, and M. H. Read, Solid State Comm. 9, 1829 (1971); L. R. Testardi, J. H. Wernick, W. A. Royer, D. D. Bacon, and A. R. Storm, J. Appl. Phys. 45, 446 (1974).
18. J. R. Gavaler, Appl. Phys. Lett. 23, 480 (1973).
19. R. A. Hein in "The Science and Technology of Superconductivity", ed. by W. D. Gregory, W. N. Mathews, Jr., and E. A. Edelsack, (Plenum Press, N.Y.), Vol. 1, 333 (1973).
20. L. R. Testardi, to be published in "Cryogenics" (1977).
21. L. R. Testardi, R. L. Meek, J. M. Poate, W. A. Royer, A. R. Storm, and J. H. Wernick, Phys. Rev. B11, 4304 (1975); R. C. Dynes, J. M. Poate, L. R. Testardi, and R. H. Hammond, in "1976 Applied Superconductivity Conference", Stanford, to be published.

22. R. D. Blaugher, A. Taylor, and M. Ashkin, Phys. Rev. Lett. 33, 292 (1974).
23. C. M. Varma, J. C. Phillips, and S. T. Chui, Phys. Rev. Lett. 33, 1223 (1974).
24. J. M. Poate, L. R. Testardi, A. R. Storm, and W. M. Augustyniak, Phys. Rev. Lett. 35, 1290 (1975).
25. A. R. Sweedler, D. G. Schweitzer, and G. W. Webb, Phys. Rev. Lett. 33, 168 (1974).
26. L. R. Testardi, J. M. Poate, and H. J. Levinstein, Phys. Rev., to be published.
27. L. R. Testardi, J. M. Poate, W. Weber, and J. H. Barrett, to be published.

SUPERCONDUCTIVITY AND MARTENSITIC TRANSFORMATIONS IN A-15 COMPOUNDS

W. L. McMillan

Department of Physics and Materials Research Laboratory
University of Illinois at Urbana-Champaign
Urbana, IL 61801, USA

I would like to discuss the current theories of the martensitic transition in A15 compounds and our understanding of the interplay between the martensitic transition and superconductivity. Labbe and Friedel[1] proposed that the martensitic transition is driven by an electronic band Jahn-Teller effect. Gorkov[2,3] proposed an alternative model in which the martensitic transition is driven by the Peierls mechanism of an energy gap opening up near the Fermi energy. Both models are based on one-dimensional or quasi-one-dimensional energy band models which are inconsistent with APW band calculations.[4] Bhatt[5] has developed a very successful Landau theory which is based on the Gorkov model and which is, of course, free of any microscopic assumptions about the nature of the band structure. Bhatt[6] has also developed a microscopic theory using a band structure model based on the APW band structure. This model includes both the band Jahn-Teller effect and the Peierls effect and is therefore a synthesis of the Labbe-Friedel and Gorkov models. The effects on superconductivity are illustrated in a calculation by Bilbro,[7] based on the Gorkov model, in which both a Peierls energy gap and the BCS energy gap are included in the theory. I want to discuss the physical assumptions that go into the various theoretical models, without presenting much of the mathematics, and then discuss the predictions of the models and the comparison with experiment.

The most widely studied A15 compounds are Nb_3Sn and V_3Si. In Nb_3Sn there is a cubic to tetragonal structural transition at 46°K; the superconducting transition is at 18°K. In V_3Si these transitions occur at 21°K and 17°K. In the cubic phase of Nb_3Sn the Sn atoms sit on body-centered cubic lattice sites and the Nb atoms are on the cube faces at, for example, (¼,½,0) and (3/4,½,0) on the xy face. The transition metal atoms form linear chains with equally spaced atoms and with the chains running in the three orthogonal directions on the three faces. This linear chain picture is central to both the Labbe-Friedel and Gorkov models. At the martensitic transition the cubic cell distorts to a tetragonal one with less than 1% distortion and the transition metal atoms on two of the three sets of linear chains pair up as in a Peierls transition. There is a drastic softening of one elastic constant $(C_{11}-C_{12})$ as one approaches the transition from above.

The corresponding phonon, the transverse (110) phonon with $(1\bar{1}0)$ polarization, goes soft at long wavelengths and is the soft mode of the transition.

The Labbe-Friedel model is based on a one-dimensional band structure for electrons moving along one linear chain. Consider one atomic orbital of a particular symmetry on each atom. The energy band for an x-direction chain is

$$E_K^x = \pm\ 2\beta_x \cos(k_x a/2)$$

where β_x is the x-chain transfer integral and "a" is the cubic lattice spacing. The zone boundary is at π/a so that the two portions of the band are degenerate at this point. In the cubic phase $\beta_x = \beta_y = \beta_z$ so the band structures of the three types of chains are identical except for a rotation of the momentum space axes. In the tetragonal phase this degeneracy of the electronic energy bands is broken. Suppose the unit cell dimension is decreased in the z direction and increased in the x and y directions to maintain constant volume, then $\beta_z > \beta_x = \beta_y$ and the bottom of the z-chain band is lower than the bottom of the x and y-chain bands. If the Fermi level lies near the bottom of the bands in the cubic phase electrons will be transferred in the tetragonal phase from the x and y chains into the z chain and the total electronic energy will be lowered. If the electronic energy decrease more than offsets the increased elastic energy the tetragonal phase will be the observed phase at low temperature. Since the electronic entropy favors the cubic phase there will be a phase transition to the cubic phase at finite temperature. Thus the Labbe-Friedel model qualitatively explains the martensitic transition in A15 compounds although the quantitative fit to experiment is not particularly good. Note the particular type of degeneracy which is broken in this band Jahn-Teller effect. The degeneracy is the equivalence of the band structure at different places in the Brillouin zone required by cubic symmetry. Breaking the cubic symmetry allows the bands to move and a redistribution of electrons can lower the electronic energy. This effect occurs with any band structure and is not a special property of one-dimensional bands. In order for this contribution to the energy to be important there must be a large density of states at the Fermi level and the "electron phonon coupling constant" for splitting these levels must be large enough.

The original Gorkov model assumed the one dimensional band structure of equation (1) but concentrated on the degeneracy of the energy bands at the zone boundary. In the tetragonal phase the pairing of the transition metal atoms introduced a potential with a periodicity of "a" and opened a Peierls energy gap at the zone boundary. If the band is half full the Fermi level lies in the Peierls gap and the electronic energy is lowered. The physics is the same as the charge density wave models applied to the layered compounds. The degeneracy of energy levels at the zone boundary (more particularly at the x point at the center of the face of the cubic Brillouin zone) is a property of the A15 crystal symmetry and is not a special property of the one-dimensional bands. Gorkov[3] has treated a band structure model including interchain coupling to produce quasi-one dimensional bands. It is necessary to assume that the bands are quite flat on the zone face in order that the density of states affected by the Peierls gap be large enough to explain the data.

Both the Labbe-Friedel and the Gorkov movels are electronic models based on particular band structure assumptions. The principal criticism of the models, it seems to me, is that the band structure models are unrealistic. Matthiess[4] has calculated APW band structures for several A15 compounds and there are no bands in Matthiess' calculation which resemble the one-dimensional or quasi-one-dimensional bands used in the models.

One way of avoiding unrealistic band structure assumptions while retaining the physical assumptions of the Gorkov model is to work with a Landau theory similar to that applied to the layered compounds. We assume that the electronic order parameters are the amplitudes of three CDW's in the (100) directions. The CDW's are locked in to the lattice with the wavelength equal to the cubic lattice spacing and there are no phase fluctuations; the order parameters are real. We write down the usual expansion of the free energy in powers of the order parameters and gradients of the order parameters. The theory is dynamical and we assume that the dominant dissipation is electronic and arises from the redistribution of electrons as the energy gap changes. The electronic order parameters are directly coupled to the amplitudes of the three optical phonons (at Γ) which modulate the transition metal atom separation. In the A15 structure these optical modes are bilinearly coupled to elastic strain and we must include the three acoustic phonon modes; we have a nonlinear dynamical problem with nine coupled modes. We use the mean field approach and find the static mode amplitudes which minimize the free energy. We then expand the free energy around this minimum to find the mode frequencies. Since dissipation is included the phonon modes have lifetimes and we actually calculate the dynamical structure factor. This type of theory can predict central peaks arising from the coupling of the phonons to the overdamped electronic mode; however, for the A15's no central peak is predicted.

The qualitative predictions of the theory fall into two categories. The first concerns the behavior of the soft mode. The model predicts that the elastic constant $C_{11}-C_{22}$ goes to zero at an extrapolated critical temperature T^* slightly below the first order structural transformation temperature. This means that the velocity of the long wavelength (110) transverse phonon with polarization $(1\bar{1}0)$ goes to zero at T*. As one moves out in momentum space the phonon mode starts to recover its stiffness, and the phonon softens dramatically only near Γ. A two-parameter fit produces quantitative agreement with the elastic constant versus temperature and with the transverse phonon frequency versus temperature and momentum. From this fit we find a correlation length $\pi\xi_0 \approx 2a$ where a is the cubic lattice spacing.

The second group of qualitative predictions concerns the behavior of the cubic terms in the free energy. Within the Gorkov model the structural transition temperature is maximum if the Fermi energy is equal to the x-point energy E_x of the relevant energy band. Since the x-point energy is strain sensitive strain (either compressive or tetragonal) changes the transition temperature T_m and there are cubic terms (proportional to the strain and to the CDW amplitude squared) in the free energy which change sign as E_F crosses E_x (say in an alloy series). The cubic terms control the sign of the tetragonality (c/a - 1) but are weak enough that they do not affect the magnitude of the tetragonality at low temperature. Thus one expects (c/a - 1) to change sign at constant magnitude as E_F crosses E_x. This is observed in $Nb_{3-x}Sb_xSn$[8] alloys. The sign of the pressure dependence of T_m is controlled by the cubic terms and the pressure dependence of the superconducting transition temperature is opposite to that of T_m. These quantities should correlate with the sign of the tetragonality and this correlation is observed to hold.

There are some quantitative cross checks of the Landau theory. One uses up several experiments in determining parameters of the theory and can only check the theory if there are more experiments than parameters. The heat capacity jump in V_3Si is predicted to be 0.7 joules/mole K and observed to be 0.4-0.5 joules/mole K,[9,10] which is satisfactory.

There are problems in trying to predict Landau theory parameters from microscopic models. From the Gorkov or Peierls models the electronic energy terms can be predicted from the electronic density of states $N_\uparrow(0)$. The change in susceptibility is also proportional to $N_\uparrow(0)$.[↑] The values of $N_\uparrow(0)$ required to explain the two results differ by a factor of three to five.

Now for Bhatt's microscopic calculation. Since the predictions of the microscopic theory depend critically on the band structure chosen, it is important to work with the most realistic band structure model available. It is not possible to superimpose the CDW calculation on the full APW calculation and one is forced to work with simplified models. Fortunately, the density of states near E_F in Matthiess' calculation is dominated by two bands of $\delta_1(x^2 - y^2)$ character. These bands appear not to hybridize strongly over much of the Brillouin zone. Bhatt's model is a tight binding model with one $\delta_1(x^2 - y^2)$ orbital per transition metal atom, with nearest and next-nearest neighbor hopping integrals which reproduce these two bands of Mattheiss' calculation. The hopping integrals are assumed to vary linearly with interatomic spacing as the lattice distorts. Bhatt calculates the band structure of the distorted lattice and computes the electronic free energy which he then minimizes with respect to distortion amplitude. There is essentially only one free parameter, the ratio of the two hopping integrals, which he chooses to fit the observed electronic density of states. The Gorkov energy term from the Peierl's energy gap near the x-point is included in the calculation; however there is insufficient phase space for this contribution to be large enough to explain the phase transition. There is, however, an electronic Jahn-Teller contribution from non-degenerate bands far from the X-point which makes up the deficit. Bhatt's model, therefore, includes both a Gorkov-Peierls contribution and a Jahn-Teller contribution to the stability of the tetragonal phase. The agreement with a wide variety of experiments on Nb_3Sn and V_3Si is nearly quantitive with discrepancies typically between 25% and 50%. The correlation lengths are large enough that the phonon entropy is not dominant but could cause substantial corrections. In addition to providing a better quantitative fit to experiment than the Labbe-Friedel or Gorkov models, Bhatt's calculation contains important contributions from both mechanisms and represents a synthesis of the earlier microscopic models.

The effects of the martensitic transition on superconductivity can be understood qualitatively very simply. As one approaches the martensitic transition by lowering the temperature, by changing pressure or by alloying, the lattice softens as the structure nears the instability. This softening is most pronounced in the transverse acoustic modes near Γ but it does extend over an appreciable fraction of the Brillouin zone and the optic modes may be affected as well. This lattice softening increases the electron-phonon coupling constant and enhances the superconducting transition temperature T_c. After passing through the martensitic transition the lattice stiffens and T_c is reduced. I do not know how important this effect is quantitatively. The second effect is that,within the Gorkov model, there is a competition between the Peierls energy gap and the BCS energy gap. Bilbro has carried out detailed calculations for the A15 compounds. He finds that when the martensitic transition occurs at higher temperature the superconducting transition temperature is reduced (but only by 0.3 K for V_3Si) and that the martensitic transformation is arrested at T_c. Both of these effects are observed for V_3Si. The theory predicts that when the superconducting transition occurs first the martensitic transition is completely suppressed and the cubic phase is stabilized. Quantitatively the effect of the Pererls gap on T_c is small and the effect of the lattice softening is unknown. The theoretical models predict a peak in T_c near structural transitions and this behavior is observed in many systems. However, it is not clear that these effects "explain" the high T_c's of the A15 compounds. At the present time it appears to be more correct to state that the structural instability and the high T_c are both produced by the large density of states. More quantitative work on this question is highly desirable.

REFERENCES

1) J. Labbe and J. J. Friedel, J. Phys. Radium 27, 708 (1966).

2) L. P. Gor'kov, JETP Lett. 17, 379 (1973; Sov. Phys. JEPT 38, 830 (1974).

3) L. P. Gor'kov and O. N. Dorokhov, J. Low Temp. Phys. 22, 1 (1976); JETP Lett. 21, 310 (1975).

4) L. F. Mattheiss, Phys. Rev. 138, A112 (1965); Phys. Rev. B12, 2161 (1975).

5) R. N. Bhatt and W. L. McMillan, Phys. Rev. B14, 1007 (1976).

6) R. N. Bhatt, Ph.D. thesis, Dept. of Physics, University of Illinois at Urbana-Champaign (1976), to be published.

7) Griff Bilbro and W. L. McMillan, Phys. Rev.

8) L. J. Vieland, J. Phys. Chem. Solids 31, 1449 (1970).

9) J. C. F. Brock, Solid State Commun. 7, 1789 (1969).

10) J. P. Maita and E. Bucher, Phys. Rev. Lett. 29,931 (1972).

p-d HYBRIDIZATION, INCIPIENT LATTICE INSTABILITIES AND SUPERCONDUCTIVITY IN TRANSITION METAL COMPOUNDS

J. HAFNER, W. HANKE and H. BILZ

Max-Planck-Institut für Festkörperforschung

D 7 Stuttgart - 80, Germany

Neutron-scattering studies /1/ revealed the existence of anomalies in the phonon dispersion curves (Fig.1) of superconducting transition metal compounds such as the carbides and nitrides, thus establishing an empirical correlation between actual or incipient lattice instabilities and high superconducting transition temperatures. These instabilities are usually associated with a rela-

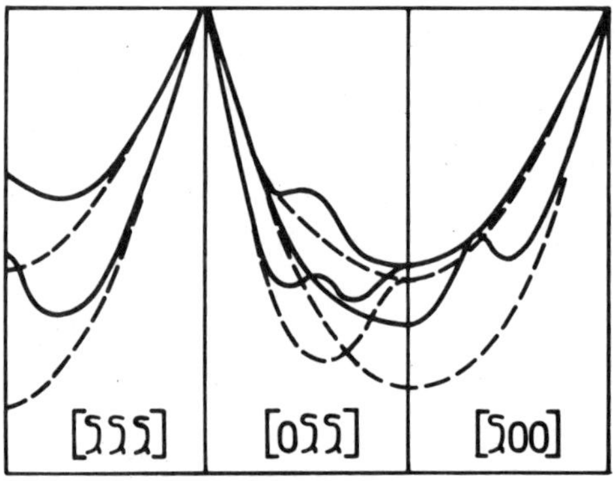

Fig.1: Acoustic phonon dispersion curves in transition metal compounds with NaCl structure. Solid lines with anomalies (NbC T_c=11°K, NbN 16.8°K, TiN 5.5°K), broken lines no anomalies (ZrC,TiC T_c <0.05°K, TiO <1°K, $NbC_{0.79}$<0.05°K).

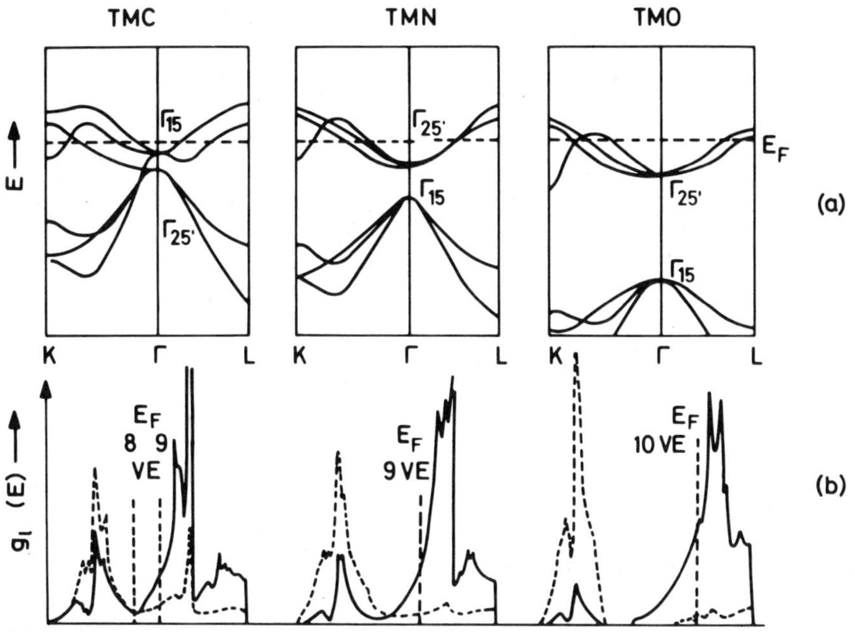

Fig.2: $E(\vec{k})$ curves (a) and partial LCAO densities of state (solid lines metal d_{xy}, broken lines nonmetal p states) for the transition metal carbides, nitrides and oxides (schematical, after Neckel et al./2/). The approximate position of the Fermi level for different numbers of valence electrons is indicated.

tively high density of d-states at the Fermi level. However, this is in direct contradiction to the fact that both the superconductivity and the phonon anomalies disappear when we go from the carbides and nitrides to the oxides, or introduce vacancies, whereas the d-electron density of states further increases.

Band-structure calculations of the superconducting carbides and nitrides of Schwarz and co-workers /2/, which are confirmed by the X-ray emission spectra /3/, show a strong p-d_{xy} hybridization near the Fermi level (Fig.2). The hybridization may be described as a covalent-ionic bonding with the most important contributions coming from planar $(p\text{-}d)_\pi$ interactions between nonmetal and metal ions. The "p"- and "d"-type band complexes (Γ_{15} and $\Gamma_{25'}$, at the center of the Brillouin zone) are constructed of linear combinations of bonding $(d+p)_\pi$ and antibonding $(d-p)_\pi$ hybrides. For one specific transition metal the band-structure calculations show a decrease in the hybridization when we go from the carbides and nitrides to the oxides, although the total density of states increases. Similarly, in the substoichiometric carbides the empty p-states near the Fermi-level which are essential for the hybridization mechanism are no longer available /4/. If we compare the carbides of the group Vb and group IVb metals (with 9 and 8 valence electrons respectively) we see that the IVb carbides have a low density of states and very weak hybridization. Correspondingly, TiC, ZrC and HfC are non-superconducting, whereas for VC $T_c=8.7^\circ K$, for NbC $T_c=11.1^\circ K$ and for TaC $T_c=10.4^\circ K$. For the nitrides of both groups, the p and d-states are strongly hybridized (TiN $T_c=5.5^\circ K$, ZrN $T_c=10.0^\circ K$, HfN $T_c=8.8^\circ K$, VN $T_c=8.5^\circ K$, NbN $T_c=16.8^\circ K$, TaC $T_c=14.3^\circ K$).

The aim of this paper is to demonstrate that in the superconducting materials the formation of covalent bonds due to the hybridization of metal-d_{xy} and nonmetal-p states leads to a resonance-like increase in the nonlocal dielectric response. The anomalous increase in the screening enhances the electron-phonon coupling, thus produces the phonon anomalies. Furthermore, it explains the high superconducting transition temperatures as resulting from a simultaneous increase of the electron-phonon coupling and a lowering of the phonon frequencies /5/.

The inverse dielectric screening matrix is the solution of the integral equation

$$\varepsilon^{-1} = 1 + v\chi\varepsilon^{-1} \tag{1}$$

where v is the electron-electron interaction. In a localized-orbital representation the polarizability χ is of a separable form /6-8/

$$\chi(\vec{q}+\vec{G},\vec{q}+\vec{G}') = \sum_{ss'} A_s(\vec{q}+\vec{G})\, N_{ss'}(\vec{q})\, A^+_{s'}(\vec{q}+\vec{G}') \tag{2}$$

hence eqn.(1) is readily solved to yield

$$\varepsilon^{-1} = 1 + v\,A\left\{N^{-1}-V\right\}^{-1} A^+ \tag{3}$$

The A_s may be interpreted as the form factor of a generalized charge density wave, the index $s=(\nu,\mu,e)$ standing for the set of quantum numbers of the localized orbitals ϕ_ν,ϕ_μ.

$$A_s(\vec{q}+\vec{G}) = \int \phi_\nu^*(\vec{r})\, e^{-i(\vec{q}+\vec{G})\vec{r}}\, \phi_\mu(\vec{r}+\vec{R}_1)\, d^3r. \qquad (4)$$

N is the bare polarizability,

$$N_{ss'}(\vec{q}) = \sum_{k,n,n'} \frac{f_n(\vec{k})-f_{n'}(\vec{k}+\vec{q})}{E_n(\vec{k})-E_{n'}(\vec{k}+\vec{q})}\, e^{-i(\vec{k}+\vec{q})\cdot(\vec{R}_1-\vec{R}_{1'})}$$

$$\times\, e_\nu^*(n,\vec{k})\, e_\mu^*(n',\vec{k}+\vec{q})\, e_\mu(n',\vec{k}+\vec{q})\, e_{\nu'}(n,\vec{k})$$

where $f_n(\vec{k})$ is the Fermi occupation factor, $E_n(\vec{k})$ the one-particle energies and $\vec{e}(n,\vec{k})$ is an eigenvector of the band-complex. N^{-1} may be viewed as the kinetic energy of the charge density waves, whereas V

$$V_{ss'}(\vec{q}) = \sum_{G''} A_s^*(\vec{q}+\vec{G}'')\, v(\vec{q}+\vec{G}'')\, A_{s'}(\vec{q}+\vec{G}'') \qquad (6)$$

represents their coulomb energy. Hence the screening matrix $S=(N^{-1}-V)$ coupling two charge density waves has the form of an energy denominator. In a metal it is convenient to separate out a diagonal (in G and G') part $\chi_o = -(\varepsilon_o-1)/v$ from the polarizability. The inverse dielectric matrix then becomes

$$\varepsilon^{-1} = \varepsilon_o^{-1} + v\varepsilon_o^{-1} A\{N^{-1} - V\}^{-1} A^+ \varepsilon_o^{-1} \qquad (7)$$

Note that the interaction between the electrons in the localized states is now screened by the diagonal part ε_o^{-1}. The solution ε^{-1} provided by equ. (7) allows for an explicit inclusion of local-field effects, which are important whenever localized electrons participate in the screening. In the approximation of a local electron-ion potential the local-field correlation, i.e. the second term in equ. (7) gives the following contribution to the dynamical matrix /6-9/

$$E_{\alpha\beta}(\vec{q}) = \sum_{ss'} F_\alpha^s(\vec{q}) \left\{N^{-1}(\vec{q}) - V(\vec{q})\right\}_{ss'}^{-1} F_\beta^{s'}(\vec{q}) \qquad (8)$$

where $F_\alpha^s(q)$ is the component of the force in direction α experienced by an ion due to the coupling to the charge density wave.

There are two different schools of thought as to the interrelation between the dips in the phonon dispersion curves and the structure of the screening matrix $S^{-1} = (N^{-1}-V)^{-1} = N(1-VN)^{-1}$. Band-theorists have pointed out that the intraband contribution to the bare susceptibility $N(\vec{q})$ from the bands crossing the Fermi level shows maxima at the positions of the phonon anomalies, attributed to a "nesting" of the Fermi surface /10,11/. However, the interband contributions are larger and to some extent anti-correlate with the intraband contributions. Furthermore, recent work /12/ has established the importance of the matrix-elements which again diminish the structure in the polarizability. On the other side it has been proposed to ascribe the phonon-dips to a resonance-like increase of the screening matrix $(N^{-1}-V)^{-1}$ which becomes large when the kinetic energy of the charge density waves is nearly equal to their Coulomb energy, $N^{-1} \approx V$ /13/. The difference between both interpretations is most easily demonstrated by considering the schematic diagram expansion in Fig. 3.

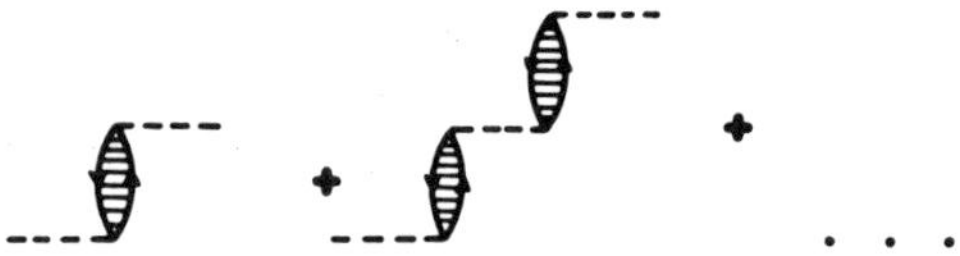

Fig. 3: Diagram expansion for the polarizability.

The band-theorists point of view corresponds to $S^{-1} \approx N$, i.e. only the simple electron-hole loop diagram is considered. Our interpretation is a many-body theorists standpoint: $S^{-1} = N/(1-VN)$ corresponds to a summation over the infinite series of polarization diagrams. This is a random-phase approximation for the local field effects, plus local corrections for exchange and correlation, which are incorporated in the electron-electron interaction v. In our theory the driving mechanism for promoting the phonon anomalies is the increase in the screening matrix. The structure in the "bare" polarizability may be helpful, but is not of decesive importance.

It is interesting to point out that, in a nonadiabatic formulation, $||N^{-1}-V||=0$ corresponds to a plasmon condition. This means that we can interpret the softening mechanism as a tendency of the localized electrons to move collectively. A coupling between a bare phonon (screened by ε_o^{-1} only) and a plasmon mode can cause an instability of the lattice. This coupling would be very effective, provided an "acoustical" plasmon /14/ exists. In our case, the plasmon frequency remains finite and corresponds to the energy required to make a rigid translation of the localized electrons.

Using metal-d on-site interactions only in a parametrized version of the screening theory, Sinha and Harmon /15/ have been able to reproduce the anomalies in the longitudinal dispersion branches. In their work the most important role is assigned to the density of d-states at the Fermi level - in striking disagreement with the trends outlined in the introduction. The transverse anomalies, which are decisive for the superconductivity and for the lattice instabilities, cannot be reproduced by d-d interactions alone /16/. In our study we consider a simple band model based in the self-consistent band calculations of Schwarz et al. /2/. The "p" and "d" band complexes are described by linear combinations of bonding $(d+p)_\pi$ and antibonding $(d-p)_\pi$ hybrids. The physical mechanism is illustrated in Fig. 4 at the example of the anomalous longitudinal phonon at (0.5,0,0). In this direction alternating Nb and C atoms form $(pd)_\pi$ bonds (Fig.4a). The phonon moves only the Nb atoms, the C atoms are at rest. The covalent bonds from the C atoms to its Nb neighbours are alternatively strengthened and weakened. Thus this displacement pattern creates an electronic superstructure (a charge density wave) which has just twice the periodicity of the lattice (Fig.4c). These charge density fluctuations are strongly coupled to the lattice and give rise to the phonon anomalies. Of course this happens only if

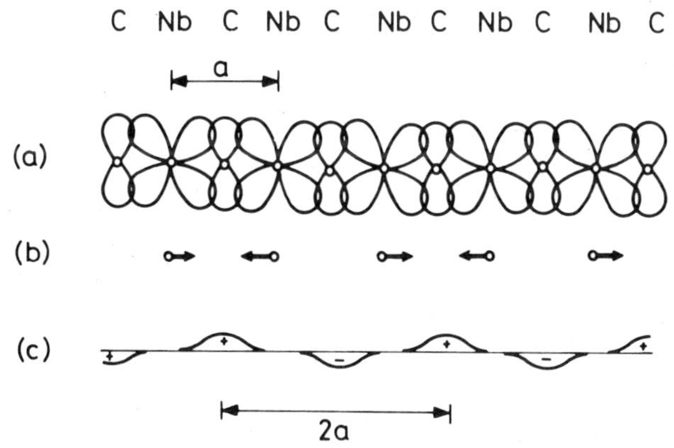

Fig. 4: (a) (p,d_{xy}) bonds along a chain of Nb and C atoms in the (100) direction.
(b) Displacement pattern of a (0.5,0,0)-L phonon.
(c) Charge fluctuation induced by the displacements.

the formation of this electronic superstructure is energetically favourable. This is the case only if the kinetic and the Coulomb energies nearly cancel, i.e. when the resonance condition $(N^{-1}-V)\approx 0$ is satisfied. The dominant p-d contribution to the polarizability stems from the diagonal term $A_s N_{ss} A_s^+$, which is proportional to the phase factor $\sin^2(\vec{q}\cdot\vec{R}/2)$ where $\vec{R}$ stands for the distance between two hybrids in the plane π .The diagonal part of the local field factor $V_{ss} \sim A_s v A_s$ has essentially the same symmetry. Hence the resonance condition $(\underline{N}^{-1}-\underline{V}) \sim 0$ may be approximated very roughly by $\Delta E_{eff} - \sin^2(\vec{q}.\vec{R}/2) \times |A_{pd}|^2 v \approx 0$, where ΔE_{eff} stands for some effective energy difference between the bonding and anti-bonding hybrides. To satisfy this condition ΔE_{eff} must be small (i.e. there must be bonding states just below and anti-bonding ones just above the Fermi level) and the form factor $|A_{pd}|$ must be large enough. Because of the phase factor, the concellation will occur at positions q_α =0.5 in reduced units. The same phase factor occurs in the corresponding contribution to the dynamical matrix (Eq.(8)), which peaks at q_α=0.5. Higher overlap tends to shift the peaks toward the zone boundary.

The validity of our concept is illustrated by a model calculation for the acoustical vibrations in NbC. This model is based on an LCAO-description of the p- and d-states, nearest-neighbour hybrid overlap, an effective-mass approximation for $N(\vec{q})$ and a local pseudopotential for the electron-ion interaction. Our model calculation, for which a quantitative agreement with experiment is hardly to be expected, yields local minima in the transverse as well as in the longitudinal dispersion branches (Fig.5). The depth of the minima is reduced by dehybridization and the non-locality of the (d-p) inter-

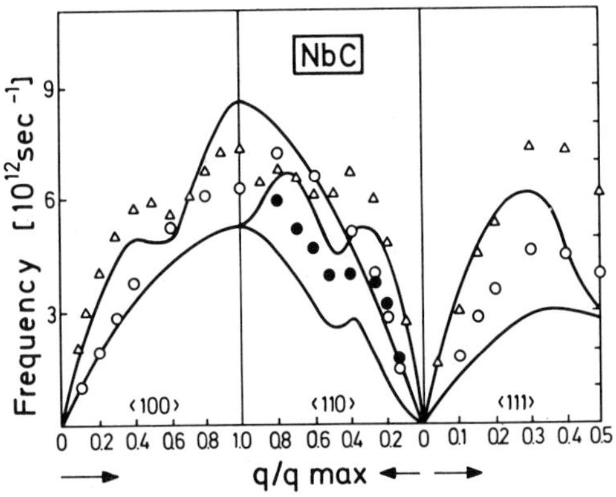

Fig.5: Dispersion curves for acoustic phonon branches in NbC. Circles and triangles show the exp.results/1,

action is essential: if in the screening matrix $(N^{-1}-V)$ the local-field factor V is neglected, the phonon dips vanish at the scale of our figure.

To study the interrelation between the phonon-anomalies and the high superconducting transition temperatures, we express the phonon-induced electron-electron attraction in the strong-coupling formulation by the nonlocal screening function. The screened electron-phonon coupling is determined by the gradient of the effective electron-ion potential, which we obtain by nonlocal screening of the bare electron-ion potential. The basic quantitiy is the electron-phonon spectral function $\alpha^2F(\omega)$ which is given by /17/

$$\alpha^2F(\omega) = \text{const.} \sum_{ll'} \sum_{tt'} \sum_{k,q} e^{i\vec{k}\vec{R}_l} e^{i(\vec{k}+\vec{q})\vec{R}_{l'}}$$

$$\times (1-VS^{-1})_{lt} (\vec{F}^t(q)\vec{e}(q)) (\vec{F}^{t'}(q)\vec{e}(q)) (1-VS^{-1})_{t'l'}$$

$$\times \delta(E(k)-E_F)\ \delta(E(k+q)-E_F)\ \delta(\omega-\omega(q)) \qquad (9)$$

Here q stands for momentum and polarization of a phonon with frequency $\omega(q)$ and polarization vector $e(q)$, k is short for electron momentum and band index. It is immediately apparent that the resonance in the screening matrix leads to a selective enhancement of the spectral function: α^2 is no longer a constant or a slowly decaying function. As a first step we have calculated the electron-phonon coupling parameter $\lambda=2\int(\alpha^2F(\omega)/\omega)d\omega$. Using our NbC model we have calculated for three different cases: (a) the local-field effects are neglected, $\lambda=0.4$, (b) they are included in evaluating the phonon frequencies, $\lambda=0.5$ and (c) they are taken into account both in the dynamical matrix and in the electron-phonon matrix-elements, $\lambda=0.6$. This corresponds (via the McMillan equation with a Coulomb repulsion of $\mu^*=0.1$ and an adjusted prefactor) to a relative change from $T_c=11.1\,^\circ$K to 6.2 K (neglecting V in the matrix elements) and further down to $T_c=2.7\,^\circ$K (neglecting V in the dynamical matrix too). Thus, the phonon anomalies are not alone responsible for the superconducting properties, the essential thing is to take the local-field effects in the spectral function into account.

In summary we have shown that covalent bond formation due to (d-p) hybridization leads to an electronically driven incipient lattice instability resulting from a resonance-like increase of the nonlocal inverse screening matrix. This incipient instability produces a softening of the phonon frequencies and an increased electron-phonon coupling. Both effects together are responsible for the high superconducting transition temperatures.

REFERENCES

1 H.G. Smith and W. Gläser, Phys. Rev. Letters 25, 1611 (1970) and 29,353(1972).
2 K. Schwarz, J. Phys.C8,809(1975); A. Neckel, P.Rastl, R. Eibler, P.Weinberger and K. Schwarz, J. Phys.C9,579 (1976).
3 L. Ramqvist,B.Ekstig, E.Källne,E.Noreland and R.Manne, J.Chem. Phys.Solids 32,149(1971)
4 K. Schwarz and N.Rösch, J. Phys. F6,L433 (1976)
5 W. Hanke,J.Hafner and H.Bilz,Phys.Rev.Lett. 37,1560(19
6 W. Hanke, Phys. Rev. B8,4585,4591(1973)
7 L.J. Sham, Phys. Rev. B6, 3581 (1972)
8 R.Pick, Phonons (Flammarion,Paris,1971),p.20.
9 S.K.Sinha,R.P.Gupta and D.L. Price,Phys. Rev. B9,2564 (1974).
10 M. Gupta and A.J.Freeman, Phys.Rev.Lett. 37,364(1976).
11 B.M.Klein,D.A.Papaconstantopoulos and L.L. Boyer, Sol. State Comm. 20, 937(1976)
12 R.P. Gupta and A.J.Freeman, Phys.Rev. B13,4316 (1976).
13 W.Hanke and H.Bilz, In Proc. of the Int.Conf.on Inelastic Scattering of Neutrons (IAEA Vienna,1971).
14 H.Fröhlich,Phys.Lett. A26,169(1968);B.N. Ganguly and R.F.Woods,Phys.Rev.Lett. 28,681 (1972)
15 S.K.Sinha and B.N.Harmon, Phys.Rev.Lett. 35,1515(1975)
16 W.Hanke,J.Hafner and H.Bilz, in Proc. of the Int.Conf. on Low-Lying Vibrational Modes and their Relation to Superconductivity and Ferroelectricity,Puerto Rico 1975 (in print).
17 J.Bar-Sagi and W.Hanke, in Proc. of the Int.Conf. on Low-Lying Vibrational Modes and their Relation to Superconductivity and Ferroelectricity, Puerto Rico 1975 (in print).

PSEUDO-SPIN APPROACH TO STRUCTURAL PHASE TRANSITIONS

R. B. Stinchcombe

Theoretical Physics Department
Oxford University, U.K.

ABSTRACT

Systems undergoing structural phase transitions and describable in a pseudo-spin formalism are reviewed, with particular emphasis on spin-phonon systems, Jahn-Teller systems and order-disorder and tunnelling ferroelectrics. Models for these systems are introduced, and their static and dynamical properties developed and compared to experiment. Mixed and diluted systems, and some aspects of the central peak problem and of critical behaviour are also discussed.

1. INTRODUCTION

This section begins with an extremely brief review of various types of lattice instability to put against a broader background the subset of inter-related systems to be discussed subsequently. Those systems will be ones for which a pseudo-spin description applies.

We list below examples of the various instabilities, roughly in order of decreasing applicability of the pseudo-spin description, giving for each type what drives the transition and the order parameter which develops below the transition temperature T_c.

A spin description obviously (partly) applies for linearly coupled spin-phonon systems; there the exchange of phonons can give rise to an effective spin-phonon interaction causing spontaneous ordering of the spin system for $T < T_c$ with possibly an accompanying distortion of the lattice. In this case the order parameter is the thermodynamic average $\langle S^{\alpha} \rangle$ of the component α of spin which couples to the phonons, or the average displacement $\langle Q \rangle$ of the phonon mode to which it couples.

A completely analogous situation occurs in the cooperative Jahn-Teller transition, where the discrete levels of different ions are coupled by phonons so that in the ordered regime the levels split and again an appropriate phonon mode becomes macroscopically occupied. The order parameter is $\langle Q \rangle$ for the phonon mode, or the related electronic energy splitting, which can be written in terms of $\langle \theta \rangle$ where θ is an appropriate operator for the electronic states of the ion, in simple cases just a (pseudo) spin operator.

Some transitions are driven by highly anharmonic phonon effects and accompanied by an ordering of some constituents among various positions of equilibrium. A simple example of such order-disorder transitions is $Na\,NO_2$ where the equilibrium positions relate to the rotational configurations of the NO_2^- ions. These can be represented by a pseudo spin operator S^z in an effective Ising Hamiltonian, with order parameter $\langle S^z \rangle$.

The hydrogen-bonded ferroelectrics, typified by KDP, share some of these characteristics: in these systems the equilibrium positions are those of the protons in their double well. The pseudo-spin states $S^z = \pm 1$ then represent the two positions for each proton, or in a more sophisticated picture (developed in §2) they can refer to some symmetrised coordinate for all the protons around a phosphate group. The order parameter is $\langle S^z \rangle$, or the average of the symmetrised coordinate, or the spontaneous electric polarisation associated with the ferroelectric transition triggered by the proton motion. Other hydrogen-bonded materials like the isomorphs of KDP and also TGS and Rochelle Salt can be described similarly.

The Peierls transition has some similarities to the Jahn-Teller transition: the continuum of conduction band electron states develops a gap at the fermi wave vector, for a half filled band, as the (one dimensional) lattice doubles its unit cell. The order parameter is the energy gap, or the distortion coordinate, or the occupation of the $q = 2k_F$ longitudinal acoustic phonon mode. In the simplest (Frohlich) model for these systems, the analogue of the pseudo-spin is the local conduction electron density. It would be inappropriate to represent this by a spin operator, but some of the formal treatments of the Frohlich Hamiltonian parallel those for pseudo-spin-phonon coupled systems. The structural transition in the A-15 high temperature superconducting systems is similar in origin to the Peierls transition but differs from it in being intrinsically three dimensional because of the role played by the weak coupling between chains.

Anharmonic phonon effects are responsible for the properties of displacive ferroelectrics such as $K\,Ta\,O_3$, in which the order parameter is the distortion coordinate corresponding to the $q = 0$ soft mode; also for the behaviour of cell-multiplying structural transitions such as the cell-doubling transition in $Sr\,Ti\,O_3$ where the distortion coordinate has $q = (\frac{1}{2},\frac{1}{2},\frac{1}{2})^{\pi}/a$; and for the incommensurate transition in for example $Ba\,Mn\,F_4$. In all these cases a pseudo-spin picture is not obviously applicable.

To the extent to which the special properties of the superionics can be regarded as due to a sublattice melting, they are in a rather special class. The pseudo-spin picture could at best apply only in some sort of lattice-gas picture of the melting sublattice.

Despite the wide variety of these systems, and whether or not the pseudo-spin picture applies, the lattice instability is normally associated with some instability in the system to which the phonons are coupled - for example, other phonons (as in for example KDP) or conduction electrons (Peierls transition) or pseudo-spins (Jahn-Teller) and so on. And the distortion coordinate (or average of the soft mode eigenvector) is always one of the possible order parameters.

In the rest of this paper the discussion is confined to spin-phonon systems, Jahn-Teller systems and order-disorder and tunnelling ferroelectrics: these are the most strongly inter-related systems and also those describable by a pseudo-spin formalism. Occasional references will be made to parallel features of some of the other systems for which, though a different formalism may be more appropriate, some basic ideas are similar.

Reviews dealing with the systems listed above and particularly those not considered further here are: Jahn-Teller system[1,2], order-disorder and hydrogen bonded ferroelectrics[3,4], structural phase transitions[5-9], improper ferroelectrics[10], Peierls transition and A15 compounds[11-18], superionics[19]. The present volume, and reports of previous Geilo conferences[20,21] give comprehensive treatments of most of these topics.

The next section discusses models for the selected systems; it is followed by a section (§3) on general static and dynamical properties of the models; section 4 discusses mixed and dilute systems, while the final section (§5) considers some aspects of the central peak problem and critical properties, in an attempt to provide some background to current developments.

2. MODELS

(i) Spin-Phonon Systems

As an illustrative example we consider a spin $\frac{1}{2}$ system with linear coupling of the z-component of spin S_i^z at site i to a particular phonon branch. Considering only that branch, the Hamiltonian has the form

$$H = \sum_q \hbar\omega_q (a_q^+ a_q + \tfrac{1}{2}) + \sum_{iq} S_i^z A_q(i)\,(a_q^+ + a_{-q}) \tag{1}$$

where $A_{-q}^*(i) = A_q(i) \equiv A_q\, e^{i\underline{q}\cdot\underline{r}_i}$.

The last term can be written in the alternative form $\sum_{iq} S_i^z \Lambda_q(i)\, Q_q$, where $\Lambda_q(i) = (2m\omega_q/\hbar)^{\frac{1}{2}} A_q(i)$, and represents a coupling of spin to lattice displacement Q. The exact canonical transformation

$$a_q \to \alpha_q = a_q + \sum_i A_q(i)\, S_i^z/\hbar\omega_q$$
$$S_i^z \to S_i^z \tag{2}$$

which corresponds to a shift of phonon coordinates, takes the Hamiltonian to the form

$$H = H_P + H_{SI}$$

where

$$H_P = \sum \hbar\omega_q (\alpha_q^+ \alpha_q + \tfrac{1}{2}) \tag{3}$$

$$H_{SI} = -\tfrac{1}{2}\sum_{ij} J_{ij}\, S_i^z S_j^z \;, \qquad J_{ij} \equiv 2\sum_q A_q(i) A_q^*(j)/\hbar\omega_q \,, \quad (i \neq j). \tag{4}$$

This separation of the spin and phonon parts leaves the spins with an effective Ising interaction, resulting from exchange of phonons. The Ising model is perhaps the simplest exhibiting a phase transition, and straightforward extensions provide descriptions of Jahn-Teller systems (§2 (ii)) and tunnelling ferroelectrics (§2(iii)).

The consequences of (3) and (4) are that the phonon frequencies are unshifted, and that $\langle S^z\rangle$ becomes non zero below the Ising transition temperature and so, since $\langle\alpha\rangle = \langle\alpha^+\rangle = 0$,

$$\langle Q_q\rangle = -\sum_i \frac{\Lambda_q(i)}{m\omega_q^2} \langle S_i^z\rangle = -\frac{\Lambda_q}{m\omega_q^2} \langle S_q^z\rangle . \tag{5}$$

This corresponds to a spin ordering and a lattice distortion, at the value of q at which the Fourier transform J(q) of J_{ij}, equation (4), has its maximum. This is normally at q = 0, though exceptions analogous to the antiferromagnet or spin spiral can occur[22,23].

An important case is when the spin coupling is to strain, ε. This is usually treated separately from the phonons because of difficulties in applying boundary conditions to a strained crystal.[24,25] For uniform strain, the corresponding strain coupling term, and elastic energy terms (replacing or supplementing the phonon terms in (1)) are

$$\sum_i S_i^z \Lambda(i)\varepsilon + \tfrac{1}{2}c\varepsilon^2 . \tag{6}$$

This leads to the development of a macroscopic strain below the transition. §3 will refer to approximate methods for obtaining quantities like $\langle S^z \rangle$, such as molecular field theory which is exact for strain coupling.

The above discussion actually applies more properly to pseudo-spin than to true spin systems : since the coupling to phonons is restricted by time-reversal invariance the linear coupling of a true spin operator to displacement as written in (1) could not actually occur, though coupling to the lattice momentum operator in principle could. A more interesting example of the spin-phonon system is single ion-lattice magnetostriction[26-28] where H is as above but with S_i^z typically a quadrupole operator. The coupling as written in (1) applies when S^z is a pseudo-spin operator, as it will be in all the subsequent discussions.

(ii) Jahn-Teller Systems[1,2]

The simplest situation occurs in Tm VO_4, and Tm As O_4, where (in zero field) a degenerate low lying doublet is linearly coupled to a non-degenerate phonon mode. All other electronic levels can be neglected as long as the Jahn-Teller splitting is sufficiently small. The two states $S^z = \pm \frac{1}{2}$ of a pseudo-spin-$\frac{1}{2}$ operator can be used to represent the two levels of the crystal field doublet for the ion at site i. The ion is coupled to a local lattice distortion through a Jahn-Teller energy term of the form $\sum_{iq} S_i^z \Lambda(i) Q_q$, which leads to a Hamiltonian of the same form (1) (or equivalently (3), (4)) as previously. As before, a lattice distortion $\langle Q_q \rangle$ develops below the transition at which $\langle S_i^z \rangle$ becomes non-zero.

In the molecular field approximation the Hamiltonian takes the form

$$H \approx -\sum_i \gamma_i S_i^z + \cdots$$

where the molecular field experienced by the 'spin' at site i is

$$\gamma_i = \sum_j J_{ij} \langle S_j^z \rangle . \tag{7}$$

This can be interpreted as the level splitting.

Care has to be taken with the J_{ii} term in (4) if molecular field theory is being used. Since $(S_i^z)^2 = \frac{1}{4}$, the J_{ii} term is a constant and without significance in an exact treatment. In molecular field approximation (which replaces $(S_i^z)^2$ by $2 \langle S_i^z \rangle S_i^z$) it would appear as a spurious contribution unless subtracted out[24,29] and similar considerations apply if random phase approximations are used. It is therefore convenient to take J_{ii} as zero or, equivalently

$$J(q) = 2\{ |A_q|^2/\hbar\omega_q - \sum_{q'} |A_{q'}|^2/\hbar\omega_{q'} \}. \tag{8}$$

In Tm VO_4 and Tm As O_4, the strain terms (6) play an important role and have to be included in a straightforward generalisation of the above discussion.

Various additional generalisations of the model are required for other Jahn-Teller systems:

Coupling to a two-dimensional distortion gives rise to an interaction term of which the simplest form is $\Sigma\, \Lambda(S^z Q_1 + S^x Q_2)$, leading to an X-Y Hamiltonian in place of (4)[25,30]. This description applies to Pr $C\ell_3$[31] and Pr Eth SO_4[32]. Outside of mean field approximation, the X-Y form of the transformed Hamiltonian is only approximate (c.f. the discussion under (10), below).

Most systems have splitting in the high temperature phase, which corresponds to an additional term $- \sum_i \Gamma_i S_i^\alpha$ in the original Hamiltonian (1). If $\alpha = z$, the splitting acts like a longitudinal field in the resulting Hamiltonian (4), and no (sharp) transition occurs. This is the situation[33,34] in Ce Eth SO_4. In general the additional term prevents the transition unless $\alpha = x$. In this case, the 'transverse field' term is

$$-\sum_i \Gamma S_i^x . \tag{9}$$

This applies to Dy VO_4 which, having two Kramers doublets, is effectively a two-level system. The same term arises in Tm VO_4 if a (true) magnetic field Γ is applied since the otherwise degenerate doublet is magnetic. Spin and phonon parts cannot be exactly separated for this case, since the transformation accomplishing (2) is, for any operator A,

$$A \to e^{iS} A e^{-iS}, \qquad S = i \sum_q \frac{A_q(i)}{\hbar\omega_q} S_i^z (a_q^+ - a_{-q}). \tag{10}$$

The transverse field term (9) transforms to $-\Sigma\,\Gamma\, e^{iS}\, S^{x}\, e^{-iS}$, which still involves phonon terms since $[S^{x}, S] \neq 0$. This can be approximately simplified[35,1], in a manner analogous to the treatment of Debye-Waller factors to yield a Ham quenched[36] effective transverse field. In molecular field approximation these complications do not appear, and the resulting Hamiltonian involves an unperturbed (shifted) phonon part and a spin part of 'transverse Ising' form

$$H_{s2} = -\sum_{i} \Gamma S_i^{x} - \frac{1}{2}\sum_{ij} J_{ij}\, S_i^{z} S_j^{z} \,. \tag{11}$$

This model, which also applies for rare-earth group V compounds with the NaCℓ structure[37] and the tunnelling ferroelectrics[3,4,38], has been rather fully treated in the theoretical literature[38-43] and its properties will be discussed in §3. The fact that the observed phonon frequencies in for example Dy VO_4 appear to be almost unaffected by the Jahn-Teller interaction[44] indicates that the molecular field approximation holds well there, implying long range effective interactions.

Other generalisations of the model are required for example for the four level system Tb VO_4, which can be represented by two pseudo-spin $\frac{1}{2}$ operators[45,44], or for triplet ground state systems[46].

The pseudo-spin method is most useful when only a limited number of low lying levels is involved. The next subsection considers its use for ferroelectrics, particularly of the tunnelling type.

(iii) Order-disorder and Tunnelling Ferroelectrics[3-6,29,38,47]

The pseudo-spin method was first used for order-disorder ferroelectrics with well-defined positions of equilibrium (eg Na NO_2[48,49]). A generalisation of this type is the KDP class of (hydrogen-bonded) ferroelectrics, for which the descriptions have evolved from simple pseudo-spin models to more sophisticated models[50,4] which include pseudo-spin phonon coupling and make evident the relationship to soft mode behaviour.

The crucial feature of all these models is the existence of a double well (for the coordinate of a single proton in simple models, or more generally for some coordinate related to the collective proton displacement which freezes in below T_c).

The simple spin model[3,47] is developed as follows. With ξ_i representing the coordinate of a single proton at site i, the simplest double well potential is

$$V = -\frac{1}{2} a \xi_i^{2} + \frac{1}{4} b \xi_i^{4} \tag{12}$$

with minima at $\delta = \pm (a/b)^{\frac{1}{2}}$. The two values $S_i^z = \pm \frac{1}{2}$ refer to which well the proton at i occupies. Tunnelling between the two wells causes a splitting Γ, which is accommodated by the following term in the Hamiltonian

$$-\sum_i \Gamma S_i^x . \tag{13}$$

Coupling between displacements of different protons, through a harmonic potential $V_{ij}(\xi_i - \xi_j)^2$ then leads to the spin interaction term

$$-\frac{1}{2}\sum_{ij} J_{ij} S_i^z S_j^z , \tag{14}$$

and we recover the transverse Ising Hamiltonian (11).

This development is inadequate in several respects. First, all but the lowest two states in the well were neglected. This approximation will fail[51] if the separation of the higher levels becomes comparable to J, Γ, where a displacive situation more correctly treated in an anharmonic phonon picture will occur.

Secondly, in the case of a specific example such as KDP, the nature of the coupling between different protons around the same PO_4 group has to be more carefully considered for two related reasons: (a) the soft mode for the system is a particular (B_2 symmetry) mode involving all the protons[52], as shown in Figure 1(a), and (b) certain configurations of the protons around a PO_4 group are energetically favoured over others[53,54]. In order of increasing energy the configurations are: the B_2 type displacement with two protons "in" (adjacent to the PO_4 group) and two "out", other displacements (E_1, E_2 symmetry) with two protons in and two out, arrangements with three in and one out or one in and three out, and least favoured of all is that (A_2 symmetry) with four in or out.

Both of these points can be overcome[29,4] by using symmetrized local normal coordinates ξ_α for the four protons, rather than a single proton coordinate, and corresponding to the B_2, A_2 and E modes shown in Figure 1(b). The $\xi_\alpha(\alpha = B_2, A_2, E_1$ or $E_2)$ are appropriate linear combinations of the four proton coordinates q_n, n = 1,...4. The effective potential of the four protons surrounding the PO_4 group in the centre of cell i can then be written as a power series in the ξ-variables. The quadratic terms are

$$\sum_\alpha A_\alpha \xi_\alpha^2(i) \tag{15}$$

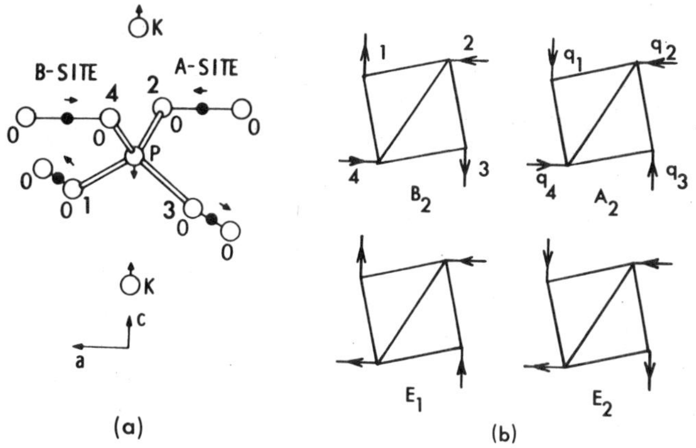

Figure 1. (a) Soft mode of KDP (after Cochran[52])
(b) Symmetrised proton displacements in KDP. The B_2 mode corresponds to the soft mode depicted in (a).

where $A_{E_1} = A_{E_2}$. By putting each q_n equal to $\pm\,\delta$ it should be possible to match the A_α to the energies of the Slater-Takagi model[53,54]. This model, however, involves one more parameter than so far appears in (15). This can be accommodated by adding to (15) symmetry - allowed quartic terms of the form

$$\frac{1}{2}\sum_{n=1}^{4} B q_n^4 - W \prod_{n=1}^{4} q_n \,. \tag{16}$$

Additional terms arise from considering the same types of quadratic and quartic coupling between protons (lying in different cells) which are adjacent to a PO_4 group at the boundary of the unit cell. These contributions, which involve the same parameters as in (15), (16) generate terms of the form ... $\xi(i)\ \xi(j)$, and ...$\xi(i)\ \xi(j)\ \xi(k)\ \xi(\ell)$, where cells i, j, k, ℓ are all adjacent to the same PO_4 group.

The energies of the Slater-Takagi model make it plausible that in the resulting combined Hamiltonian the coefficient of $\xi_{B_2}^2(i)$ is negative, and all other $\xi_\alpha^2(i)$ terms have large positive coefficients.

Since (16) gives rise to a term

$$\frac{1}{8}(B - \frac{1}{2}W)\ \xi_{B_2}^4(i) \tag{17}$$

a double well then arises for the ξ_{B_2} coordinate provided $B > \frac{1}{2}W$, while all other ξ_α are stable and do not play an important role in the phase transition. A pseudo-spin S_i can then be associated with the ξ_{B_2} (i) of each unit cell and the tunnelling term (13) is recovered. Of the quadratic and quartic intercell terms referred to earlier, the quadratic terms in ξ_{B_2} (i) ξ_{B_2} (j) then give the usual Ising coupling (14), where i and j label adjacent cells; the quartic terms ξ_{B_2} (i)...ξ_{B_2} (ℓ) give an additional four-spin coupling proportional to

$$S_i^z S_j^z S_k^z S_\ell^z . \tag{18}$$

The above discussion depends on arguing that coupling terms between ξ_{B_2} and the other ξ_α are either small by symmetry (e.g. for low k modes) or can be accommodated as small correction terms to the coefficients by 'decouplings' of the random-phase type[4].

In addition to the near-neighbour pseudo-spin couplings arising above, longer range pair interactions arise from the dipole-dipole interactions with more distant protons.

Only proton motion was so far considered. The spontaneous polarisation arises from the resulting largely harmonic motion of other atoms in the unit cell. Their coupling to the proton soft mode will give additional terms in the Hamiltonian of the form[50,4]

$$\Sigma \left(S_i^z \Lambda_q^z(i)\, Q_q + S_i^x \Lambda_q^x(i)\, Q_q \right) \tag{19}$$

where the Λ^x term is small and unimportant for KDP, and will be neglected in the remainder of this discussion. In addition, strain coupling terms can arise, as in (6).

The transformation (10) can again be applied to approximately uncouple 'spin' and phonon parts, resulting in a transverse Ising model with a quartic term, coming from (18) and a quadratic term arising from three sources: (a) the ξ_{B_2} (i) ξ_{B_2} (j) terms, (b) the dipole-dipole interaction, and (c) the phonon exchange term (analogous to (8)).

(iv) Displacive Ferroelectrics[5,6,55-57]

Up to a point, the development would be similar for a displacive ferroelectric. The crucial difference occurs at the introduction of the pseudo-spin, which is appropriate provided a double well arises and that the mean square fluctuation in the soft mode local coordinate ξ_α (i) remains large even above the transition. If this is not the case, an anharmonic picture would seem more suitable. The following simple discussion[55,56] gives the main characteristics of this picture. The lattice vibrational energy is expanded in terms of displacements from the high temperature configuration, using appropriate normal coordinates (and, for simplicity, considering only fourth order anharmonic terms):

$$H = \tfrac{1}{2}\Sigma\,\Omega\,\xi^2 + \Sigma\,W\,\xi\xi\xi\xi\,. \tag{20}$$

The effective Hamiltonian resulting from averaging pairs of ξ's in the anharmonic term is

$$H_{eff} = \tfrac{1}{2}\Sigma\,\omega_q^2\,\xi_q^2\,,$$

$$\omega_q^2 = \Omega_q^2 + \sum_k V(q,k)\langle\xi_k\xi_k\rangle = \Omega_q^2 + \sum_k \frac{V(q,k)}{2\omega_k}\coth\tfrac{1}{2}\beta\omega_k \tag{21}$$

where V is a coupling constant appropriately related to W.

If, for a particular mode and wave vector (q_c), $\Omega^2_{q_c}$ is so negative that zero point fluctuations alone are not sufficient to stabilize it ($\Omega^2_{q_c} < -\frac{1}{2}\,\Sigma\,V/\omega$), the displacive transition occurs at $T = T_o$ such that $\omega^2_{q_c} = 0$.

For $T > T_o$, $\omega^2 > 0$ and the reference configuration is stable.

For $T < T_o$, a non-zero $\langle\xi_{q_c}\rangle$ freezes in to stabilize a distorted lattice having $\omega^2 > 0$, and the averaging used to obtain H_{eff} has to be modified to include the additional non-vanishing averages.

In not requiring spin-like behaviour for the local soft mode coordinate, this is more general than the pseudo-spin approach (but less easy to approximate to obtain spin-model behaviour where it is appropriate).

(v) Hamiltonians

This section gathers together the basic Hamiltonians for the principal pseudo-spin systems discussed above, omitting strain terms

(a) $$H = \sum_q \hbar\omega_q\,(a_q^+ a_q + \tfrac{1}{2}) + \sum_q A_q\,S_q^z\,(a_q^+ + a_{-q}) \tag{22}$$

(simple spin-phonon, or simplest Jahn-Teller system: $TmVO_4$ in zero field). (22) becomes, after exact canonical transformation (10),

$$H = H_p + H_{S1} \equiv \sum_q \hbar\omega_q (\alpha_q^+ \alpha_q + \tfrac{1}{2}) - \tfrac{1}{2}\sum_{ij} J_{ij}\, S_i^z S_j^z \tag{23}$$

with the Ising interaction given by (4) and (8).

(b) $$H = \sum_q \hbar\omega_q (a_q^+ a_q + \tfrac{1}{2}) - \sum_i \Gamma S_i^x + \sum_q A_q S_q^z (a_q^+ + a_{-q}) \tag{24}$$

(cooperative Jahn-Teller systems Dy VO_4, or Tm VO_4 with applied field). Canonical transformation (10) takes (24) approximately to the form

$$H \approx H_p + H_{S2} \equiv \sum_q \hbar\omega_q (\alpha_q^+ \alpha_q + \tfrac{1}{2}) - \sum_i \Gamma S_i^x - \tfrac{1}{2}\sum_{ij} J_{ij}\, S_i^z S_j^z \tag{25}$$

which involves the Transverse Ising Hamiltonian with exchange again given by (4) and (8).

(c) $$H = \sum_q \hbar\omega_q (a_q^+ a_q + \tfrac{1}{2}) - \sum_i \Gamma S_i^x - \tfrac{1}{2}\sum_{ij} L_{ij}\, S_i^z S_j^z$$

$$- \sum_{ijkl} K_{ijkl}\, S_i^z S_j^z S_k^z S_l^z + \sum_q A_q S_q^z (a_q^+ + a_{-q}) \tag{26}$$

(Hydrogen-bonded ferroelectrics. The Ising pair interaction L_{ij} includes the dipole-dipole part). The canonical transformation again approximately uncouples spin and phonon terms:

$$\begin{aligned} H \approx H_p + H_{S3} \equiv \sum \hbar\omega_q (\alpha_q^+ \alpha_q + \tfrac{1}{2}) - \sum_i \Gamma S_i^x - \tfrac{1}{2}\sum_{ij} J_{ij}\, S_i^z S_j^z \\ - \sum_{ijkl} K_{ijkl}\, S_i^z S_j^z S_k^z S_l^z \end{aligned} \tag{27}$$

where

$$J_q = L_q + 2\left\{ |A_q|^2/\hbar\omega_q - \sum_{q'} |A_{q'}|^2/\hbar\omega_{q'} \right\}. \tag{28}$$

The properties of all these models are considered in the next section (see also the papers by H. Thomas, R.T. Harley, and J.K. Kjems in this volume).

3. PROPERTIES OF MODELS

The simplest model listed is (22), applying to Tm VO_4 and Tm As O_4, in zero field. From (23) and (7) the order parameter $\langle S^z \rangle$ is that for an Ising model and it should show up in experiment as the splitting of the ground state electronic doublet. For Tm VO_4 the measured splitting[58] agrees well with Ising model molecular field theory for $\langle S^z \rangle$, as shown in Figure 2, and the specific heat anomaly (Figure 3) also fits extremely well to the Ising model molecular field theory curve[59]. The validity of molecular field theory is due to the dominance of the very long

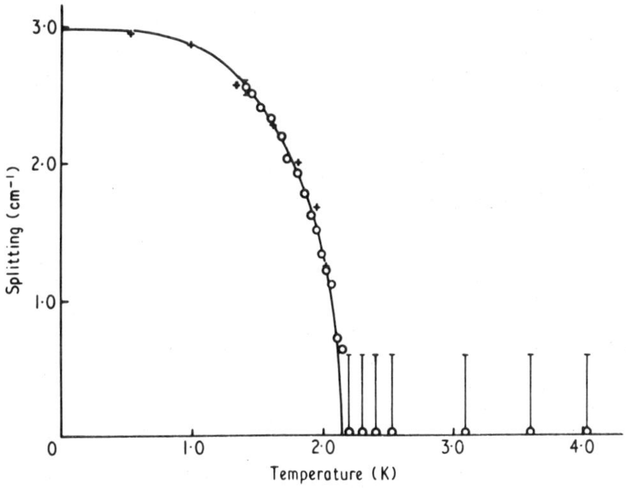

Figure 2. Ground state splitting γ in Tm VO_4 (Γ = 0), and a comparison with molecular field theory (solid line)[58].

range exchange interaction produced by strain coupling. No interesting dynamics is left in the Hamiltonian (23).

A related example is completely deuterated KDP. To the extent

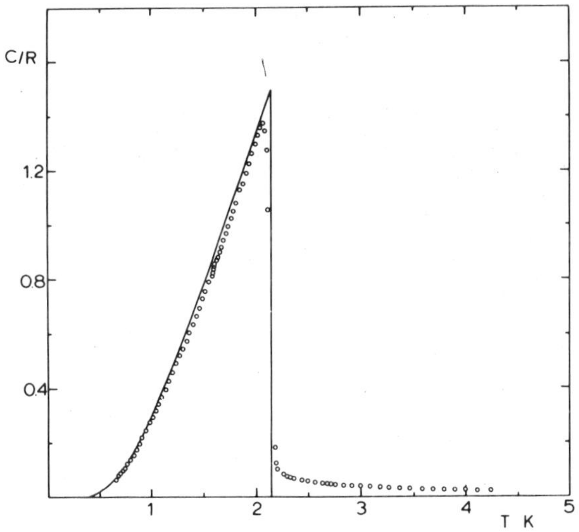

Figure 3. Experimental and mean-field (solid line) values for Specific Heat of Tm VO_4 (Γ = 0)[59].

to which it is described by (26), it is usual to place $\Gamma \sim 0$, since the heavier deuteron will tunnel less than a proton. The transformation to (27), without the tunnelling term, is then exact and the dynamics again become trivial. In particular[3] the pseudo-spin wave mode observed in KDP below the transition (§3 (iii)) should not be seen in deuterated KDP. The models (24), (26) with non-zero Γ have both interesting static and dynamic behaviour. The next three subsections briefly consider these aspects.

(i) Statics

The simplest approximation is molecular field theory[3,40,47]. Since molecular field theory for (24) is the same as for (25), it is sufficient to apply it to (25), concentrating on the effective spin Hamiltonian

$$H_{eff} = H_{S2} = -\sum_i \Gamma S_i^x - \tfrac{1}{2}\sum_{ij} J_{ij}\, S_i^z S_j^z \,. \tag{29}$$

Each spin experiences a mean field

$$\underline{\gamma} = \Gamma \hat{\underline{x}} + J(0)\langle S^z\rangle \hat{\underline{z}} \tag{30}$$

as depicted in Figure 4. The average spin vector is of magnitude $R = \frac{1}{2}\tanh\frac{1}{2}\beta\gamma$ and makes an angle θ with the z-direction. The

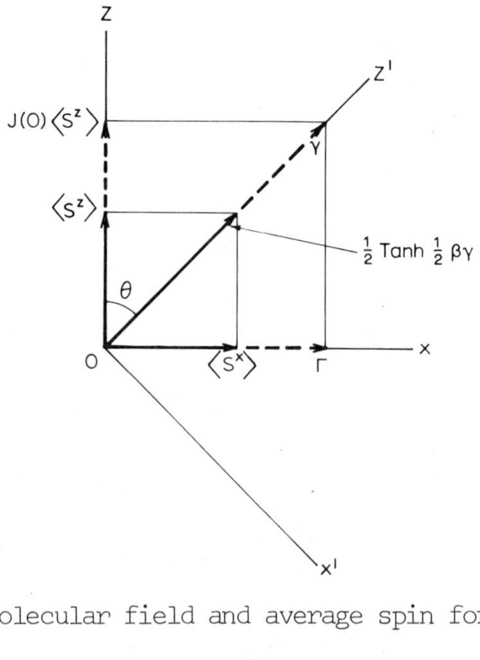

Figure 4. Molecular field and average spin for the transverse Ising model.

equation of state is therefore given by

$$\langle S^z \rangle = \tfrac{1}{2} \cos\theta \tanh \tfrac{1}{2}\beta\gamma = \gamma \cos\theta / J(0)$$

$$\sin\theta = \Gamma/\gamma , \qquad \gamma^2 = \Gamma^2 + J(0)^2 \langle S^z \rangle^2 . \tag{31}$$

Provided the transverse field is sufficiently small ($\Gamma < J(0)$) the order parameter $\langle S^Z \rangle$ will become non-zero at temperatures less than the transition temperature $T_c(\Gamma)$ given by

$$\Gamma/J(0) = \tfrac{1}{2} \tanh \tfrac{1}{2}\beta_c \Gamma . \tag{32}$$

This relationship between T_c and Γ is shown in Figure 5. For $T < T_c$, γ (and hence $\langle S^Z \rangle$) and $\langle S^X \rangle$ are given by

$$\gamma = \tfrac{1}{2} J(0) \tanh \tfrac{1}{2}\beta\gamma , \qquad \langle S^x \rangle = \Gamma / J(0) . \tag{33}$$

For $T > T_c$

$$\langle S^z \rangle = 0 , \qquad \langle S^x \rangle = \tfrac{1}{2} \tanh \tfrac{1}{2}\beta\Gamma . \tag{34}$$

The molecular field specific heat[42] is of the form of a usual $\Gamma = 0$ molecular field contribution superimposed on a Schottky anomaly. The experiments on Tm VO_4 and Tm As O_4 with no applied field ($\Gamma = 0$) agree well with the molecular field form[59,60]. Experiments[59] with an applied field ($\Gamma \neq 0$) agree satisfactorily with the relationships (33), (34) for γ and $\langle S^X \rangle$. In addition (32) accounts well for the observed dependence[59,61] of T_c on Γ.

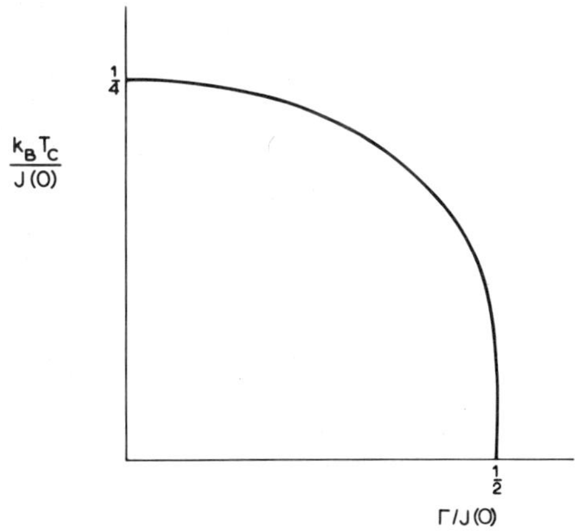

Figure 5. Mean field relationship between critical temperature and transverse field.

The decrease of T_c with Γ can also be seen in KDP, since Γ can be varied by applying pressure[62], which increases overlap and therefore the tunnelling frequency. The observed dependence of transition temperature on pressure[63] is shown in Figure 6 and has a form similar to part of the curve (Figure 5) resulting from (32). Though the coupling (28) also changes slightly with pressure (enough to cause T_c to decrease slightly with pressure for deuterated KDP, as is also shown in Figure 6), the result for KDP appears to provide strong support[64] for the tunnelling picture, and is difficult to explain in an anharmonic phonon approach.

In discussing KDP with the transverse Ising model, we have ignored the four spin coupling term in the Hamiltonian H_{S3} of (27). Its main effects, in a molecular field discussion[4], are that it yields the first order transition actually seen[65]; it also allows a satisfactory fit to the data[65] for the spontaneous polarization $P_o \langle S^z \rangle$ with roughly the same values of the parameters (e.g. P_0) as are required for the high temperature Curie-Weiss law, which supports the picture of one pseudo-spin per unit cell.

Static aspects of the transverse field Hamiltonian (29) have also been treated by series methods[66-68] or by controlled extensions[42,43,69] of molecular field theory. The series method is

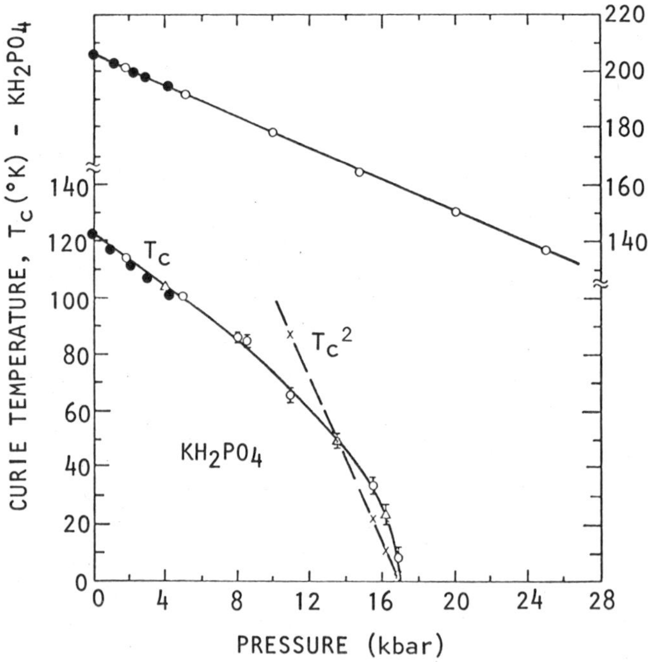

Figure 6. Relationship between critical temperature and pressure for KDP and DKDP[63].

well suited to the treatment of critical properties, which are discussed in §5. The other method is useful away from the critical region. In this approach the leading corrections can be seen to arise from spin wave and spin fluctuation effects[42]. These include zero point and linear spin wave effects, as well as effects arising (in for example the low temperature specific heat) from the temperature and field dependence of the spin wave energies (§3 (iii)). These last effects are important near $T_C(\Gamma)$ where the spin wave energy gap vanishes. Similar considerations apply to the Hamiltonian H_{S3} of (27). It should, however, be noted that, to the extent to which corrections to molecular field theory are being considered, if $\Gamma \neq 0$, models (24) and (25) (or (26) and (27)) are not equivalent.

(ii) Formalism for Dynamics

The models are all included in the following general form

$$\begin{aligned} H &= \sum \hbar\omega_q (a_q^+ a_q + \tfrac{1}{2}) + \sum_q A_q S_q^z (a_q^+ + a_{-q}) + H_S^{(0)} \\ &\equiv \sum_q \left[\frac{1}{2m} P_q P_{-q} + \tfrac{1}{2} m\omega_q^2 Q_q Q_{-q}\right] + \sum_q \Lambda_q S_q^z Q_q + H_S^{(0)} \end{aligned} \tag{35}$$

(compare (22), (24), (26)) where the spin Hamiltonian $H_S^{(0)}$ is in the case of Jahn-Teller systems the crystal field Hamiltonian for the undistorted phase or, in the case of KDP, the Hamiltonian $-\sum \Gamma S^x - \frac{1}{2}\sum L S^z S^z - \sum L S^z S^z S^z S^z$. Only phonon coupling to the z-component of pseudo-spin has been considered, but the subsequent discussion applies to the more general Jahn-Teller case involving coupling to a single electronic operator not represented as pseudo-spin $\frac{1}{2}$, or to the Frohlich Hamiltonian for the Peierls transition problem where S_q^z is to be replaced by the electron density operator $\rho_q = \sum_K a_{K+q}^+ a_K$ and $H_S^{(0)}$ becomes the electron Hamiltonian.

The use of

$$\omega^2 \langle\langle QQ\rangle\rangle = \frac{\omega}{m}\langle\langle PQ\rangle\rangle = \frac{\hbar}{m} + \omega_q^2 \langle\langle QQ\rangle\rangle + \frac{\hbar}{m}\Lambda_q \langle\langle S^z Q\rangle\rangle \,, \tag{36}$$

$$\omega^2 \langle\langle S^z Q\rangle\rangle = \omega^2 \langle\langle Q S^z\rangle\rangle = \omega_q^2 \langle\langle Q S^z\rangle\rangle + \frac{\hbar}{m}\Lambda_q \langle\langle S^z S^z\rangle\rangle \tag{37}$$

leads to the following exact result[70-72] relating the phonon and spin Green functions (momentum labels omitted, for simplicity):

$$\langle\langle QQ\rangle\rangle = \frac{\hbar/m}{\omega^2 - \omega_q^2}\left[1 + \frac{\frac{\hbar}{m}\Lambda_q^2}{\omega^2 - \omega_q^2}\langle\langle S^z S^z\rangle\rangle\right]. \tag{38}$$

There now remains the calculation of $\langle\langle S^z\, S^z \rangle\rangle$, hereafter called $\chi(q\omega)$.

For the special case where $H_S^{(0)}$ commutes with the generator S of the canonical transformation (10) (i.e. where $\Gamma = 0$ in the ferroelectric or simple Jahn-Teller cases, or where $H_S^{(0)}$ contains only the electronic operator to which the lattice couples in more complicated Jahn-Teller cases) the transformation takes the Hamiltonian H to the form

$$H \to H_p + H_S \equiv H_p + H_S^{(0)} - \sum_q \left\{ \frac{|A_q|^2}{\hbar\omega_q} - \sum_{q'} \frac{|A_{q'}|^2}{\hbar\omega_{q'}} \right\} S_q^z S_{-q}^z . \qquad (39)$$

Here, H_p is the free phonon form, and H_S (= H_{S1}, or H_{S2} or H_{S3},...) is an effective spin Hamiltonian containing no phonon term, as in §2(v). Since the spin operator is unchanged in the transformation, the special case has

$$\chi = \chi_S \qquad (40)$$

where χ_S is $\langle\langle S^z\, S^z \rangle\rangle$ evaluated with Hamiltonian H_S. In the situation (e.g. model (22)) where this procedure is exact, χ is, however, static.

The more interesting cases, discussed for the remainder of this subsection, are where $[S, H_S^{(0)}] \neq 0$. These cases are normally dealt with by introducing approximations within Green function formalisms[71,73,44] or in diagrammatic formalisms[40,41,74,75,34].

If $\chi(q\ \omega)$ is represented by diagrams using $\sum_q \Lambda_q S_q^z Q_q$ as the interaction, the diagrams represent the processes of production and absorption of phonons. Some diagrams have intermediate states corresponding to the free propagation of a single phonon with wave vector and frequency having the external values, q, ω. All other diagrams constitute the "irreducible" part $\Pi(q\ \omega)$ of $\chi(q\ \omega)$. χ is given in terms of Π by the Dyson equation[8,34,74]

$$\chi = \Pi - \Pi D^0 \Lambda_q^2 \chi = \frac{\Pi}{1 + \Pi D^0 \Lambda_q^2} \qquad (41)$$

where D^0 is the free phonon propagator

$$D^0 = \langle\langle \varphi\varphi \rangle\rangle_0 = \frac{\hbar/m}{\omega^2 - \omega_q^2} . \qquad (42)$$

The obvious approximation (of RPA type) is now to replace Π by the value corresponding to $\langle\langle S^z\, S^z \rangle\rangle$ evaluated with $H_S^{(0)}$. This, however, is not adequate for the more delicate aspects of the dynamics[71,76] essentially because the resulting χ does not reduce to the correct limit when $[H_S^{(0)}, S] \to 0$, where χ_S should be recovered.

Instead, a satisfactory approximation for χ can be obtained using the effective spin Hamiltonian H_S. The approximation is obtained by exploiting the contact between χ and χ_S when $[H_S^{(0)}, S] \to 0$, and the fact that χ_S can be written in a form analogous to (41): if H_S is separated into a 'tractable part' H_0, and a remainder of pair interaction form,

$$H_S = H_0 - \frac{1}{2}\sum_q v(q)\, S_q^z S_{-q}^z \quad , \tag{43}$$

χ_S can be developed in terms of diagrams[77,40,41] having the same topology as the diagrams for χ, but with the phonon propagators replaced by interactions v. Thus[40,70,77-80]

$$\chi_S = \frac{\Pi_S}{1 - v(q)\Pi_S} \tag{44}$$

where Π_S denotes the irreducible part of χ_S containing no v(q)'s with external label q. In the limit $[H_S^{(0)}, S] \to 0$ (denoted by asterisk) where χ_S becomes static and $\chi \to \chi_S$,

$$\frac{1}{\Pi^*(q,0)} = \frac{1}{\Pi_S^*(q,0)} - v(q) + \frac{\hbar}{m}\frac{\Lambda_q^2}{\omega_q^2} \quad . \tag{45}$$

No approximations have been made so far. The approximation now introduced is that, away from the limit, the same relationship holds between the frequency-dependent irreducible parts, so that $\Pi(q, \omega)$ can be written in terms of the simpler quantity $\Pi_S(q, \omega)$. That results in[71,76]

$$\chi(q,\omega) = \frac{\Pi_S(q,\omega)}{1 - \left[v(q) - \dfrac{2\omega^2 A_q^2}{\omega_q(\omega^2-\omega_q^2)}\right]\Pi_S(q,\omega)} \quad . \tag{46}$$

Π_S is evaluated using H_S. The separation indicated in (43) is obvious for the H_S arising from (22) and from (24) provided K = 0. Then

$$\begin{aligned} H_0 &= -\sum_i \Gamma S_i^x \\ v(q) &= J(q) \end{aligned} \tag{47}$$

with J(q) given in (8), (28) respectively. The case $K \neq 0$ can be accommodated in the scheme by approximating the four spin term by an effective pair interaction so that in (28), $L \to L + 6\,\Sigma\, K \langle S^z S^z \rangle$.

Quite crude approximations for $\Pi_S(q, \omega)$ are satisfactory in some circumstances. Among the simplest is mean field approximation. (This will give the correct mixed modes, but will not be adequate for subtler effects like damping, or the Peierls transition at T = 0). Π_S is then the single site susceptibility[76,73,72], in mean field theory for the system (43), (47). This approximation yields[40,72,70,78]

$$\Pi_S(q,\omega) = \frac{\Gamma \langle S^x \rangle}{\gamma^2 - \omega^2} + \delta(\omega)\cos^2\theta\left(\tfrac{1}{4} - R^2\right) \tag{48}$$

with R, γ, cos θ, $\langle S^x \rangle$ as in §3 (i). The zero-frequency part of Π_S vanishes at and above the transition but has important consequences below. It corresponds to longitudinal spin fluctuations and arises because below the transition $\hat{\underline{z}}\, S^z$ has a component along the molecular field direction.

For more complicated situations, e.g. Jahn-Teller systems with many levels, the corresponding approximation for Π_S is obtained by diagonalising the full single ion Hamiltonian H_S treating in the mean field approximation the term bilinear in the electronic operator[76,82,81]. In cases with degenerate levels, the zero frequency part of Π_S need not vanish at the transition if S^z (or more generally the operator to which the distortion mode couples) has non-zero matrix elements between the degenerate levels, as in Tb VO_4 or in the singlet ground state system Pr_3Tl. In such cases the zero frequency part of Π_S can cause a divergence as the transition is approached[76,83,84], leading to a "central peak" at discussed in Section 5.

(iii) Dynamic Properties

In a Raman experiment in which the coupling to the light is through the operator

$$P_1 S^z + P_2 Q \tag{49}$$

(with P_1, P_2 polarizability coefficients) the scattered intensity is proportional to

$$I(\omega) = \frac{1}{1 - e^{-\beta\omega}}\,\mathrm{Im}\left[P_1^2 \langle\langle S_q^z S_q^z \rangle\rangle + P_1 P_2 \{\langle\langle S_q^z Q_q \rangle\rangle + \langle\langle Q_q S_q^z \rangle\rangle\} + P_2^2 \langle\langle Q_q Q_q \rangle\rangle\right].$$

In the Raman case, q is effectively zero. Similar Green functions, but at finite q, occur for inelastic neutron scattering where in the Jahn-Teller case the corresponding P_1 is very small unless the pseudo-spins have an associated magnetic moment.

In a coupled system all these Green functions share the same poles (where the denominator of (46) vanishes) which determine the mode frequencies and Raman and neutron response. Inserting Π_S as given by (48) gives zero frequency modes below the transition (associated with order-parameter fluctuations) and also coupled spin-phonon modes[44,50,35] whose frequencies satisfy

$$(\omega^2-\omega_q^2)[\omega^2-(\gamma^2-\Gamma\langle S^x\rangle J(q))] = 2\Gamma\langle S^x\rangle\,\omega^2 A_q^2/\omega_q\;. \tag{50}$$

If the mixing is small, one mode is phonon-like and the other mode (the 'vibron' in Jahn-Teller systems) is close to the uncoupled 'spin wave' solution

$$\omega_S(q)^2 = \gamma^2 - \Gamma\langle S^x\rangle J(q) \tag{51}$$

(except where the modes cross). The wave-vector, and temperature-dependence of the 'spin wave' frequency[38] is shown schematically in Figure 7. The q ~ 0 vibron for Dy VO_4, Tb VO_4 has been identified in Raman scattering[44]. If only coupling to acoustic phonons is considered ω_q goes to zero at small q and the vibron solution of (50) is, for q = 0

$$\omega^2 = \gamma^2 - \lambda\Gamma\langle S^x\rangle \tag{52}$$

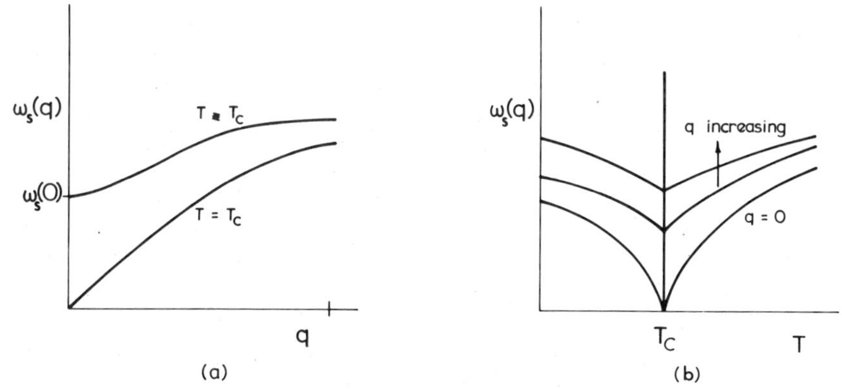

Figure 7. The (uncoupled) spin wave or tunnelling mode frequency ω_S as a function of (a) wave vector, (b) temperature.

where λ arises from the self-energy term in J(q) (the other part being cancelled by $2A_q^2/\omega_q$ from the right hand side); optic phonon contributions can also occur in λ. The temperature-dependence resulting from $\langle S^x \rangle$ above the transition is seen in the experiments.

Inelastic neutron scattering[85,83,84] on Pr Al O_3, Tb VO_4 and Tm VO_4 shows important characteristics of the dispersion relations given by (50), including the vibron-phonon mixing at finite q, and the acoustic mode softening near the transition, as depicted for Pr Al O_3 in Figure 8.

The Raman spectra[87-92] for KDP and its isomorphs show the corresponding coupled spin-phonon modes (at $q \sim 0$), and the coupled spin-phonon model allows a satisfactory fit[4] to the peak positions below T_C. In deuterated KDP ($\Gamma \sim 0$), the spin (proton) mode is not seen: from (46) and (48) it can be seen that it should have zero intensity for $\Gamma \sim 0$.

The influence of pressure allows a clear identification of the spin mode in tunnelling ferroelectrics[93-97]. Since Γ increases with pressure in a roughly linear way[62], the pressure-dependence of the Raman spectra also allows a test of results such as (50). For KDP it has been shown[96,98] that the observed pressure-dependence is in rough agreement with the model (allowing also for the fact that J decreases slightly with pressure). In particular the uncoupled spin mode frequency at fixed $T < T_c$ falls with increasing Γ (i.e. decreasing $T_c(\Gamma)$) in the way expected from (51). However, the pressure-dependence of the spin mode frequency in both phases appears to be less than is required for consistency

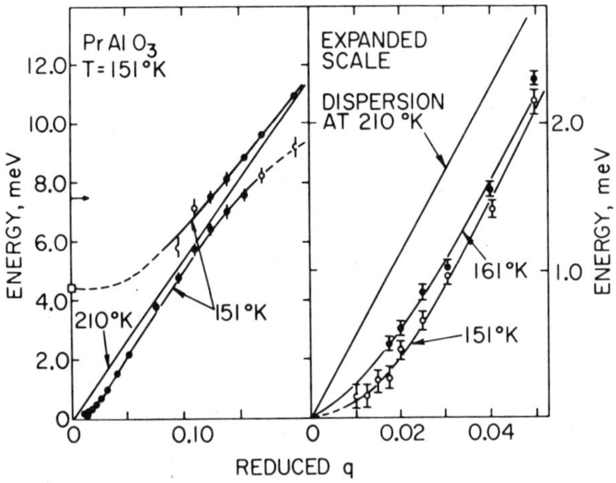

Figure 8. Inelastic neutron scattering results[85] for Pr Al O_3, showing the mixed modes and the acoustic mode softening as $T \to T_c$

with static measurements. The temperature-dependence of the uncoupled spin mode frequency in the paraelectric regime also agrees with the model result (51), as can be seen from Figure 9.

An important aspect of the pressure-dependence which makes it possible to obtain accurate spin mode frequencies in the paraelectric phase is that while the response is overdamped in the paraelectric phase for all KDP-type crystals at atmospheric pressure, application of moderate pressures makes the spin mode underdamped in KDP. Similar behaviour is seen in Rb D P[93,94] and in KDA[99]. The reason for the reduction of damping with increased Γ will be briefly discussed in §3 (iv).

The application of a magnetic field to Tm VO_4 or Tm As O_4 gives rise to the otherwise absent transverse field term. The field-dependence of the frequency and intensity of the resulting coupled mode response (analogous to its pressure-dependence in KDP) has been investigated by Raman and neutron methods[100,86].

One important effect of the coupling is on the elastic constants. From (46) and (44) it is straightforward to obtain

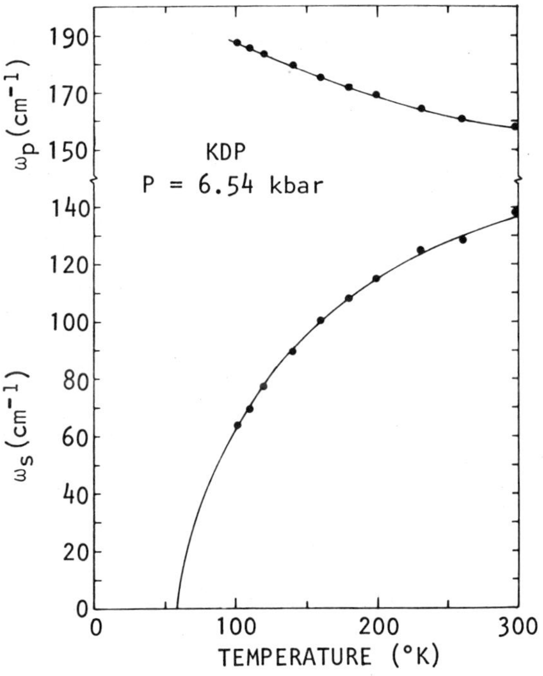

Figure 9. Raman results[96] for the temperature-dependence of the uncoupled mode frequencies in KDP. The lower full line is a fit using (51).

$$\frac{\omega^2}{\omega_q^2} = \frac{1 - v(q)\Pi_s}{1-(v(q)-\mu_q)\Pi_s} = \frac{1}{1+\mu_q\chi_s} \tag{53}$$

where $\mu_q = 2A_q^2/\omega_q$, and χ_S is the dynamic susceptibility for the effective spin system. Assuming that A_q represents coupling to an acoustic phonon such that μ is the limit of μ_q as q goes to zero, $[1 + \mu\,\chi_S(q,\omega)]^{-1}$ then gives the modification[101] and frequency dependence[102] of the elastic anomaly. This has been used to fit the elastic anomaly both in KDP[103,4] and in Dy VO_4 and Tb VO_4[102]. In the case of Tb VO_4 the (damped) zero-frequency part of Π_S appears to distinguish its behaviour from that of Dy VO_4, though this explanation is not supported by experiments[104] on Pr $Al\,O_3$.

For simple considerations it is sometimes adequate to treat the dynamics starting from the approximately transformed Hamiltonians (25), (27). The excitations are then the uncoupled phonon and pseudo-spin wave modes, whose frequencies ω_q and $\omega_S(q)$ can be obtained directly from the approximate Hamiltonians or, as was used to obtain (51) above, by ignoring the right hand side of (50). The approximate Hamiltonians also give rise to the zero frequency longitudinal spin fluctuations referred to below (48).

(iv) Damping

The theoretical discussion so far given has led to undamped modes. Though some damping effects are lost in the approximation used to get (46), the more severe approximation was the use of mean field approximation for Π_S. A proper evaluation of Π_S can include all the dominant damping effects other than those that rely on resonant spin-phonon aspects. For the simpler Jahn-Teller cases, or for KDP when the four spin term is approximated by an effective pair interaction, H_S is the transverse Ising Hamiltonian. For this model, Im Π_S, which gives the damping, has been considered using the Blume-Hubbard technique[43], by diagrammatic methods[40,41,78], or within a phenomenological approach[3]. In situations where mean field theory is a suitable first approximation the leading corrections which yield damping are associated with the break up of an excitation into (i) a pseudo-spin wave and a spin fluctuation or (ii) a pair of spin waves[41,78]. Only the first process contributes to Π_S above the transition, but both contribute below because of the mixed character of the excitations.

The resulting expressions are very complicated but simplify in various regimes. For example, for $T > T_C$ and ω near the spin wave frequency $\omega_S(q)$, the Green function χ_S for the transverse Ising model (that is to say, neglecting the mixing term in A_q^2 in (46)) becomes[41]

$$\chi_s(q,\omega) \propto [(\omega^2 - \omega_s(q)^2)^2 + (2\omega\Delta_q)^2]^{-1} \tag{54}$$

where

$$\Delta_q = \frac{\pi}{4} J(q)\, \mathrm{sech}^2 \tfrac{1}{2}\beta\Gamma \frac{\Gamma^2}{2\omega_s(q)^2} \sum_{q'} J(q')\,\delta(\omega_s(q') - \omega_s(q)). \tag{55}$$

Δ_q is a complicated function of ω for ω not near $\omega_s(q)$.

In the underdamped situation where these results apply, the ratio of damping Δ_q to frequency is, above the transition, of order $(J(o)/z\,\Gamma)\coth\frac{1}{2}\beta\Gamma$ (with z the coordination number), so that underdamping occurs for large Γ, in agreement with the pressure dependences[87,87,92] referred to in §3 (iii). The reason is that for Γ large 'spin precession about the external field' (i.e. tunnelling) dominates the interaction processes produced by J. The increase of damping with temperature (from the $\coth\frac{1}{2}\beta\Gamma$ factor) is due to the increase with temperature of the density of states into which the excitation can decay: that density of states is determined by the inverse of the spin wave band width, and at high temperatures the band width is proportional to $\langle S^x \rangle$, from (51).

In the more strongly damped situations which occur at higher T or lower Γ it is necessary to allow for the damping of the excitations into which the mode decays, and the self-consistent treatment of such effects[43] leads to a high temperature width typically of order $J(0)/4\,z^{\frac{1}{2}}$ when Γ and $J(0)$ are comparable. The first principles theory of overdamping in the pseudo-spin model is still at a somewhat primitive stage but it would clearly be desirable to compare a more detailed analysis with the most reliable experimental parameter $\tau^{-1} \equiv \omega^2/\Delta$.

For situations where the mode mixing has important (e.g. resonant) effects on the damping it is not possible to consider the damping within the transverse Ising model. Phenomenological discussions have been given[102,105], and also perturbation analyses[74,75,34] starting from the Hamiltonian (24). The damping for the Hamiltonian (26) has been treated with the Blume-Hubbard method[106] and also by a diagrammatic perturbation method[78], omitting the four spin term.

The above discussion applies only to damping intrinsic in the model. Damping can also arise from random strains, from domain boundaries, or from various types of substitutional disorder. We briefly discuss the first and last of these sources in the next section.

4. MIXED AND DILUTE SYSTEMS

Substitutionally disordered ferroelectrics and cooperative Jahn-Teller systems have been the subject of recent study. Consideration has been given to the effects of both dilution (partial substitution of an active component by an inactive one) and of mixing (partial substitution by another active component, with a different transverse field and interaction). Examples[107,108] of the dilute systems are $Tb_c\ Gd_{1-c}\ VO_4$, $Dy_c\ Y_{1-c}\ VO_4$, and an example[109,110] of the mixed system is $K(H_c\ D_{1-c})_2\ PO_4$ (partly deuterated KDP).

The simplest approach to such systems is the virtual crystal approximation, which is the disorder analogue of mean field theory for thermodynamics. For the transverse Ising model that allows an immediate treatment of static quantities in mean field theory (§3 (i)) by the replacements

$$(\Gamma/\gamma)^2 \rightarrow \langle(\Gamma/\gamma)^2\rangle_c \,, \qquad J\langle S^z\rangle \rightarrow \langle J\langle S^z\rangle\rangle_c \tag{56}$$

where the configurational average $\langle\ \rangle_c$ denotes the weighted mean, with probabilities c, 1-c of the values for pure systems composed solely of one or other of the two components. Quite simple prescriptions allow virtual crystal approximation to be extended to discuss the dynamic behaviour[108,111]. These results can be recovered within a simple scheme[72] for the transverse Ising model which automatically includes the dynamics and which can be readily generalised to deal with the coupled spin phonon system, or to allow more sophisticated treatments, such as CPA, of the disorder.

For simplicity we consider the case $T > T_c$ (deferring until the end the generalisation to $T \leqslant T_c$).

In the Hamiltonian (29) Γ is now taken to be site dependent ($\Gamma \rightarrow \Gamma_i$) and the exchange to depend in the following way on the types of spins occupying sites i, j (i.e. through μ_i, μ_j only):

$$J_{ij} \rightarrow \mathcal{J}_{ij} \equiv \mu_i J_{ij} \mu_j \,. \tag{57}$$

This replacement is exact for the dilute case, and appears to be adequate[98] for partially deuterated KDP. μ_i is like an effective dipole moment in the ferroelectric case.

The equation of motion for the Green function $\chi \equiv \langle\langle \mu_i S_i^z ; \mu_j S_j^z \rangle\rangle$ can be decoupled in random phase approximation[38] above the transition to yield

$$\chi_{ij} = \phi_i \delta_{ij} - \phi_i \sum_k \mathcal{J}_{ik} \chi_{kj} \tag{58}$$

where

$$\phi_i = \frac{\Gamma_i \mu_i^2 \langle S_i^x \rangle}{\omega^2 - \Gamma_i^2} . \tag{59}$$

This is precisely equivalent to (44), with Π_S replaced by $-\phi_i$, which is the mean field approximation (48), $T > T_c$, with the allowance of site-dependence. The disorder now occurs entirely through the site-diagonal parameter ϕ_i. The replacement of this by its mean value ϕ is virtual crystal approximation:

$$\begin{aligned}\phi_i \to \phi(\omega) &= \langle \phi_i \rangle_c \\ &= \frac{c\Gamma_A \mu_A^2 \frac{1}{2}\tanh\frac{1}{2}\beta\Gamma_A}{\omega^2 - \Gamma_A^2} + \frac{(1-c)\Gamma_B \mu_B^2 \frac{1}{2}\tanh\frac{1}{2}\beta\Gamma_B}{\omega^2 - \Gamma_B^2}\end{aligned} \tag{60}$$

where c is the concentration of species A, with parameters μ_A, Γ_A and (1 - c) is the concentration of species B. This gives

$$\chi \to \chi_q = \frac{\phi}{1 + \phi J(q)} . \tag{61}$$

(i) Statics

The transition is located by the divergence of the q = 0 static susceptibility, that is, where[72,111-113]

$$0 = 1 + J(0)\phi(0) \tag{62}$$

with ϕ given by (60). The shape of the resulting curve of transition temperature versus concentration depends on whether ordering is possible in both or just one of the two pure systems (pure A system, pure B system)[72]. The theory gives a satisfactory fit to the concentration-dependence of the transition temperature of $Dy_c Y_{1-c} VO_4$, apart from its neglect of percolation effects, and the appropriate generalisation of the theory to the four level system appropriate to $Tb VO_4$ accounts well[108] for $Tb_c Gd_{1-c} VO_4$, as shown in Figure 10. In these cases $\mu_B = 0$ since the Gd and Y ions are not Jahn-Teller active.

In the corresponding description of the static behaviour of partially deuterated KDP, Γ_B can be taken as effectively zero. The theoretical results[98] for $T_c(c)$ and C(c) (where C is the Curie constant) agree satisfactorily with experiment[109].

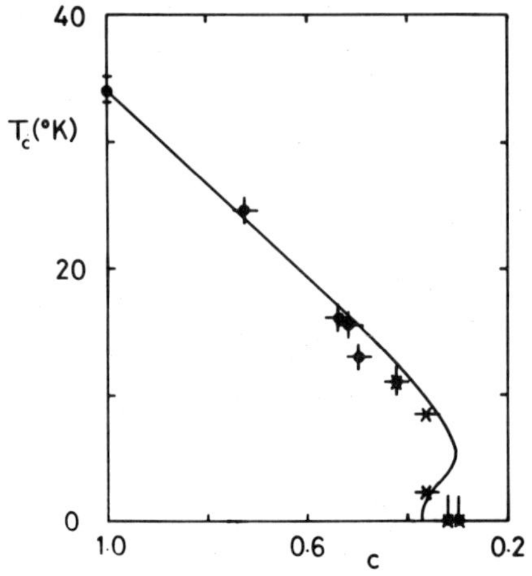

Figure 10. Measured concentration dependence of the transition temperature in $Tb_c\, Gd_{1-c}\, VO_4$. The full line is from virtual crystal theory[108].

(ii) Dynamics

Virtual crystal approximation as set out above also gives the dynamic behaviour: the excitation frequencies are given by

$$1 + \phi(\omega) J(q) = 0. \tag{63}$$

Provided none of c, (1 - c), μ_A, μ_B, Γ_A, Γ_B are zero, (63) is a quadratic equation in ω^2 with solutions $\omega_q^{(+)2}$ and $\omega_q^{(-)2}$ (with $\omega_q^{(+)} > \omega_q^{(-)}$). In virtual crystal approximation these excitations are undamped, and Im χ is then a sum of δ- functions centred at $\pm\, \omega_q^{(+)}$, $\pm\, \omega_q^{(-)}$. Defining the integrated intensities $I_q\, \pm$ of these modes to be the residues at the poles gives

$$I_q^{\pm} = \frac{\pm\pi\, (\Gamma_A^2 - \omega_q^{(\pm)2})(\omega_q^{(\pm)2} - \Gamma_B^2)}{2J(q)\, \omega_q^{(\pm)2}\, (\omega_q^{(+)2} - \omega_q^{(-)2})}\,. \tag{64}$$

It is easy to check that in the limit $c \to 0$ the mode frequencies become $\omega \sim \omega_B, \omega \sim \Gamma_A$ where ω_B is the pseudo-spin wave frequency for the pure B system, and that their respective intensities are O(1), O(c); analgous results apply near c = 1.

For $\Gamma_B = 0$ or $\mu_B = 0$ (as apply for deuterated KDP, or in diluted cases) only the A-like excitation occurs, and its intensity varies roughly like c. Related behaviour is seen in the experiments[108,110].

(iii) Generalisations

For a more complete understanding particularly of the dynamic behaviour[110] in partially deuterated KDP, the phonon coupling has to be introduced into (58) in a manner analogous to the way (46) generalises (44). The resulting generalisation[98,111] gives a more complete understanding of the observed concentration dependences of the Raman data[110], including the combined pressure and deuteration dependences. An example of this is shown in Figure 11 where the resulting theoretical expressions are compared to the concentration dependences of the observed logarithmic pressure derivatives of the low temperature mixed mode frequencies.

Intrinsic damping of the excitations can also be included[72], through a generalisation of $\phi_i(\omega)$ in (58), (59).

In order to treat damping caused by the disorder the CPA generalisation[72,114] of virtual crystal approximation is required. That leads to the replacement of $\phi(\omega)$ in the preceding discussion by $\tilde{\phi}(\omega)$ where

$$\tilde{\phi} = \langle \phi_i / [1 + (\phi_i - \tilde{\phi}) g] \rangle_c$$
$$g(\tilde{\phi}) \equiv \frac{1}{N} \sum_q J(q) / [1 + \tilde{\phi} J(q)]. \qquad (65)$$

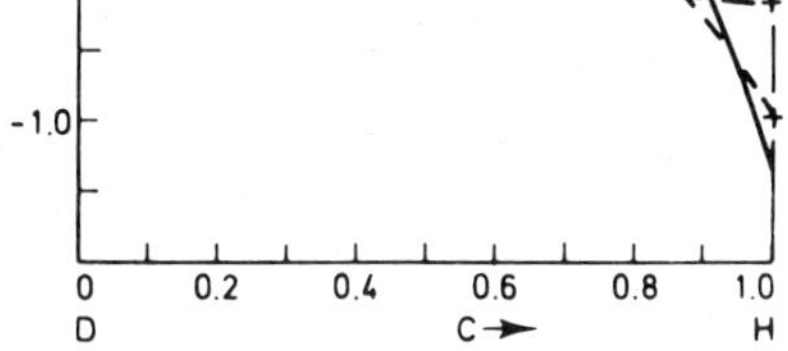

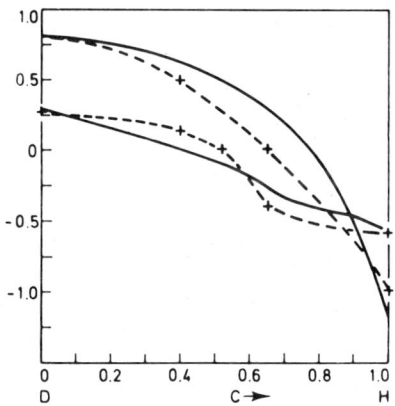

Figure 11. Experimental[110], and theoretical[98] results (full lines), for the logarithmic pressure-derivatives $\partial \ell n\, \omega_\pm / \partial p$ of the frequencies of the low temperature q = 0 mixed modes in $K(H_c D_{1-c})_2 PO_4$.

g is small $(0(^1/z))$ in mean field-like systems, and its neglect gives back the virtual crystal approximation (60). Expanding $\tilde{\phi}$ to first order in g yields the leading contribution of configurational fluctuations to the damping, through

$$\operatorname{Im}\tilde{\phi} \sim -\langle(\phi_i-\phi)^2\rangle_c \operatorname{Im} g(\phi) \tag{66}$$

(As was noted earlier, the damping arises from Im Π, which is here - Im $\tilde{\phi}$). The result (66) is proportional to the density of virtual crystal states (whose energies $\omega^{(\pm)}$ were discussed above). This simple extension of the virtual crystal approximation can give a satisfactory account of disorder damping where it is small. In other cases the self consistent CPA equations (65) have to be solved.

The above discussion considered the disorder resulting from substitution. Random strains also have similar effects, in that they lead to a transverse field Γ varying from site to site, and a resulting damping. That can be treated as above, but with $\langle\ \rangle_c$ interpreted as an averaging over the probability distribution for Γ. In particular (66) will then yield the damping arising from this source.

To generalise the results of this section to $T \leqslant T_c$, ϕ_i has to be replaced by[98,115]

$$\phi_i = \Gamma_i \mu_i^2 \langle S_i^x\rangle / [\omega^2 - \gamma_i^2] \tag{67}$$

where

$$\gamma_i^2 = \Gamma_i^2 + z_i^2 , \qquad z_i = \sum_j \mu_i J_{ij} \langle \mu_j S_j^z\rangle \tag{68}$$

(compare (48)). Virtual crystal approximation is then

$$\phi_i \to \phi = \langle\phi_i\rangle_c , \qquad z_i \to z = \langle z_i\rangle_c \tag{69}$$

which is equivalent in the static case to (56).

5. THE CENTRAL PEAK; CRITICAL BEHAVIOUR

(i) The Central Peak

In the discussion after (48) it was stated that when degenerate levels contribute to Π_S, a zero frequency response can arise; that implies[76,81,83,84] an elastic peak in the Raman or neutron response, or a dynamic one if the levels are weakly broadened by residual interactions. Such a central peak appears to have been seen in Tb VO_4[83,84].

Another mechanism for the occurrence of a central component is exhibited by the zero-frequency part of (48), which is due to the z-component of spin having a longitudinal component (with respect to the mean field direction) below the transition. Beyond molecular field theory that longitudinal component will relax and give rise to a central peak of finite width. However, this central peak only occurs in χ below the transition. (It could also appear in the Raman response above the transition if the operator S^x were also involved in the Raman operator (49) or in the phonon coupling in the Hamiltonian).

There remains the question how narrow such a peak would be. The pseudo-spin longitudinal mode just considered actually relaxes through decay into two pseudo-spin waves (§3 (iv)), which at low frequency and wave vector corresponds to a fluctuation in spin wave density. This process is analogous to the origin of the relaxational mode proposed for the anharmonic phonon model by Cowley[116-118] which relies on coupling to fluctuations in phonon density and can occur in piezoelectric non-pyroelectric crystals above T_C. In both models damping and dispersion of the decay modes (spin waves or phonons respectively) will normally give a relatively large relaxational mode damping, though this appears to be consistent with the broader peaks sometimes observed[119].

A central mode can also occur in the pseudo-spin model from non-linear fluctuations in the order parameter[3] (analogously to proposals for the anharmonic phonon model[120,121]), and indeed from most mechanisms applying in the anharmonic phonon model. Among the most clearly established of these appears to be entropy fluctuations, which is consistent with the magnitude and wave-vector dependence of the width of some very narrow peaks[119], and appears to be responsible for the narrow but resolved central peak recently seen in KDP[122]. In addition, some types of defect structure[123,124] seem to play a role in some central peaks[119,125] (both static and dynamic) at low temperature and near T_C.

(ii) Critical Behaviour

The dependence of, for example, the order parameter $\langle S^z \rangle$ on $T - T_C$ near a second order phase transition is usually described by critical indices[126] such as β in

$$\langle S^z \rangle \propto (T-T_c)^\beta . \tag{70}$$

For the transverse Ising model, the static indices are known from series expansions[127,68]. Some exact results are also known. For example, for the transverse Ising model the critical behaviour with respect to the field Γ of the thermodynamic functions at $T = 0$ in a lattice of dimensionality d, is the same as the critical behaviour of the corresponding functions with respect to T in the d + 1 dimensional Ising model : this was proved[128-130] for d = 1, and indicated for d = 2, 3 by series[66,67] and subsequently confirmed[131] using standard renormalization group methods[132]. Thus for T = 0 the transverse Ising model has mean field exponents for d > 3. In addition it can be seen that for T ≠ 0, the transverse Ising model has the same critical indices as the Ising model with the same dimensionality[131].

The crossover near T = 0 is a quantum effect arising for non-commuting operators from the usual thermodynamic frequency sum becoming an integral for T = 0, and acting like an additional wave-vector component integral in the theory. It would be interesting to see some aspects of the crossover behaviour exhibited experimentally. The above discussion applies, however, only to the transverse Ising model with short range interactions, while most real transverse Ising-like systems have also long range interactions produced by strain coupling or dipole-dipole effects. This would result in mean field exponents sufficiently close to T_C[133], as appears to have been verified by observations of β for $Pr\,Al\,O_3$[134] and for $Tb\,VO_4$[135], though not for $Dy\,VO_4$[135].

REFERENCES

1. G.A. Gehring and K.A. Gehring, 1975, Rept. Prog.Phys. 38,1.
2. H. Thomas, 1974, Anharmonic Lattices, Structural Transitions and Melting, Ed. T. Riste (Noordhoff, Leiden) p. 213.
3. R. Blinc and B. Žekš, 1972, Adv.Phys. 21, 693
4. R.J. Elliott and A.P. Young, 1974, Ferroelectrics 7, 23.
5. J.F. Scott, 1974, Rev.Mod.Phys. 46, 83.
6. G. Shirane, 1974, Rev.Mod.Phys. 46, 437.
7. W. Rehwald, 1973, Adv.Phys. 22, 721.
8. P.A. Fleury, 1972, Comments Sol.St.Phys. 4, 149, 167.
9. P.A. Fleury, 1972, Proc.Int.Conf. Light Scattering in Solids, Ed. M. Balkanski, R.C.C. Leite, and S.P.S. Porto, (Flammarion, Paris) p. 747.
10. Y. Ishibashi and Y. Takagi, 1976, J. Appl.Phys. (Japan) 15,1621.
11. H.R. Zeller, 1973, Festkörperprobleme 13, 31.
12. H.R. Zeller and S. Strassler, 1975, Comments Sol.St.Phys. 7,17.
13. W. Dietrich, 1976, Adv.Phys. 25, 615.
14. L.R. Testardi, 1975, Rev.Mod.Phys. 47, 637.
15. L.R. Testardi, 1975, Comments Sol.St.Phys. 6, 131.
16. L.N. Bulaevskiĭ, 1975, Sov.Phys.: Uspekhi 18, 131.
17. Yu.A. Izyumov and E.Z. Kurmaev, 1976, Sov.Phys.: Uspekhi 19, 26.
18. S.P. Ionov, G.V. Ionova, V.S. Lubimov and E.F. Makarov, 1975, Phys. Stat. Solidi B71, 11.
19. B.A. Huberman, 1976, Comments Sol.St.Phys. 7, 75.
20. E.J. Samuelsen, E. Andersen, and J. Feder (eds.), 1971, Structural Phase Transitions and Soft Modes (Oslo, Universitetsforlaget).
21. T. Riste (ed.), 1974, Anharmonic Lattices, Structural Transitions and Melting (Noordhoff, Leiden).
22. S. Hirotsu, 1977, J. Phys. C. 10, 967.
23. Y. Yamada, 1977, "Electron-Phonon Interaction and Charge Ordering in Insulators" - in this volume.
24. J. Kanamori, 1960, J. Appl. Phys. 31, 145.
25. E. Pytte, 1971, Phys.Rev. B3, 3505.
26. E. Callen and H.B. Callen, 1965, Phys.Rev. A139, 455.
27. H.H. Chen and P.M. Levy, 1971, Phys.Rev.Lett. 27, 1383.
28. J. Sivardière and M. Blume, 1972, Phys.Rev. B5, 1126.
29. R.J. Elliott, 1971, in Structural Phase Transitions and Soft Modes, ed. E.J. Samuelsen, E. Andersen and J. Feder (Oslo, Universitetsforlaget).
30. E. Pytte, 1973, Phys.Rev. B8, 3954.
31. J.P. Harrison, J.P. Hessler and D.R. Taylor, 1976, Phys. Rev. B14, 2979.
32. J.T. Folinsbee, J.P. Harrison, D.B. McColl and D.R. Taylor, 1977, J. Phys. C10, 743.
33. J.R. Fletcher and F.W. Sheard, 1971, Solid St. Comm. 9, 1403.
34. N.E. Buttery, 1974, D.Phil.Thesis, Oxford.

35. R.J. Elliott, 1971, Proc.Int.Conf.Light Scattering in Solids, ed. M. Balkanski (Flammarion Press, Paris) p.354.
36. F.S. Ham, 1972, in Electron Paramagnetic Resonance, ed. S. Geschwind (Plenum Press) p. 1.
37. Y.L. Wang and B. Cooper, 1968, Phys.Rev.172, 539.
38. R. Brout, K.A. Müller and H. Thomas, 1966, Solid St.Comm. 4, 507.
39. P. Pfeuty, 1971, Ph.D. Thesis, Paris.
40. R.B. Stinchcombe, 1973, J. Phys. C6, 2459.
41. R.B. Stinchcombe, 1973, J. Phys. C6, 2484.
42. R.B. Stinchcombe, 1973, J. Phys. C6, 2507.
43. M.A. Moore and H.C.W.L. Williams, 1972, J. Phys. C5, 3168,3185.
44. R.J. Elliott, R.T. Harley, W. Hayes and S.R.P. Smith, 1972, Proc.Roy.Soc. A328, 217.
45. E. Pytte and K.W.H. Stevens, 1971, Phys.Rev. Lett. 27, 862.
46. D.K. Ray and A.P. Young, 1973, J. Phys. C6, 3353.
47. P.G. de Gennes, 1963, Solid St. Comm. 1, 132.
48. Y. Yamada and T. Yamada, 1966, J.Phys.Soc. Japan 21, 2167.
49. Y. Yamada, Y. Fujii and I. Hatta, 1968, J. Phys. Soc. Japan, 24, 1053.
50. K.K. Kobayashi, 1968, J. Phys. Soc. Japan 24, 497.
51. A.R. Bishop and J.A. Krumhansl, 1975, Phys. Rev. B12, 2824.
52. W. Cochran, 1961, Adv.Phys. 10, 401.
53. J.C. Slater, 1941, J. Chem. Phys. 9, 16.
54. Y. Takagi, 1948, J. Phys. Soc. Japan 3, 271.
55. E. Pytte, 1973, Comments Sol.St.Phys. 5, 41.
56. E. Pytte, 1973, Comments Sol.St.Phys. 5, 57.
57. E. Pytte, 1972, Phys.Rev.Lett. 28, 895.
58. P.J. Becker, M.J.M. Leask and R.N. Tyte, 1972, J.Phys. C5, 2027.
59. A.H. Cooke, S.J. Swithenby and M.R. Wells, 1972, Solid St. Comm. 10, 265.
60. J.H. Colwell and B.W. Mangum, 1972, Solid St. Comm. 11, 83.
61. B.W. Mangum, J.N. Lee, and H.W. Moos, 1971, Phys.Rev.Lett. 27, 1517.
62. G.A. Samara, 1970, J.Phys.Soc. Japan 28, Suppl.,Proc. Second Int. Meeting on Ferroelectricity p. 399.
63. G.A. Samara, 1971, Phys.Rev.Lett. 27, 103.
64. R. Blinc, S. Svetina and B. Žekš, 1972, Solid St.Comm. 10, 387.
65. W.J. Benepe and W. Reese, 1971, Phys.Rev. B3, 3032.
66. R.J. Elliott and P. Pfeuty, 1971, J.Phys. C4, 2370.
67. R.J. Elliott and C. Wood, 1971, J. Phys. C4, 2359.
68. R.J. Elliott and I.D. Saville, 1974, J.Phys. C7, 4293.
69. L. Lam and A. Bunde, 1977, J.Phys. C10, 693.
70. S. Takada, I. Ohnari, H. Kurosawa, and Y. Ohmura, 1975, Prog. Theor. Phys. 53, 936.
71. A.P. Young, 1975, J.Phys. C8, 3158.
72. E.J.S. Lage and R.B. Stinchcombe, 1976, J. Phys. C9, 3295.
73. W.J.L. Buyers, T.M. Holden and A. Perreault, 1975, Phys. Rev. B11, 266.
74. F.W. Sheard and G.A. Toombs, 1971, J.Phys. C4, 315.

75. G.A. Toombs and F.W. Sheard, 1973, J.Phys. C6, 1467.
76. A.P. Young, 1976, Proc. Int. Conf. Light Scattering in Solids, ed. M. Balkanski, R.C.C. Leite, and S.P.S. Porto (Flammarion, Paris) p. 817.
77. V.G. Vaks, A.I. Larkin, and S.A. Pikin, 1968, Sov.Phys. JETP 26, 188.
78. S. Takada, I. Ohnari, H. Kurosawa and Y. Ohmura, 1976, Prog.Theor.Phys. 55, 989
79. B. Žekš and F.C. de Sá Barreto, 1976, Proc.Int.Conf. Light Scattering in Solids, ed. M. Balkanski, R.C. Leite and S.P.S. Porto (Flammarion, Paris) p. 822.
80. F. Englert, 1963, Phys.Rev. 129, 567.
81. S.R.P. Smith, 1976, Proc. Int. Conf. Light Scattering in Solids, ed. M. Balkanski, R.C.C. Leite, and S.P.S. Porto (Flammarion, Paris) p. 329.
82. E. Pytte, 1973, Phys.Rev. B8, 3954.
83. M.T. Hutchings, R. Scherm, S.H. Smith and S.R.P. Smith, 1975, J.Phys. C8, L393.
84. M.T.Hutchings, R. Scherm, and S.R.P. Smith, 1975, AIP Conference Proceeding No. 29, 372.
85. J.K. Kjems, G. Shirane, R.J. Birgenau, and L.G. Van Uitert, 1973, Phys.Rev.Lett. 31, 1300.
86. J.K. Kjems, W. Hayes and S.H. Smith, 1975, Phys.Rev.Lett. 35, 1089.
87. C.M. Wilson and H.Z. Cummins, 1971, Proc.Int.Conf. Light Scattering in Solids, ec. M. Balkanski (Flammarion, Paris) p. 420.
88. R.S. Katiyar, J.F. Ryan, and J.F. Scott, 1971, Phys.Rev. B4, 2835.
89. I.P. Kaminov, and T.C. Damen, 1968, Phys.Rev.Lett. 20, 1105.
90. R.A. Cowley, G.J. Coombs, R.S. Katiyar, J.F. Ryan, and J.F. Scott, 1971, J.Phys. C4, L203.
91. C.Y. She, T.W. Broberg, L.S. Walland, and D.F. Edwards, 1972, Phys.Rev. B6, 1847.
92. N. Lagakos, and H.Z. Cummins, 1974, Phys.Rev. B10, 1063.
93. P.S. Peercy, and G.A. Samara, 1973, Phys.Rev. B8, 2033.
94. P.S. Peercy, 1973, Phys.Rev.Lett. 31, 379.
95. P.S. Peercy, 1975, Comments Sol.St.Phys. 7, 37.
96. P.S. Peercy, 1975, Phys.Rev. B12, 2725.
97. P.S. Peercy, 1976, Proc.Int.Conf. Light Scattering in Solids, ed. M. Balkanski, R.C.C. Leite, and S.P.S. Porto, (Flammarion, Paris) p. 782.
98. E.J.S. Lage and R.B. Stinchcombe, 1976, J.Phys. C9, 3681.
99. R.C. Leung, W.B. Spillman, N.E. Tornberg, and R.P. Lowndes, 1976, Proc.Int.Conf. Light Scattering in Solids, ed. M. Balkanski, R.C.C. Leite,and S.P.S. Porto (Flammarion,Paris) p. 796.
100. R.T. Harley, W. Hayes and S.R.P. Smith, 1972, J.Phys. C5,1501.
101. V. Dvorak, 1970, Czech. J. Phys. B20, 1.

102. J.R. Sandercock, S.B. Palmer, R.J. Elliott, W. Hayes, S.R.P. Smith and A.P. Young, 1972, J.Phys. C5, 3126.
103. E.M. Brody, and H.Z. Cummins, 1974, Phys.Rev. B9, 179.
104. P.A. Fleury, P.D. Lazay, and L.G. Van Uitert, 1974, Phys.Rev. Lett. 33, 492.
105. J. Feder and E. Pytte, 1973, Phys.Rev. B8, 3978.
106. M.A. Moore, and H.C.W.L. Williams, 1972, J.Phys. C5, 3222.
107. R.T. Harley, W. Hayes, A.M. Perry and S.R.P. Smith, 1974, Solid St.Comm. 14, 521.
108. R.T. Harley, W. Hayes, A.M. Perry, S.R.P. Smith, R.J. Elliott, and I.D. Saville, 1974, J.Phys. C7, 3145.
109. G.A. Samara, 1973, Ferroelectrics 5, 25.
110. P.S. Peercy, 1976, Phys.Rev. B13, 3945.
111. R. Blinc, R. Pirc, and B. Žekš, 1976, Phys.Rev. B13, 2943.
112. E. Shiles, G.B. Taggart, and R.A. Tahir-Kheli, 1974, J. Phys. C7, 1515.
113. U.Schmidt, 1974, Z. Physik. 267, 277.
114. R. Pirc and P. Prelovšek, 1976, Proc.Int.Conf. Light Scattering in Solids, ed. M. Balkanski, R.C.C. Leite, and S.P.S. Porto (Flammarion, Paris) p. 829.
115. R. Pirc and P. Prelovšek, 1977, J.Phys. C10, 861
116. R.A. Cowley, 1970, J.Phys.Soc.Japan Suppl. 28, 239.
117. G.J. Coombs and R.A. Cowley, 1973, J.Phys. C6, 121.
118. R.A. Cowley and G.J. Coombs, 1973, J.Phys. C6, 143.
119. P.A. Fleury and K.B. Lyons, 1976, Phys.Rev.Lett. 37, 161, 1088.
120. J. Feder, 1974, Anharmonic Lattices, Structural Transitions and Melting, ed. T. Riste (Noordhoff, Leiden) p.113.
121. R. Silberglitt, 1972, Solid St. Comm. 11, 247.
122. M.D. Mermelstein and H.Z. Cummins, to be published.
123. B.I. Halperin and C.M. Varma, 1976, Phys.Rev. B14, 4030.
124. J.A. Krumhansl and J.R. Schrieffer, 1976, Solid St.Comm. 17, 1515.
125. R.A. Cowley, J.D. Axe and M. Iizumo, 1976, Phys.Rev.Lett. 36, 806.
126. M.E. Fisher, 1967, Rept.Prog. Phys. 30, 615.
127. C. Domb, 1974, in Phase Transitions and Critical Phenomena, ed. C. Domb and M.S. Green (Academic Press, New York) p.357.
128. P. Pfeuty, 1970, Ann. Phys. 27, 79.
129. M. Suzuki, 1971, Phys.Lett. 34A, 94.
130. P. Pfeuty, 1976, J.Phys. C9, 3993.
131. A.P. Young, 1975, J. Phys. C8, L309.
132. K.G. Wilson and J. Kogut, 1974, Physics Reports 12, 75.
133. R.A. Cowley, 1976, Phys.Rev. B13, 4877.
134. R.J. Birgenau, J.K. Kjems, G. Shirane, and L.G. Van Uitert, 1974, Phys.Rev. B10, 2512.
135. R.T. Harley and R.M. MacFarlane, 1975, J.Phys. C8, L451.

THEORY OF JAHN-TELLER TRANSITIONS

H. Thomas

Institut für Physik, Universität Basel

Klingelbergstrasse 82, CH-4056 Basel, Switzerland

1. INTRODUCTION

In crystals containing ions with orbitally degenerate ground state, the interactions of the electron orbitals with the ligand displacements have a destabilizing effect on the ionic configuration: Jahn-Teller (JT) effect. This provides a mechanism for structural phase transitions in crystals which would be stable in the absence of electron-ion interaction. We give an introduction into the theory of such JT-induced transitions, with particular emphasis on a classification of the various types of JT-coupling and their dynamic characteristics. The dynamics of the cooperative JT effect manifests itself in bands of collective vibronic modes arising from transitions between low-lying vibronic levels of the JT complexes, and an analysis of the collective dynamic behaviour requires a thorough understanding of the vibronic excitations of a single JT complex.

Surveys of experimental and theoretical work on the JT effect can be found in Refs /1-4/. As a mechanism for structural phase transitions, the JT effect was first considered by Dunitz and Orgel /5/. Transitions of this type have been found in a number of compounds containing transition-metal or rare-earth JT ions. References to earlier work, notably on spinels and perovskites containing Cu^{2+}, Ni^{2+}, Cr^{2+} and Mn^{3+} ions, are given in Refs /6,7/. As more recent examples we mention the transitions observed in the Cu^{2+}-hexanitro compounds /8-11/, in $CsCu\ell_3$ /12-14/, in the rare-earth vanadates and arsenates with zircon structure /15-19/, in $PrA\ell O_3$ /19-21/, and in the intermetallic compound TmCd /22/. The theory of such JT transitions has been treated by various authors /6,7,16,18,23-29/.

Electron config.	d^1	d^2	d^3	d^4	d^5	d^6	d^7	d^8	d^9
Splitting of free ion term ^{2S+1}L by cubic crystal field (intermediate field)	^2D: E, T_2	^3F: A_2, T_2, T_1	^4F: T_1, T_2, A_2	^5D: T_2, E	^6S: A_1	^5D: E, T_2	^4F: A_2, T_2, T_1	^3F: T_1, T_2, A_2	^2D: T_2, E
Filling of single-electron crystal-field levels e_g, t_{2g} (strong field) High S: upper row Low S: lower row	e_g —, t_{2g} ↑; T_2	—, ↑↑; T_1	—, ↑↑↑; A_2	↑, ↑↑↑; E —, ↑↑↑↑; T_1	↑↑, ↑↑↑; A_1 —, ↑↑↑↑↑; T_2	↑↑, ↑↑↑↑; T_2 —, ↑↑↑↑↑↑; A_1	↑↑, ↑↑↑↑↑; T_1 ↑, ↑↑↑↑↑↑; E	↑↑, ↑↑↑↑↑↑; A_2	↑↑↑, ↑↑↑↑↑↑; E
Examples of 3d-ions	V^{4+} Ti^{3+}	Cr^{4+} V^{3+} Ti^{2+}	Mn^{4+} Cr^{3+} V^{2+}	Mn^{3+} Cr^{2+}	Co^{4+} Fe^{3+} Mn^{2+} Cr^{+}	Co^{3+} Fe^{2+} Mn^{+}	Ni^{3+} Co^{2+} Fe^{+}	Cu^{3+} Ni^{2+} Co^{+}	Cu^{2+} Ni^{+}

Table 1: Orbital ground states of d-ions in octahedral crystal field. The ground states in tetrahedral crystal field can be obtained by the following procedure:
Second row: Reverse order of levels, or interchange $d^n \leftrightarrow d^{10-n}$
Third row: Reverse order of levels, interchange $d^n \leftrightarrow d^{10-n}$, interchange electrons $\leftrightarrow$ holes.

2. DYNAMICS OF JT-SYSTEMS

2.1. Electronic Configuration

Our starting point is the electronic structure of a JT ion in the crystal field of the ligands fixed in their symmetric reference configuration. We consider specifically the case of a transition metal ion in a crystal field of cubic symmetry with

crystal-field energy > spin-orbit coupling

(intermediate or strong crystal field). Table 1 shows the orbital ground states for octahedral coordination. In tetrahedral crystal field, the order of the levels is reversed, and the ground states are related to those of the octahedral case by

$$|d^n, \text{tetrahedral}> = |d^{10-n}, \text{octahedral}>.$$

For intermediate crystal field (crystal-field energy $\lesssim$ intraatomic Coulomb energy), the ground state is found from the splitting of the free-ion term ^{2S+1}L by the crystal field (Table 1, second row).

Spin state	Symmetry of crystal field	E doublet	T_1 or T_2 triplet
High-Spin	Octahedral	d^4, d^9	d^1, d^2, d^6, d^7
	Tetrahedral	d^1, d^6	d^3, d^4, d^8, d^9
Low-Spin	Octahedral	d^7, d^9	d^1, d^2, d^4, d^5
	Tetrahedral	d^1, d^3	d^5, d^6, d^8, d^9

Table 2: JT configurations of d-ions in cubic crystal field (small spin-orbit coupling).

JT ions are expected for the configurations listed in the first two rows of Table 2.
For strong crystal field (crystal-field energy > intraatomic Coulomb energy), the crystal-field splitting of the single-electron d level into a t_{2g} triplet and an e_g doublet has to be taken into account before the Coulomb interaction. These levels are then filled successively under observation of Hund's maximum-spin rule: For (crystal-field energy < intraatomic exchange energy), Hund's rule holds for the whole d shell: High-spin case (Table 1, upper part of third row). In this case, one finds the same ground states as for intermediate crystal field (first two rows of Table 2). For (crystal-field energy > intraatomic exchange energy), Hund's rule holds only within each subshell: Low-spin case (Table 1, lower part of third row). In this case, JT-ions are expected for the configurations listed in the last two rows of Table 2.

In crystal fields of lower than cubic symmetry, the levels are generally split further. Twofold orbital degeneracy still occurs in tetragonal, trigonal and hexagonal symmetry. In lower symmetry, all orbital states are split into singlets.

We disregard spin-orbit interaction and consequently spin degeneracy, which is justified if (JT-coupling > spin-orbit interaction). In the opposite case (JT-coupling < spin-orbit interaction), the spin-orbit splitting of the crystal-field ground-state multiplet has to be taken into account before the coupling to the ligand motion. The situation is then qualitatively similar to the case of weak crystal field (crystal-field energy < spin-orbit interaction): For non-Kramers ions (even number of electrons) one can apply the same considerations to the total-angular momentum ground-state, and one finds the same types of ground-state degeneracies as in the orbital-momentum case. For Kramers ions (odd number of electrons), on the other hand, the ground-state degeneracy in the crystal field is usually reduced to that of a Kramers doublet, which shows no JT-coupling. The only exception is the case of a $G_{3/2}$ quartet state in cubic crystal field.

2.2. Vibronic Coupling

Next, we consider the coupling of the electrons orbitals at the JT-ion with the distortions of the ligand configuration. The dynamics of the complex are described by the electron-ion Hamiltonian:

$$H = H_{e\ell}\cdot 1_{ion} + H_{ion}\cdot 1_{e\ell} + V(x,Q) \tag{2.1a}$$

$$H_{e\ell} = \sum p^2/2m + V_{e\ell}(x) \,, \qquad H_{ion} = \sum P^2/2M \,. \tag{2.1b}$$

Here, (x,p) stand for the coordinates and momenta of the electrons in unfilled shells, and (Q,P) for the ionic normal coordinates of the complex and their conjugate momenta, with effective mass M.

The electronic operators are expressed as matrices in a fixed basis consisting of states $|\psi_i\rangle$ belonging to the reference configuration Q = 0:

$$\langle\psi_i|H_{e\ell}\cdot 1_{ion} + V(x,Q)|\psi_j\rangle = V_{ij}(Q) \,. \tag{2.2}$$

The vibrational amplitudes $\chi_i(Q)$ of the total vibronic wave function

$$\psi(x,Q) = \sum_i \chi_i(Q)|\psi_i\rangle \tag{2.3}$$

are approximately determined by the matrix Schrödinger equation

$$\left(H_{ion}\, \underline{\underline{1}}_{e\ell} + \underline{\underline{V}}(Q)\right)\cdot\underline{\chi}(Q) \equiv \underline{\underline{H}}_{JT}\cdot\underline{\chi}(Q) = E\,\underline{\chi}(Q) \tag{2.4}$$

in which $\underline{\underline{V}}(Q)$ serves as an effective potential-energy matrix for the ionic motion (generalized Born-Oppenheimer approximation). The electronic motion can in good approximation be restricted to the n-dimensional subspace spanned by the states $|\psi_i\rangle$ with excitation energies $\lesssim kT_c$. Usually, the cubic crystal-field splitting is larger than kT_c (for d-ions typically of the order $10^4 cm^{-1}$), such that in cubic symmetry only the ground-state multiplet has to be taken into account. In crystal fields of lower than cubic symmetry, on the other hand, the further splitting of the ground state can be comparable to kT_c, especially if the distortion from cubic symmetry is small. In such cases, the lowest excited states have to be included.

We introduce a complete set of n x n matrices

$$\underline{\underline{\xi}}_0 = \underline{\underline{1}}_{e\ell}\ ,\ \underline{\underline{\xi}}_1\ ,\ \underline{\underline{\xi}}_2 \tag{2.5}$$

with

$$\mathrm{tr}_{e\ell}\underline{\underline{\xi}}_\nu = 0 \quad (\nu\neq 0) \tag{2.6}$$

acting on the electronic states (electronic pseudospin operators), and express the electronic operators as linear combinations of the $\underline{\underline{\xi}}_\nu$,

$$\underline{\underline{V}}(Q) = \sum_\nu V_\nu(Q)\underline{\underline{\xi}}_\nu = V_0(Q)\ \underline{\underline{1}}_{e\ell} + \underline{\underline{V}}_{JT}(Q) \tag{2.7}$$

such that

$$\mathrm{tr}_{e\ell}\ \underline{\underline{V}}_{JT}(Q) = 0\ . \tag{2.8}$$

We obtain thus a separation of $\underline{\underline{V}}(Q)$ into a part $V_0(Q)\underline{\underline{1}}_{e\ell}$ which is independent of the electronic state (vibrational part), and a part $\underline{\underline{V}}_{JT}(Q)$ which vanishes on averaging over the electronic states (JT-part).

An expansion in powers of Q gives rise to the following classification of terms:
Vibrational part $\nu = 0$:

$$V_0(Q) = V_0^{(0)} - F\cdot Q + V_{harm}(Q) + V_{anh}(Q)\ . \tag{2.9}$$

The zeroth-order term is a constant and may be dropped. The second-order term is the harmonic part, and the higher-order terms are the anharmonic parts of the vibrational potential of the JT-complex. We further include a linear coupling to an external field F which will later be identified with the molecular field due to lattice-dynamical interactions with neighbouring complexes.

JT-parts $\nu \neq 0$:

$$\underline{\underline{V}}_{JT}(Q) = \underline{\underline{V}}_\Delta + \underline{\underline{V}}_{JT}^{(1)}(Q) + \underline{\underline{V}}_{JT}^{nonlin}(Q)\ . \tag{2.10}$$

The zeroth-order term describes the electronic ground-state splitting; it occurs only when excited electronic states have to be taken into account, and is thus not a JT term in the strict sense. The first-order term is the linear JT effect which plays the decisive role for the transitions considered, because it favours a distorted ligand configuration. The JT theorem states that for every reference configuration with a degenerate ground state (except for linear molecules and for Kramers degeneracy), there exists a vibrational mode Q for which symmetry permits

$$\underline{\underline{V}}_{JT}^{(1)}(Q) \neq 0 \,. \tag{2.11}$$

The higher-order terms describe nonlinear JT interactions.

We introduce symmetry-adapted vibrational coordinates Q_γ^Γ transforming as the irreducible representations Γ of the symmetry group of the JT complex, and the corresponding electronic operators $\underline{\underline{\mathcal{E}}}_\gamma^\Gamma$, such that

$$V_{harm}(Q) = \sum_\Gamma \frac{1}{2} M \omega_\Gamma^2 \sum_\gamma (Q_\gamma^\Gamma)^2 \tag{2.12a}$$

$$\underline{\underline{V}}_{JT}^{(1)}(Q) = -\sum_\Gamma A_\Gamma \sum_\gamma Q_\gamma^\Gamma \underline{\underline{\mathcal{E}}}_\gamma^\Gamma \tag{2.12b}$$

According to the JT theorem, at least one coupling constant A_Γ is allowed by symmetry. The possible types of JT coupling which can occur in crystals are listed in Table 3. Figure 1 shows the ligand configurations corresponding to the symmetry-adapted coordinates appearing in the linear JT effect. In the following, only the JT-active modes with a nonvanishing JT coupling will be taken into account.

Site Symmetry	Electronic State	Vibrational Coordinate	Vibronic Coupling
tetragonal	2-fold: E	1-fold: β_{1g}, β_{2g}	$E \otimes (\beta_1 + \beta_2)$
trigonal hexagonal cubic	2-fold: E	2-fold: ε_g	$E \otimes \varepsilon$
cubic	3-fold: T_1 or T_2	2-fold: ε_g 3-fold: τ_{2g}	$T \otimes (\varepsilon + \tau_2)$
cubic	4-fold: $G_{3/2}$ (spin-orbit)	2-fold: ε_g 3-fold: τ_{2g}	$G_{3/2} \otimes (\varepsilon + \tau_2)$

Table 3: The four types of linear Jahn-Teller coupling. The representations of the electronic states and of the vibrational coordinates are denoted by upper-case Roman letters and lower-case Greek letters, respectively.

The eigenvalues of $\underline{\underline{V}}(Q)$ form a potential-energy surface in Q-space consisting of n sheets, and the vibrational motion can be visualized as a motion in this surface. From the linear JT term together with the harmonic vibrational term, one obtains minima at

$$Q_{JT} \sim A/M\omega^2 \tag{2.13}$$

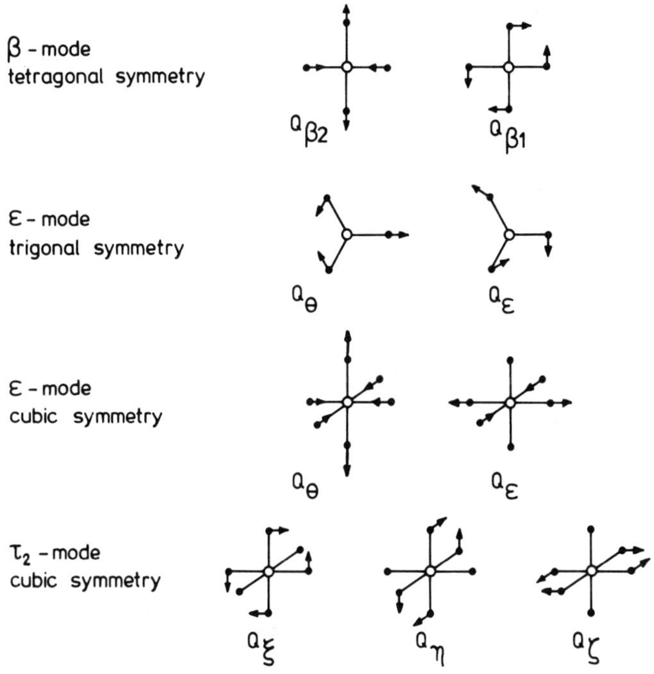

Figure 1: Ligand modes coupling to JT ions

with depth

$$E_{JT} \sim A^2/2M\,\omega^2. \qquad (2.14)$$

Thus, the JT effect favours a distorted state $Q = Q_{JT}$ over the undistorted state $Q = 0$ by an energy E_{JT} (JT destabilization energy). We are interested here in the case of strong JT effect where the JT destabilization energy is large compared with the vibrational zero-point energy, or equivalently, where the JT distortion is large compared with the vibrational zero-point amplitude:

$$E_{JT} \gg \frac{1}{2}\,\hbar\omega \; ; \qquad Q_{JT} \gg (2\hbar/M\omega)^{1/2} \; . \qquad (2.15)$$

In all cases, the vibronic ground state has the same degeneracy as the orbital crystal-field multiplet: The JT interaction cannot split the ground state, but can only mix electronic and vibrational motion because it has the full symmetry of the crystal field.

There exist two cases with basically different dynamic behaviour: If $\underline{\underline{V}}(Q)$ is simultaneously diagonal for all values of Q, vibrational motion leaves the electronic orbitals fixed. This occurs for $E \otimes \beta$ and $T \otimes e$ coupling. If the eigenstates of $\underline{\underline{V}}(Q)$ depend on Q, then vibrational motion will cause the electronic orbitals to follow, and one obtains dynamically coupled vibronic motion. This occurs for $E \otimes \varepsilon$ and $T \otimes \tau_2$ coupling.

The eigenvalues E_n of $\underline{\underline{H}}_{JT}$ determine the vibronic excitations of the JT complex. The dynamic properties are conveniently described by the dynamic susceptibility $\underset{\approx}{\chi}_s(\omega)$ to an external force $\underset{\sim}{F}$ at frequency ω,

$$\underset{\approx}{\chi}_s(\omega) = \frac{1}{Z} \sum_{mn} \frac{\langle m|\underset{\sim}{Q}|n\rangle\langle n|\underset{\sim}{Q}|m\rangle}{E_n - E_m - \hbar\omega - i0^+} \left(e^{-\beta E_n} - e^{-\beta E_m}\right) \tag{2.16}$$

It shows with which strength the various vibronic transitions can be excited by a given force $\underset{\sim}{F}$.

2.3. Collective Behaviour, Mean-Field Approximation (MFA)

If coupling to all other modes is neglected, the crystal is described by a model Hamiltonian of the form

$$\underline{\underline{H}} = \sum_{\ell} \underline{\underline{H}}_{JT,\ell} + \underline{\underline{H}}^{int} \,. \tag{2.17}$$

We assume ordinary lattice-dynamical bilinear interactions between the coordinates $\underset{\sim}{Q}_\ell$ in different cells:

$$\underline{\underline{H}}^{int} = -\frac{1}{2} \sum_{\ell\ell'}{}' \underset{\sim}{Q}_\ell \cdot \underset{\approx}{v}_{\ell\ell'} \cdot \underset{\sim}{Q}_{\ell'} \, \underline{\underline{1}}_{e\ell} \,. \tag{2.18}$$

The matrix $\underset{\approx}{v}_{\ell\ell'}$ together with the force constants of V_{harm} of the single cell (Eq. 2.9) form that part of the dynamic matrix which describes the phonon bands constructed from the local normal coordinates $\underset{\sim}{Q}_\ell$. If the ligands of a given JT ion belong to different lattice cells, there arise problems with the orthogonality of the Q_ℓ for different ℓ /30/. These problems have been discussed in detail in Ref. /31/; they will be disregarded in the present context. The collective response to an external field can be decomposed into the single-cell response to the molecular field and the response of the molecular field,

$$\delta\underset{\sim}{Q}_\ell = \underset{\approx}{\chi}_s(\omega) \cdot \delta\underset{\sim}{F}^{mol} \,, \quad \delta\underset{\sim}{F}_\ell^{mol} = \sum_{\ell'} \underset{\approx}{v}_{\ell\ell'} \cdot \delta\underset{\sim}{Q}_{\ell'} + \delta\underset{\sim}{F}_\ell^{ext} \,, \tag{2.19}$$

and one obtains after Fourier transformation the usual feedback result of MFA /30,32/

$$\underset{\approx}{\chi}_q(\omega) = \underset{\approx}{\chi}_s(\omega) \cdot \left(\underset{\approx}{1} - \underset{\approx}{v}_q \cdot \underset{\approx}{\chi}_s(\omega)\right)^{-1} \tag{2.20}$$

where

$$\underset{\approx}{v}_q = \sum_{\ell'}' \underset{\approx}{v}_{\ell\ell'} \exp(iq \cdot R_{\ell\ell'}) \tag{2.21}$$

is the Fourier transform of the interaction $\underset{\approx}{v}_{\ell\ell'}$. The collective modes are found from the poles of $\underset{\approx}{\chi}_q(\omega)$, i.e. from the zeros of

$$\underset{\approx}{v}_q \cdot \underset{\approx}{\chi}_s(\omega) = \underset{\approx}{1} . \tag{2.22}$$

Since

$$\sum_q \underset{\approx}{v}_q = \underset{\approx}{v}_{\ell\ell} = 0 , \tag{2.23}$$

the eigenvalues of $\underset{\approx}{v}_q$ take on positive and negative values in the Brillouin zone. As Figure 2 illustrates, the single-cell excitations give thus rise to bands of collective modes consisting of linear superpositions of single-cell excitations with phases $\exp(iq \cdot R_\ell)$. Of particular interest is the possibility of critical slowing down of a collective mode due to the feedback effect of the molecular field. One of the main objects of the theory is the study of the nature of the soft mode associated with the transition. This requires a thorough understanding of the underlying single-cell excitations. It is important to note that the collective modes originating from vibronic excitations of the single complex are not optical-phonon modes of the crystal, but are modes of vibronic character occurring in addition to the phonon modes in JT crystals.

The type of phase transition is determined by that mode which yields the largest feedback enhancement, i.e. by the maximum eigen-

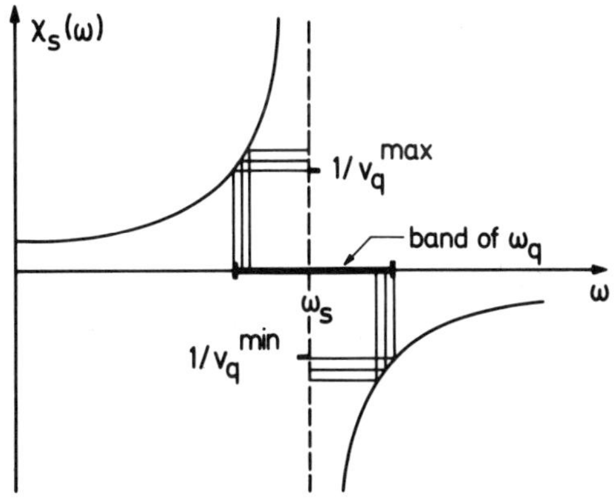

Figure 2: Single-cell excitation at frequency ω_s giving rise to band of collective excitations with frequencies ω_q.

value of $v_{\underset{\approx}{q}}$. If this occurs at q = 0, the transition will be ferrodistortive; for q = 1/2 x (reciprocal lattice vector), it will be antiferrodistortive; and for other values in the Brillouin zone, one finds transitions to more complicated commensurate or incommensurate phases. In the ferrodistortive case, the only candidate for a continuous phase transition is the $E \otimes \beta$ coupling, as is easily seen from the Landau criteria: Both the ε and the τ_2 mode have third-order invariants, and yield therefore necessarily discontinuous transitions. In the antiferrodistortive case, on the other hand, no third-order invariants exist, and continuous transitions are possible for all types of coupling.

In the low-symmetry phase, there exists a static molecular field. For this reason, it is important that we study the behaviour of the JT complex in the presence of a static field,

$$V_1^{(0)} = - \underset{\sim}{F} \cdot \underset{\sim}{Q} \tag{2.24}$$

as indicated in Eq. (2.9).

It should be noted that in MFA in its proper sense only the correlations between coordinates in one cell and the molecular field arising from the interactions with different cells are neglected, but the vibronic correlations between the electron orbitals and the ligand modes of the same complex, which play an important role in JT transitions /7,26-28,33/, are fully taken into account. Neglect of these intracell correlations by a factorization of correlation functions $\langle Q \underline{\underline{\boldsymbol{\xi}}} \rangle$ /6,24/ yields unphysical results such as a phase transition for non-interacting JT complexes and a decreasing T_c with increasing interaction strength unless spurious so-called "self-energy" terms are subtracted out /6,16, 25/. In the MFA as used here, no such ad hoc corrections are required.

2.4. Coupling to Elastic Strain

The model considered so far does not contain elastic deformations because the JT-active coordinates $\underset{\sim}{Q}_\ell$ leave the center of gravity of each cell at rest. Coupling of the $\underset{\sim}{Q}_\ell$ to elastic strain will however play an important role, especially in the ferrodistortive case: There exist strain components of the same symmetry $\beta_{1,2}$, ε, and τ_2 as the JT-active modes for all types of JT coupling, which gives rise to bilinear piezodistortive coupling.

We assume a crystal structure in which the ligand modes Q_ℓ are true intracell distortions. The elastic deformation is described by a displacement field consisting of rigid translations $\underset{\sim}{s}_\ell$ of cells ℓ which are assumed to vary slowly with ℓ. Then, the elastic strain tensor $\underset{\approx}{\varepsilon}_\ell$ can be defined in analogy to the continuum case as the symmetrized gradient of the displacement vector, i.e.

$$\underset{\sim}{s}(\underset{\sim}{R}_\ell+\underset{\sim}{\xi}) = \underset{\approx}{A} \cdot \underset{\sim}{s}(\underset{\sim}{R}_\ell) + \underset{\approx}{\varepsilon}_\ell \cdot \underset{\sim}{\xi} \; , \tag{2.25}$$

where $\underset{\approx}{A}$ describes an infinitesimal rotation. In terms of symmetry-adapted strain components $\varepsilon^\Gamma_{\gamma,\ell}$, the elastic Hamiltonian has the form /24/

$$H^{elast} = \sum_\ell \{\frac{1}{2} M_s \dot{s}^2_\ell + \sum_\Gamma c_\Gamma \sum_\gamma (\varepsilon^\Gamma_{\gamma,\ell})^2\} \tag{2.26}$$

where the c_Γ are the bare elastic constants per cell volume for strain of symmetry Γ. The strain is coupled to the distortions Q_ℓ of the JT complexes by a piezodistortive coupling term /26,28,29/

$$H' = \sum_\ell \sum_\Gamma g_\Gamma \sum_\gamma Q^\Gamma_{\gamma,\ell} \varepsilon^\Gamma_{\gamma,\ell} \tag{2.27}$$

assumed to be local. Equations (2.26) and (2.27) are reasonable descriptions in the long-wavelength limit. Since the elastic energy (2.26) together with compatibility requirements for the strain field makes short-wavelength strain components unfavourable, piezodistortive coupling is most effective for ferrodistortive and small q modes.

This coupling has important consequences even in the absence of JT coupling: It leads to an intracell distortion

$$Q^{ind}_\Gamma = \chi_\Gamma \, g_\Gamma \, \varepsilon_\Gamma \tag{2.28}$$

induced by a uniform strain ("internal strain"), and gives rise to a renormalization of elastic constants

$$c^*_\Gamma = c_\Gamma - \chi_\Gamma g^2_\Gamma \quad . \tag{2.29}$$

Here, χ_Γ is the ferrodistortive static susceptibility of mode Q_Γ in the absence of piezodistortive coupling. On the other hand, a ferrodistortive distortion Q_Γ induces a uniform strain of magnitude

$$\varepsilon^{ind}_\Gamma = c^{-1}_\Gamma \, g_\Gamma \, Q_\Gamma \tag{2.30}$$

and leads to a renormalized ferrodistortive susceptibility

$$\chi^*_\Gamma = \chi_\Gamma \left(1 - c^{-1}_\Gamma \chi_\Gamma g^2\right) \; . \tag{2.31}$$

From Equations (2.29) and (2.31) there follows the interesting relation

$$c_\Gamma^* \chi_\Gamma^* = c_\Gamma \chi_\Gamma \tag{2.32}$$

between the renormalized and the unrenormalized quantities /29/.

For a ferrodistortive transition, Eq. (2.29) shows that there occurs an elastic instability ($c_\Gamma^*=0$) before the phase transition of the uncoupled system ($\chi_\Gamma=\infty$) is reached: Because of piezodistortive coupling, the soft vibronic Q-mode pushes the acoustic mode of the appropriate symmetry down until the sound velocity vanishes at the elastic instability $c_\Gamma^* = 0$. In the ordered phase, there exists a strain ε_Γ given by Eq. (2.30) proportional to the order parameter Q_Γ. Such elastic instabilities associated with a soft acoustic mode have been observed in a number of rare-earth compounds /15,17,22/, even for systems where the structural transition coincides with a magnetic transition /34,35/, as well as in nickel chromite /36/. The coupling strength g_Γ can be determined by fitting the experimental results to an equation of the form (2.29). The results show that strain coupling is very important in the rare-earth systems investigated.

By qualitative arguments it may be expected that the piezodistortive coupling in transition-metal compounds is generally smaller than in rare-earth compounds. Equation (2.28) shows that the coupling constant g_Γ measures the amount of distortion of the ligand configuration induced by a given strain. In transition-metal compounds, covalent bonding by the d-orbitals makes the ligand complex very stable, thus preventing large distortions by strain. In rare-earth compounds, on the other hand, covalent bonding by the f-orbitals in much weaker, the ligand complexes will be less stable and will therefore more readily adjust to strain.

For antiferrodistortive transitions, the piezodistortive coupling (2.27) still has important effects if the distortion has a ferrodistortive component (secondary order parameter). This is to be expected in the case of ε and τ_2 modes for which configurations $\underset{\sim}{Q}$ and $-\underset{\sim}{Q}$ are not equivalent. Consequences of this coupling for antiferrodistortive $E \boxtimes \varepsilon$ systems are discussed in Refs. /28,29/ (see Section 3.2). In a purely antiferrodistortive structure as it may occur in the $E \boxtimes \beta$ case, on the other hand, bilinear coupling to uniform strain is ineffective, and one has to consider higher-order piezodistortive couplings like

$$H'' = - \sum g_{\gamma\gamma'\gamma''}^{\Gamma\Gamma'\Gamma''} Q_\gamma^\Gamma Q_{\gamma'}^{\Gamma'} \varepsilon_{\gamma''}^{\Gamma'} \quad . \tag{2.33}$$

Up to this point it was assumed that JT coupling occurs only with intracell distortions Q_Γ of the ligand configuration. In principle, there exists also a direct JT coupling to strain of the form

$$\underline{\underline{V}}'_{JT} = -\sum_{\Gamma} A'_{\Gamma} \sum_{\gamma} \varepsilon_{\gamma}^{\Gamma} \underline{\underline{\xi}}_{\gamma}^{\Gamma} \quad , \tag{2.34}$$

as has been considered in Refs. /6,24/. This coupling is due to the crystal field at the JT ion produced by displacements of neighbouring complexes, with all ligands kept fixed in their undistorted symmetric configuration. Since this contribution to the crystal field will be much smaller than the contribution produced by the distortion of the ligand configuration, the intracell coupling (2.12b) is expected to be the dominant one if an intracell ligand distortion exists at all in the crystal structure under consideration. If, on the other hand, the distortion of the ligand configuration is the strain coordinate itself, as is the case in the perovskite structure /6/ and in the $CsC\ell$ and $NaC\ell$ structures /22,35/, then only the strain-JT coupling (2.34) exists, and the model of Ref. /24/ which takes only acoustic modes into account applies without modification. In the general case, it will however be difficult to distinguish experimentally a direct strain-JT coupling (2.34) from an indirect coupling via the piezodistortive effect (2.27), i.e. to distinguish A' from $A\chi g$.

3. SPECIFIC CASES

3.1. E x β Coupling

This coupling which occurs in tetragonal symmetry represents the simplest case. Disregarding anharmonic and nonlinear-JT terms, the single-cell JT Hamiltonian has the form

$$\underline{\underline{H}}_{JT} = \left(\frac{1}{2M} P^2 + \frac{1}{2} M \omega_\beta^2 Q^2\right) \underline{\underline{1}}_{e\ell} - AQ\,\underline{\underline{\xi}}_3 - FQ\underline{\underline{1}}_{e\ell} \tag{3.1}$$

where $\underline{\underline{\xi}}_3$ is the Pauli matrix $\begin{pmatrix} 1 & 0 \\ 0 & -1 \end{pmatrix}$. The potential energy "surface" consists of two displaced parabola

$$V(Q) = \frac{1}{2} M \omega_\beta^2 Q^2 - (\pm A + F)\, Q \tag{3.2}$$

displayed in Figure 3a. Since there is only one electronic operator, the eigenvalue problem separates into two independent harmonic oscillators ($\zeta = \pm 1$), and one finds the eigenvalues

$$E_{n,\zeta} = -E_{JT} - \frac{1}{2} F^2/M\omega_\beta^2 + (n+\tfrac{1}{2})\hbar\omega_\beta - F\,Q_{JT}\,\zeta \tag{3.3}$$

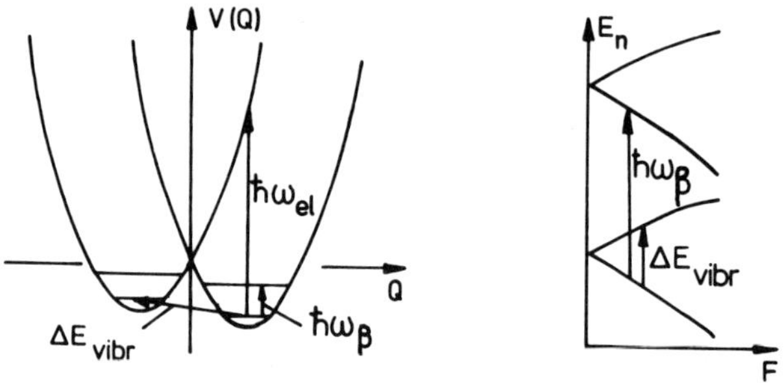

Figure 3: E ⊠ β coupling
a: Potential energy V(Q)
b: Shift of vibronic energy levels with field

shown in Figure 3b. Here, $Q_{JT} = A/M\omega_\beta^2$ is the JT displacement and $E_{JT} = AQ_{JT}/2$ is the JT stabilization energy.

We now turn to the important question, by what type of experiment can the various transitions be induced, i.e. which type of coordinate will show the various excitation frequencies. Since $\underline{\underline{\xi}}_3$ commutes with P and Q, the vibrational motion does not couple at all to the motion of the electron orbital, and we find for the deviation $\delta Q = Q - Q_0$ from the minima $Q_0 = \pm Q_{JT} - F/m\omega_\beta^2$ the equation of motion

$$\delta\ddot{Q} + \omega_\beta^2 \,\delta Q = 0 \tag{3.4}$$

with unperturbed vibrational frequency ω_β: A motion of the electron orbital has no effect on the ionic motion. The low-frequency vibronic transition ΔE_{vibr} can thus certainly not be excited by a force acting on the coordinate Q. This result is due to the fact that no tunnelling occurs between the two parabola. The equations of motion for $\underline{\underline{\xi}}_\pm$, on the other hand,

$$\hbar \dot{\underline{\underline{\xi}}}_\pm = 2i\, A\, Q\, \underline{\underline{\xi}}_\pm \tag{3.5}$$

couple to the vibrational coordinate Q. However, for strong JT-effect, Q fluctuates only little about the minima $Q_0^\pm$, and the main part of the motion of the electronic pseudospin $\underline{\underline{\xi}}$ consists in a precession with frequency

$$\hbar\, \omega_{e\ell} = 4\, E_{JT} \pm 2\, F\, Q_{JT} \tag{3.6}$$

corresponding to the vertical (Franck-Condon-like) transition in Figure 3a: The ionic motion has only a small effect on the electron state, consisting in a small modulation of the precession frequency. The vibronic transition ΔE_{vibr} can thus neither be excited by a force acting on the electron orbitals. An operator associated with the vibronic transition ΔE_{vibr} is obtained by coupling the pseudospin-flip operator $\underline{\underline{\xi}}_{\pm}$ with the vibrational translation operators ("ionic configuration-flip operators")

$$T_{\pm} = \exp(\pm i\, Q_{JT} P) \tag{3.7}$$

One finds indeed

$$\hbar(\underline{\underline{\xi}}_{\pm} T_{\pm})^{\bullet} = \pm\, 2i\; F\; Q_{JT}\; (\underline{\underline{\xi}}_{\pm} T_{\pm}) \tag{3.8}$$

i.e. a precession with the vibronic frequency $\Delta E_{vibr} = 2\,F\,Q_{JT}$.

In the cooperative system, the vibrational excitation gives rise to an optical-phonon band, which in this model is not influenced at all by the transition. The electronic excitation remains local in character, and is split by the molecular field F. The vibronic excitation will for $T < T_c$ give rise to a vibron band which becomes soft as $T \to T_c$. For $T > T_c$ the molecular field vanishes, yielding $\Delta E_{vibr} = 0$. In a realistic system, there will occur thermally activated hopping between the two degenerate states $\zeta = \pm 1$, giving rise to a vibronic relaxation band which becomes soft as $T \to T_c$. The decoupling of the phonons from the electronic motion can be formally taken into account by a transformation to displaced phonon coordinates δQ and pseudospin variables $S_{\pm} = \underline{\underline{\xi}}_{\pm} T_{\pm}$, $S_3 = \underline{\underline{\xi}}_3$ /37,38/. One then obtains an interacting Ising system completely decoupled from the optical-phonon band /38/, and the vibronic excitations become the spinwaves of this model.

In order to excite the vibronic mode, one would have to couple an external probe to the operators $\underline{\underline{\xi}}_{\pm} T_{\pm}$, i.e. the coupling would have to have matrix elements between the ground states in each of the two parabola. This appears impossible in the strong JT case considered here, where the two wave functions are spatially well separated. There will therefore probably exist no method for direct observation of this vibronic mode. But indirect methods can be found: the splitting ΔE_{vibr} has been observed in $TmVO_4$ by optical transitions from the split doublet ground state to a high-lying singlet /39/. The results displayed in Figure 4 show clearly the expected softening. Another method consists in splitting the electronic degeneracy by a static magnetic field, which changes the system into a pseudo-JT system treated in Section 3.5. Then tunnelling takes place between the two parabola, and the vibronic transition can be excited by EPR /19/.

In crystals of this type, the JT ion can couple both to the β_1 mode and the β_2 mode, and it depends on the coupling constants and the interactions which coupling determines the phase transi-

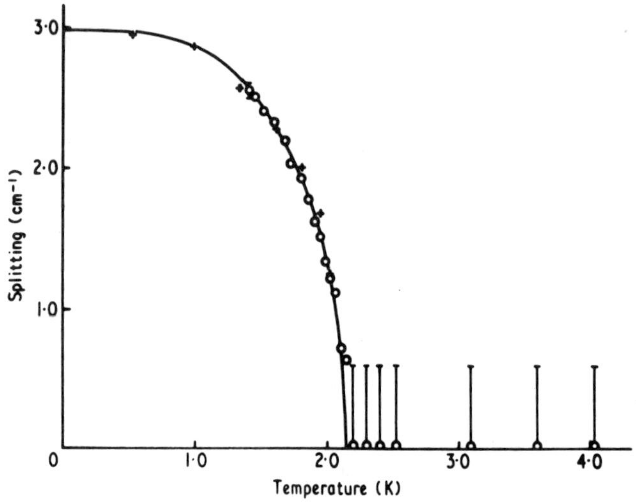

Figure 4: Ground-state splitting of $TmVO_4$ as function of temperature. Circles from optical absorption, crosses from magnetic measurements. (After /39/).

tion. In $TmVO_4$ there seems to occur an interesting competition between these two coupling schemes /15b/.

3.2. E x ε Coupling

The linear JT coupling has the form

$$\underline{\underline{V}}_{JT}^{(1)}(Q) = -A_{\varepsilon}\,(Q_{\theta}\,\underline{\underline{\mathcal{E}}}_{\theta} + Q_{\varepsilon}\,\underline{\underline{\mathcal{E}}}_{\varepsilon}) \qquad (3.9)$$

where

$$\underline{\underline{\mathcal{E}}}_{\theta} = -\underline{\underline{\mathcal{E}}}_{3} = \begin{pmatrix} -1 & 0 \\ 0 & 1 \end{pmatrix}, \qquad \underline{\underline{\mathcal{E}}}_{\varepsilon} = \begin{pmatrix} 0 & 1 \\ 1 & 0 \end{pmatrix} \qquad (3.10)$$

are electronic operators of E symmetry acting on the two-dimensional E-Hilbert space, with respect to a basis transforming as

$$\psi_{\theta} = |2z^2-x^2-y^2\rangle \;; \qquad \psi_{\varepsilon} = |\sqrt{3}(x^2-y^2)\rangle \;. \qquad (3.11)$$

The potential energy surface of $\underline{\underline{V}}_{JT}^{(1)}(\underset{\sim}{Q}) + V_{harm}(\underset{\sim}{Q})\underline{\underline{1}}_{e\ell}$ is the well-known Mexican hat (Figure 5a). In this case, there exists a strong dynamic coupling between the ionic configuration $\underset{\sim}{Q}$ and the electronic state: When $\underset{\sim}{Q} = Q\,(\cos\phi,\ \sin\phi)$ is turned by an angle $\Delta\phi$, the electronic state vector in the two-dimensional Hilbert space rotates by $-\Delta\phi/2$. Thus, by a full rotation of $\underset{\sim}{Q}$, the electronic wave function changes sign, and so does the vibrational amplitude $\chi(Q)$ in order to make the total vibronic wave function invariant.

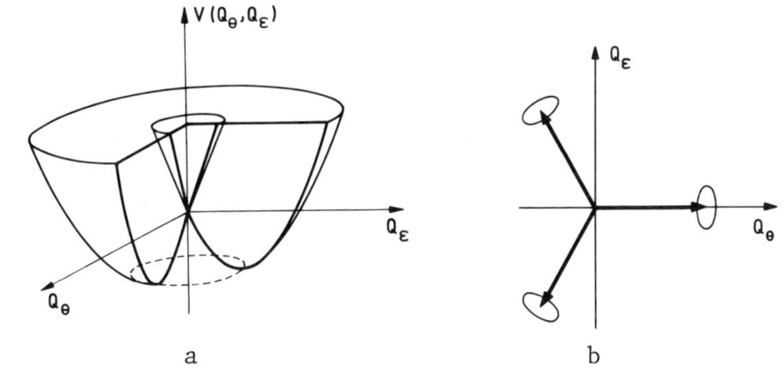

Figure 5: Potential-energy surface $V(Q_\theta, Q_\varepsilon)$ for $E \otimes \varepsilon$ coupling.
a: $\underline{\underline{V}}_{JT}^{(1)} + V_{harm}\, \underline{\underline{1}}_{e\ell}$ ("Mexican hat")
b: Warping by $\underline{\underline{V}}_{JT}^{(2)} + V_{anh}^{(3)}\, \underline{\underline{1}}_{e\ell}$ ("tricorn")

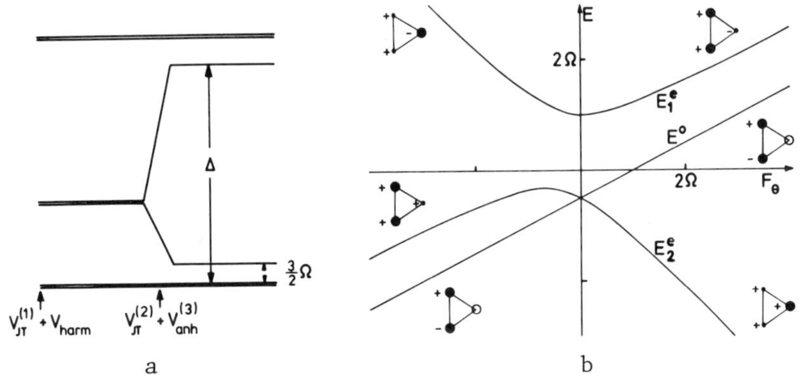

Figure 6: Vibronic energy levels for $E \otimes \varepsilon$ coupling.
a: Splitting of the rotator levels of the Mexican hat by $\underline{\underline{V}}_{JT}^{(2)} + V_{anh}^{(3)}\underline{\underline{1}}_{e\ell}$.
b: Level shift with field in θ-direction

This "anticyclic" boundary condition $\chi(\phi+2\pi) = -\chi(\phi)$ has important consequences for the ordering of the energy levels of $\underline{\underline{H}}_{JT}$: One obtains doubly degenerate rotator eigenvalues for the motion in the brim of the hat corresponding to half-integer quantum numbers $m = \pm 1/2, \pm 3/2, \ldots$ /27/.

Second-order JT coupling $\underline{\underline{V}}_{JT}^{(2)}(\underset{\sim}{Q})$ and third-order anharmonic terms $V_{anh}^{(3)}(\underset{\sim}{Q})$ gives rise to a warping of the brim, changing the hat into a tricorn with three minima (Figure 5b). Correspondingly, the doubly degenerate rotator eigenvalues for the vibronic motion in the brim change as shown in Figure 6a into triplets corresponding to states localized in these three minima, which are split into doublets and singlets because of tunnelling between the minima. Contrary to ordinary potential problems, we obtain on account of the anticyclic boundary condition a ground-state doublet and an excited singlet, separated by a tunnel splitting 3 $\Omega/2$.

We assume the excitation energy Δ to the next higher triplet to be large compared to kT_c, and describe the vibronic motion in terms of S = 1 pseudospin operators acting on the lowest vibronic triplet /26/. The Hamiltonian of the JT complex takes the form

$$H_{JT} = -\frac{1}{2}\Omega\left[\sqrt{2}\,S_x-(S_x^2-S_y^2)\right] - F_\theta Q_\theta - F_\varepsilon Q_\varepsilon \ , \qquad (3.12)$$

where the coordinates Q_θ, Q_ε are in the strong localization limit given by

$$Q_\theta = \frac{1}{2}(2-3\,S_z^2), \qquad Q_\varepsilon = \frac{1}{2}\sqrt{3}\,S_z \qquad (3.13)$$

The shift of the eigenvalues with a field in the θ-direction is shown in Figure 6b. The distribution of the vibrational amplitude over the three valleys is schematically indicated for the different states. In the $\Omega \to 0$ limit, this model goes over into the three-states Potts model /29/.

In cooperative systems, the transitions between these levels

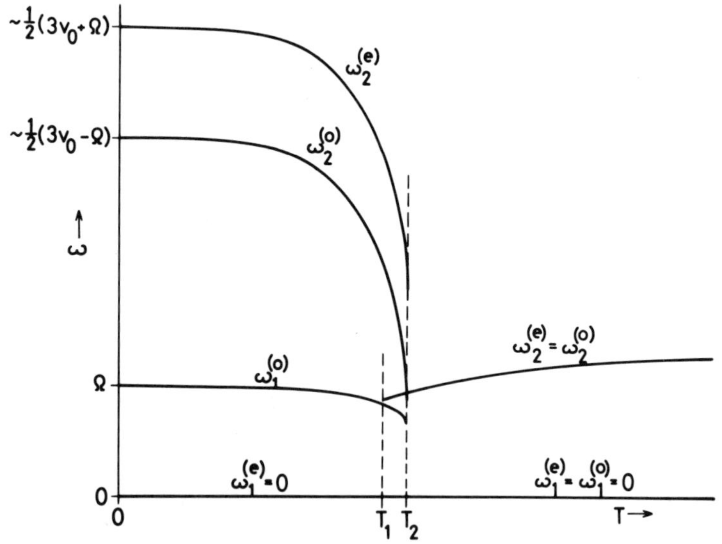

Figure 7: Collective vibronic modes for ferrodistortively coupled $E \otimes \varepsilon$ complexes (After /26/).

give rise to strongly temperature-dependent vibronic bands. This behaviour was studied for the phase transitions occurring in a system consisting of a simple cubic arrangement of JT complexes /26-29/. Figure 7 shows the results for the case of a ferrodistortive transition /26/ which is necessarily discontinuous (see Section 2.3). The bands obtained are easily associated with transitions between the levels of Figure 6b, for $T > T_c$ at $F = 0$ and for $T < T_c$ at $F = F^{mol}(T) > 0$. An important result is the fact that none of these finite-frequency modes is the soft mode associated with the stability limits T_1, T_2 of the two phases. In the high-temperature phase, it is the zero-frequency transition within the ground state doublet which will give rise to a soft relaxational mode. In the low-temperature phase, on the other hand, coupling to the energy will yield a soft heat-diffusion mode, because of the discontinuous nature of the transition /40,41/.

In the case of antiferrodistortive interactions, there exist two competing low-temperature phases of tetragonal symmetry, the θ-phase with the sublattice distortions in one of the three equivalent θ directions, and the ε phase with sublattice distortions which close to T_c are along one of the ε directions, but cant towards the two closest θ directions with decreasing T (see the inserts in Figure 8) /28,29/. The θ phase is favoured by tunnelling and at $T \neq 0$ also by entropy. Therefore, in the absence of further couplings, the θ phase is stable at all temperatures up to the

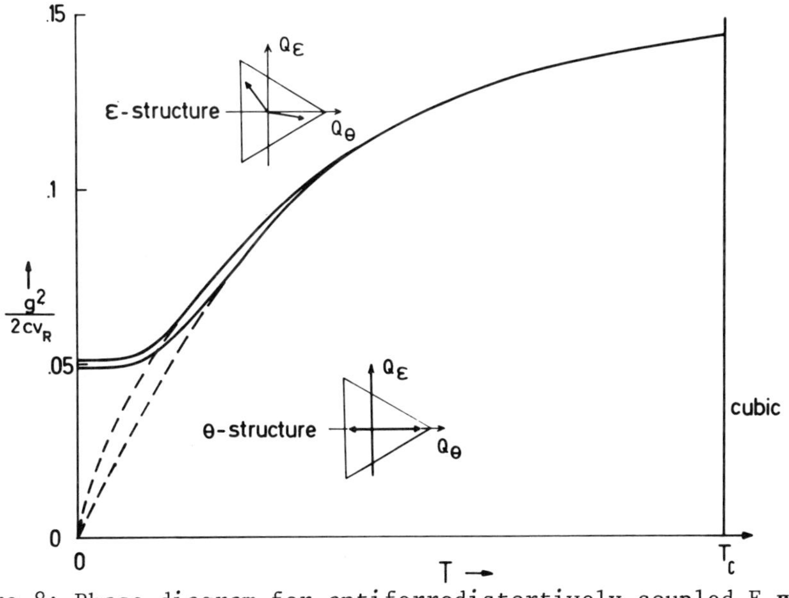

Figure 8: Phase diagram for antiferrodistortively coupled $E \otimes \varepsilon$ complexes with piezodistortive coupling. (Full line: $\Omega = v_R/4$, dashed line: $\Omega=0$) (After /28,29/).

critical temperature T_c where a continuous phase transition occurs to the cubic phase. However, both phases have a ferrodistortive component and show therefore piezodistortive coupling (2.27) to strain which favours the ε phase. The resulting phase diagram is displayed in Figure 8. (Preliminary results for the $\Omega = 0$ case reported in Ref. /27/ contain an error). One finds that the θ phase is stable for low values, and the ε phase is stable for high values of the coupling constant g. The two phases are separated by continuous phase transitions to an intermediate phase of orthorhombic symmetry existing in a very narrow region. Thus, below a critical value of the coupling constant, the θ structure is stable up to T_c where it undergoes a continuous transition to the cubic phase. In an adjacent narrow range of g, one finds the intermediate phase at low temperatures. For larger values of g, the ε structure is stable at low temperatures, and up to a second critical value of g there occurs with increasing temperature first a continuous transition to the intermediate phase and at a slightly higher temperature another continuous transition to the θ phase. For still larger values of g, the ε phase remains stable up to the continuous transition to the cubic phase at T_c.

The temperature dependence of the collective modes is shown in Figure 9. The number of modes is now twice as large as in the ferrodistortive case, because the two sublattices experience dif-

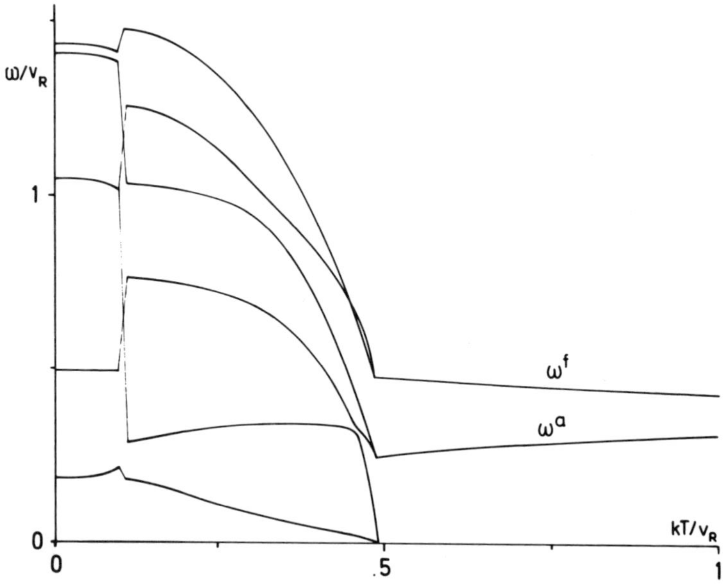

Figure 9: Collective vibronic modes for antiferrodistortively coupled E ⊗ ε complexes with piezodistortive coupling ($\Omega = v_R/4$, $g^2/2c = 0.075\ v_R$). (After /28/).

ferent molecular fields. For $T > T_c$, the soft mode is again a relaxation mode originating from the zero-frequency transition within the ground state doublet. For $T < T_c$, we now find a dynamic soft mode which hybridizes with other modes already at temperatures close to T_c.

All modes consist of coupled vibrational and electronic motions. The temperature dependence of their oscillator strengths and of the relative sublattice amplitudes is discussed in detail in Ref. /28/. In addition to the vibronic excitations, there will occur interesting elastic anomalies on account of the piezodistortive coupling. Experimental investigations of the dynamics of such $E \otimes \varepsilon$ systems appear highly desirable.

3.3. $T \otimes \varepsilon$ Coupling

In this case, the linear JT coupling has the form

$$\underline{\underline{V}}^{(1)}_{JT} = -A_\varepsilon \, (Q_\theta \underline{\underline{\mathcal{E}}}_{T,\theta} + Q_\varepsilon \underline{\underline{\mathcal{E}}}_{T,\varepsilon}) \tag{3.14}$$

where

$$\underline{\underline{\mathcal{E}}}_{T,\theta} = \begin{pmatrix} -1/2 & 0 & 0 \\ 0 & -1/2 & 0 \\ 0 & 0 & 1 \end{pmatrix} \quad , \quad \underline{\underline{\mathcal{E}}}_{T,\varepsilon} = \begin{pmatrix} \sqrt{3}/2 & 0 & 0 \\ 0 & -\sqrt{3}/2 & 0 \\ 0 & 0 & 0 \end{pmatrix} \tag{3.15}$$

are electronic operators of E symmetry acting on the threedimensional T-Hilbert space. The potential energy surface consists of

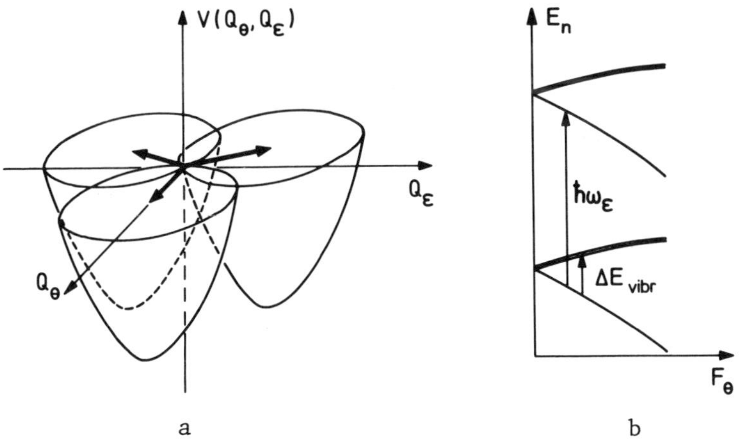

Figure 10: $T \otimes \varepsilon$ coupling.
a: Potential-energy surface $V(Q_\theta, Q_\varepsilon)$
b: Shift of vibronic energy levels with field in θ-direction

three shifted paraboloids, each belonging to a fixed electronic state (Figure 10a). Since the two electronic operators commute, the vibronic eigenvalue problem separates into three independent harmonic oscillators ($\zeta = 1,2,3$), with eigenvalues

$$E_n = -E_{JT} - \frac{1}{2} F^2/m\omega_\varepsilon^2 + (n+\frac{1}{2})\hbar\omega_\varepsilon - Q_{JT} \cdot \begin{cases} F_\theta \\ (-\frac{1}{2}F_\theta+\frac{1}{2}\sqrt{3}F_\varepsilon) \\ (-\frac{1}{2}F_\theta-\frac{1}{2}\sqrt{3}F_\varepsilon) \end{cases} \quad (3.16)$$

displayed in Figure 10b.

This case is thus very similar to the E ⊠ β case discussed in Section 3.1. One finds the same types of transitions: Vibrational transitions $\hbar\omega_\varepsilon$ giving rise to an optical-phonon band, electronic Franck-Condon like transitions of local character, and vibronic transitions ΔE_{vibr} between the ground states of the three parabola which constitute the soft modes, and which will be as difficult to excite directly as in the E ⊠ β case. By a transformation to displaced phonon coordinates, the system can be transformed into a three-states Potts model completely decoupled from the optical-phonon band.

3.4. T ⊠ τ_2 Coupling

This case is described by a linear JT coupling

$$\underline{\underline{V}}_{JT}^{(1)} = -A_\tau \, (Q_\xi \underline{\underline{\boldsymbol{\Sigma}}}_\xi + Q_\eta \underline{\underline{\boldsymbol{\Sigma}}}_\eta + Q_\zeta \underline{\underline{\boldsymbol{\Sigma}}}_\zeta) \quad (3.17)$$

with electronic operators

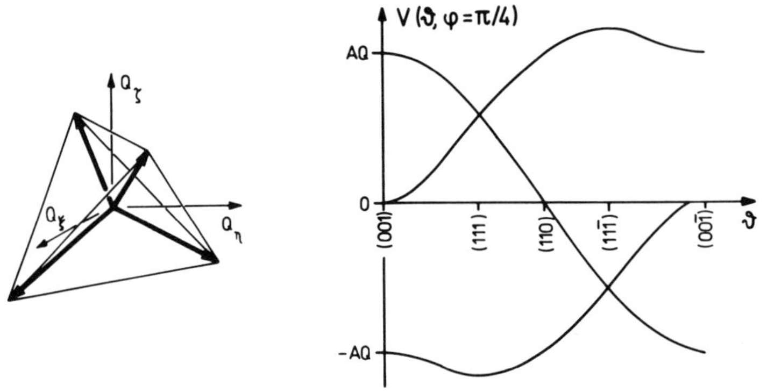

a b

Figure 11: Potential-energy surface $V(Q_\xi,Q_\eta,Q_\zeta)$ for T ⊠ τ_2 coupling
a: Minima of $V(Q_\xi,Q_\eta,Q_\zeta)$
b: The three sheets of $V(Q_\xi,Q_\eta,Q_\zeta)$ as function of ϑ at $\phi = \pi/4$

$$\underline{\underline{\Sigma}}_{\xi} = \begin{pmatrix} 0 & 0 & 0 \\ 0 & 0 & 1 \\ 0 & 1 & 0 \end{pmatrix} , \quad \underline{\underline{\Sigma}}_{\eta} = \begin{pmatrix} 0 & 0 & 1 \\ 0 & 0 & 0 \\ 1 & 0 & 0 \end{pmatrix} , \quad \underline{\underline{\Sigma}}_{\zeta} = \begin{pmatrix} 0 & 1 & 0 \\ 1 & 0 & 0 \\ 0 & 0 & 0 \end{pmatrix} \quad (3.18)$$

of T symmetry acting on the threedimensional T-Hilbert space. Because they do not commute, this case is similar to the $E \otimes \varepsilon$ case discussed in Section 3.2. However, the potential-energy surface is anisotropic already in the absence of higher-order coupling. It has minima for $\tilde{Q}$ along (1,1,1), (1,-1,-1), (-1,1,-1), (-1,-1,1) in τ_2 configuration space (Figure 11a), which correspond to trigonal distortions of the ligand configuration. The potential-energy surface consists of three sheets, two of which penetrating each other in any of the eight (111) directions. Figure 11b shows a cross section through the surface along the meridian $\phi = \pi/4$.

This case has not been studied in much detail. The nature of the lowest vibronic modes is discussed in Ref. /42/. If the minima of $V(\tilde{Q})$ are sufficiently deep, one expects a ground state quartet corresponding to wave functions localized in the four minima, which is split into a low-lying triplet and an excited singlet due to tunnelling between the minima, in a manner analogous to the $E \otimes \varepsilon$ case (Figure 6a). If all other states have excitation energies $> kT_c$, one may construct an $S = 3/2$ pseudospin model for the vibronic motion. In the cooperative system, the transitions within this quartet will again give rise to bands of strongly temperature dependent collective modes. In the limit of zero tunnel splitting, the model goes over into a four-states Potts model.

In a T-type JT crystal, the JT ions can couple both to the ε mode and the τ_2 mode, and it depends on the coupling parameters and the interactions which of the two coupling schemes determines the transition. For suitable values of the parameters, there may occur an interesting competition between these two cases.

3.5. (A+B) $\otimes$ β Pseudo-JT Coupling

One may visualize the two electronic states as the components of an E doublet split by a small distortion of the crystal field from higher symmetry. The single-cell pseudo-JT Hamiltonian

$$\underline{\underline{H}}_{JT} = \left(\frac{1}{2M} P^2 + \frac{1}{2} M\omega_\beta^2 Q^2\right)\underline{\underline{1}}_{e\ell} - \Delta\cdot\underline{\underline{\Sigma}}_1 - AQ\,\underline{\underline{\Sigma}}_3 - FQ\underline{\underline{1}}_{e\ell} \quad (3.19)$$

then contains a term $\underline{\underline{V}}_\Delta = -\Delta\underline{\underline{\Sigma}}_1$ describing this electronic ground-state splitting 2Δ, in addition to the linear JT term $\underline{\underline{V}}_{JT}^{(1)} = -AQ\,\underline{\underline{\Sigma}}_3$. The potential energy "surface" is shown in Figure 12a. Now, there occurs tunnelling between the two minima, and the vibronic states are split already in zero field (Figure 12b). If all higher exci-

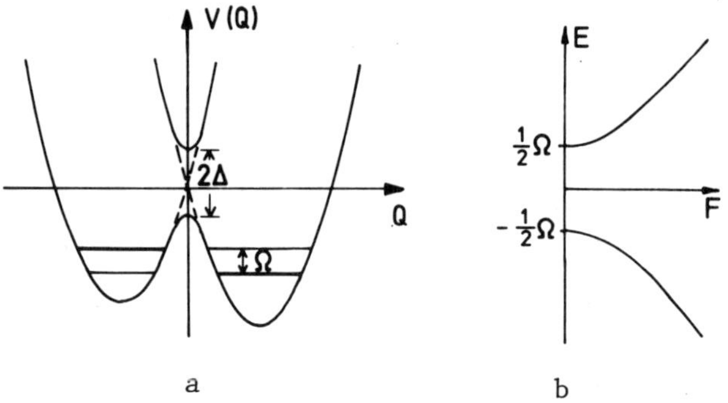

Figure 12: (A+B) ⊠ β coupling (Pseudo-JT effect)
a: Potential energy V(Q)
b: Shift of vibronic energy levels with field

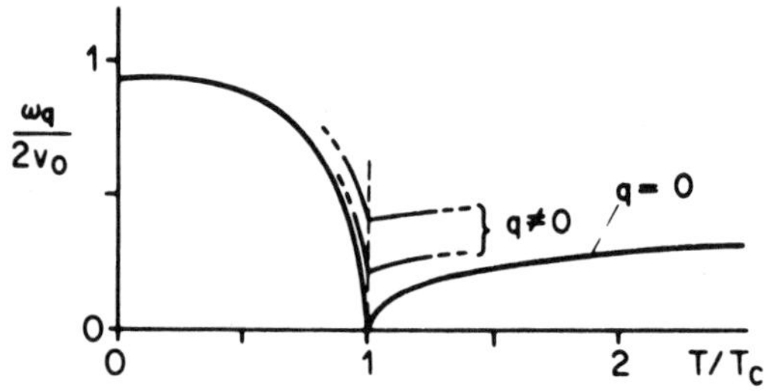

Figure 13: Collective vibronic modes for (A+B) ⊠ β pseudo-JT effect

ted states have excitation energies > kT_c, the vibronic motion can be described in terms of S = 1/2 pseudospin operators acting on the vibronic ground state doublet with tunnel splitting Ω. The pseudo-JT Hamiltonian then takes the form

$$H_{JT} = -\Omega\, S_x - 2F\, Q_0\, S_z \qquad (3.20)$$

which is the same as that used for hydrogen-bonded ferroelectrics /43/ and for magnetic singlet-singlet crystal-field transitions /44/.

For the cooperative system, this model becomes the Ising model with a transverse field, which shows a continuous phase transition. The tunnelling excitations give rise to a band of collective tunnelling modes (pseudospin waves) which become soft at the transition (Figure 13).

REFERENCES

/1/ M.D. Sturge, Solid State Phys. 20, 95 (1967)
/2/ W. Gebhardt, Der Jahn-Teller Effekt. In: Festkörperprobleme IX, edited by O. Madelung, p. 99. Pergamon Vieweg 1969
/3/ F.S. Ham, Jahn-Teller Effects in Electron Paramagnetic Resonance Spectra, Plenum Press (1971)
/4/ R. Englman, The Jahn-Teller Effect in Molecules and Crystals, Wiley-Interscience (1972)
/5/ J.D. Dunitz, L.E. Orgel, J. Phys. Chem. Solids 3, 20 (1957)
/6/ J. Kanamori, J. Appl. Phys. 31, 14S (1960)
/7/ R. Englman, B. Halperin, Phys. Rev. B2, 75 (1970)
B. Halperin, R. Englman, Phys. Rev. B3, 1698 (1971)
/8/ D. Reinen, C. Friebel, K.P. Reetz, J. Solid State Chem. 4, 103 (1972)
/9/ C. Friebel, Z. anorg. allg. Chemie 417, 197 (1975)
/10/ D. Reinen, C. Friebel, to be published
/11/ Y. Yamada, this volume
/12/ C.J. Kroese, W.J.A. Maaskant, Chem. Phys. 5, 224 (1974)
/13/ J. Fernandez et al., Mater. Res. Bulletin 11, 1161 (1976)
/14/ S. Hirotsu, J. Phys. C10, 967 (1977)
/15a/ R.L. Melcher, B.A. Scott, Phys. Rev. Letters 28, 607 (1972)
/15b/ R.L. Melcher, E. Pytte, B.A. Scott, Phys. Rev. Letters 31, 307 (1973)
/16/ R.J. Elliott, R.T. Harley, W. Hayes, S.R.P. Smith, Proc. R. Soc. A328, 217 (1972)
/17/ J.R. Sandercock et al., J. Phys. C5, 3126 (1972)
/18/ G.A. Gehring, K.A. Gehring, Rep. Prog. Phys. 38, 1 (1975)
/19/ R.T. Harley, this volume
/20/ R.T. Harley et al., J. Phys. C6, 2382 (1973)
/21/ R.J. Birgeneau et al., Phys. Rev. B10, 2512 (1974)
/22/ B. Lüthi et al., Phys. Rev. B8, 2639 (1973)
/23/ P.J. Wojtowicz, Phys. Rev. 116, 32 (1959)
/24/ E. Pytte, Phys. Rev. B3, 3503 (1971)
/25/ E. Pytte, Phys. Rev. B8, 3954 (1973)
/26/ H. Thomas, K.A. Müller, Phys. Rev. Letters 28, 820 (1972)
/27/ H. Thomas, The Jahn-Teller Effect as a Mechanism for Structural Phase Transitions, In: Anharmonic Lattices, Structural Transitions and Melting, edited by T. Riste, p. 213, Leiden: Noordhoff (1974)
/28/ G. Schröder, H. Thomas, Z. Phys. B25, 369 (1976)
/29/ K.-H. Höck, G. Schröder, H. Thomas, to be published
/30/ H. Thomas, In: Structural Phase Transitions and Soft Modes, edited by E.J. Samuelsen, E. Andersen and J. Feder, p.15. Oslo: Universitetsforlaget (1971)
/31/ J.C.M. Tindemans-van Eijndhoven, C.J. Kroese, J. Phys. C8, 3963 (1975)

/32/ H. Thomas, Mean Field Theory of Phase Transitions. In: Local Properties at Phase Transitions, Proceedings of the International School of Physics "Enrico Fermi", Course LIX, edited by K.A. Müller and A. Rigamonti, p. 3, Bologna: Società Italiana di Fisica (1976)

/33/ B. Halperin, Phys. Rev. B7, 897 (1972)

/34/ M.E. Mullen et al., Phys. Rev. B10, 186 (1974)

/35/ P. Morin et al., Phys. Rev. B14, 2972 (1976)

/36/ Y. Kino et al., Solid State Comm. 12, 275 (1973)

/37/ J. Feder, E. Pytte, Phys. Rev. B8, 3978 (1973)

/38/ R.B. Stinchcombe, this volume

/39/ P.J. Becker, M.J.M. Leask, R.N. Tyte, J. Phys. C5, 2027 (1972)

/40/ E. Pytte, H. Thomas, Solid State Comm. 11, 161 (1972)

/41/ H. Thomas, Soft Modes at First- and Second-Order Phase Transitions. In: Anharmonic Lattices, Structural Phase Transitions and Melting, edited by T. Riste, p. 231. Leiden: Noordhoff (1974)

/42/ M.J. Shultz, R. Silbey, J. Chem. Phys. 65, 4375 (1976)
M.J. Shultz, this volume

/43/ R. Brout, K.A. Müller, H. Thomas, Solid State Comm. 4, 507 (1966)

/44/ P. Fulde, I. Peschel, Adv. Phys. 21, 1 (1972)

LOCAL JAHN-TELLER EFFECT AT A STRUCTURAL PHASE TRANSITION

K.-H. Höck and H. Thomas

Institut für Physik, Universität Basel

Klingelbergstrasse 82, CH-4056 Basel, Switzerland

ABSTRACT

We consider a Jahn-Teller (JT) impurity ion with a doublet electronic ground state of E_g-symmetry at a lattice site of cubic symmetry and investigate the critical enhancement of the local JT-effect caused by the vibronic coupling to the soft mode of a structural phase transition of the host crystal. Furthermore, we study the temperature-dependence of the strain field around the JT-centre induced by the local JT-effect.

1. INTRODUCTION

Commonly, the JT-effect of JT-impurities in crystals is described in terms of a phenomenological cluster model /1/, which takes into account only the coupling of the electronic state of the JT ion to the vibrational modes of the cluster consisting of its ligands, but disregards lattice-dynamical interactions with the rest of the crystal. Recently, progress has been made by treating the more realistic problem of coupling to the continuum of phonon modes of the host crystal (multimode JT-effect) /2,3/. It was shown, that the properties of the JT-impurity become modified due to the vibronic coupling to such a continuous phonon spectrum of the host crystal as compared with the coupling to the single mode of the simplified quasi-molecular description of the cluster model. Within the framework of harmonic lattice theory these effects are expected to be ordinarily rather small, because the induced JT-distortion extends only over a few lattice distances. However, if the JT-distortion couples to the soft mode of a structural phase transition of the host crystal, the range of distortion increases as the critical temperature T_c is approached, and one obtains a strong, temperature-dependent enhancement of the local JT-effect. We show that due to this enhancement an initially weak JT-effect changes into a strong JT-effect close to the transition temperature. This behaviour is reflected in a characteristic temperature-dependence of the reduction factors p and q of the Ham-effect /1/, which measure the influence of the JT-effect on the EPR-spectrum.

2. MULTIMODE JT-EFFECT

In order to study these effects, we consider a cubic crystal composed of separate octahedral ionic complexes, the distortions of which are described in terms of local normal coordinates $\underset{\sim}{Q}_\ell$ transforming according to the irreducible representations of the cubic point group associated with the lattice-sites $\underset{\sim}{\ell}$. In the centre of the octahedron at $\underset{\sim}{\ell} = 0$, we assume the JT-impurity with a doublet electronic ground state $(\psi_{\ell=0,\theta}, \psi_{\ell=0,\varepsilon})$ of E_g-symmetry, which is linearly coupled to the local normal coordinates $(Q_{\ell=0,\theta}, Q_{\ell=0,\varepsilon})$ of the octahedron at lattice site $\underset{\sim}{\ell} = 0$ only. $Q_{\ell=0,\theta}$ and $Q_{\ell=0,\varepsilon}$ which transform in the same way as the electronic wave functions, on the other hand, interact lattice-dynamically with local normal coordinates of neighbouring octahedra, so that the localized electronic state is indirectly coupled to the whole lattice. In the following, we assume that the host crystal undergoes a displacive structural phase transition with a soft optical phonon branch composed of local normal coordinates of type $\underset{\sim}{Q}_\ell = (Q_{\ell,\theta}, Q_{\ell,\varepsilon})$ at every lattice site. We take only the soft phonon branch into account, and restrict the discussion to the high temperature phase $T > T_c$. In the basis spanned by the electronic wave functions, the Hamiltonian we then consider reads,

$$H = \frac{1}{2} \sum_{\ell} \{\underset{\sim}{P}_\ell^2 + \Omega_s^2(T)\underset{\sim}{Q}_\ell^2 - \sum_{\alpha} v\underset{\sim}{Q}_\ell \cdot \underset{\sim}{Q}_{\ell+\xi\alpha}\}\cdot\underset{\sim}{1} + V\underset{\sim}{Q}_{\ell=0} \cdot \underset{\sim}{\sigma} \ . \tag{1}$$

$\underset{\sim}{P}_\ell = (P_{\ell,\theta}, P_{\ell,\varepsilon})$ are the momenta conjugate to the coordinates $\underset{\sim}{Q}_\ell$, and $\underset{\sim}{\sigma} = (\sigma_\theta, \sigma_\varepsilon)$ are the Pauli spin matrices with

$$\sigma_\theta = \begin{pmatrix} -1 & 0 \\ 0 & 1 \end{pmatrix} \qquad \sigma_\varepsilon = \begin{pmatrix} 0 & 1 \\ 1 & 0 \end{pmatrix}.$$

Here, we have described the host crystal by an effective harmonic Hamiltonian with a temperature-dependent effective single-particle Einstein frequency $\Omega_s(T)$, an approximation usually applied in the theory of displacive structural phase transition /4/ and well justified for the study of the local JT-effect in the high-temperature phase. v and V are, respectively, the interaction coefficient of a bilinear, isotropic nearest-neighbour interaction and the local

JT-coupling constant. Due to the local JT-interaction, the translational symmetry of the system is broken and the Hamiltonian (1) cannot be diagonalized directly with the help of a Fourier transformation to phonon coordinates. The only symmetry operations which leave the system invariant are those belonging to the cubic point group at the JT-impurity site. In order to take account of the reduced symmetry, we introduce so-called symmetry-adapted coordinates /5,6/ which are irreducible E_g-linear combinations of the degenerate local normal coordinates $(Q_{\ell,\theta}, Q_{\ell,\varepsilon})$ belonging to the star of a lattice vector $\underset{\sim}{\ell}$. One finds that in general the star of a lattice vector $\underset{\sim}{\ell}$ is associated with four pairs of degenerate symmetry-adapted coordinates. In the case of isotropic nearest-neighbour interaction, however, only one pair plays a rôle in the study of the local JT-coupling. This pair is given by the totally symmetric linear combination of local normal coordinates of type $Q_{\ell,\theta}$ and $Q_{\ell,\varepsilon}$, respectively, at each member of the star. Performing a final Fourier transformation to symmetry-adapted phonon coordinates $\underset{\sim}{q}_q$ associated with the star of the wave vector q, the vibronic Hamiltonian (1) is then transformed into the so-called multimode JT-Hamiltonian, where the localized electronic doublet is coupled to the continuum of the $\underset{\sim}{q}_q$,

$$H = \sum_q \hbar\omega_q(T)\{\tfrac{1}{2}(\underset{\sim}{p}_q^2+\underset{\sim}{q}_q^2)\cdot\underset{\approx}{1} + K_q(T)\underset{\sim}{q}_q\cdot\underset{\sim}{\sigma}\}$$
$$K_q^2(T) = (N^{-1}V^2)/(\hbar\omega_q^3) \qquad (2)$$

We consider a ferrodistortive transition and assume a Debye dispersion

$$\omega_q^2(T) = \omega_0^2(T) + \alpha q^2$$

for the soft optical phonon branch with a MFA-type temperature dependence

$$\omega_0^2(T) \propto T-T_c \quad .$$

3. CRITICAL ENHANCEMENT

To treat the multimode JT-problem (2), we apply the method described by Englman and Halperin /3/. The Hamiltonian (2) is separated by an orthogonal transformation $\hat{q}_{q',\kappa} = \Sigma_q A_{q'q} q_{q,\kappa}$ into three parts: An effective-single-frequency JT-Hamiltonian $\hat{H}^{JT}$, an uncoupled harmonic part H_{harm}, and a part H' which for a proper choice of the transformation $\underset{\approx}{A}$ is coupled only weakly to the electronic states,

$$H = \hbar\Omega_{eff}(T)\{\tfrac{1}{2}(\hat{p}_1^2+\hat{q}_1^2)\cdot 1 + K_{eff}(T)\hat{q}_1\cdot\sigma\} + H_{harm} + H' . \qquad (3)$$

The transformed mode $\hat{q}_1 = \Sigma_q A_{1q} q_q$, represents some kind of interaction mode strongly affected by the JT-coupling. Its spacial extension is no longer restricted to the immediate nearest neighbours of the JT-impurity ion as in the simple cluster description. The effective frequency and the effective JT-coupling constant, associated with the interaction mode are given by

$$\Omega_{eff} = \sum_q \omega_q A_{1q}^2 \quad , \qquad K_{eff} = \Omega_{eff}^{-1} \sum_q \omega_q K_q A_{1q}$$

The transformation A is determined by minimizing the free energy calculated with a density matrix $\rho = Z^{-1}\exp\{-\beta(\hat{H}^{JT}+H_{harm})\}$. It is further assumed that the first excited state of $\hat{H}^{JT}$ has excitation energy $\gg kT$. Then, the coefficients A_{1q} are found by minimizing the ground state energy $\varepsilon_{JT}(\Omega_{eff},K_{eff})$ of $\hat{H}^{JT}$. Working out the variational calculation, we obtain the following result /Fig. 1/. For high temperatures, we assume a weak JT-effect: the minimum of the harmonic potential energy is only slightly lowered by the JT-interaction, and the gain in JT-stabilization energy E_{JT} is smaller then the zero-point energy $\frac{1}{2}\hbar\omega_{eff}$. As the temperature is lowered, the JT-energy increases while the zero-point energy of oscillation decreases, and below a certain temperature T_1 the gain in JT-stabilization energy will become larger than the zero-point energy of

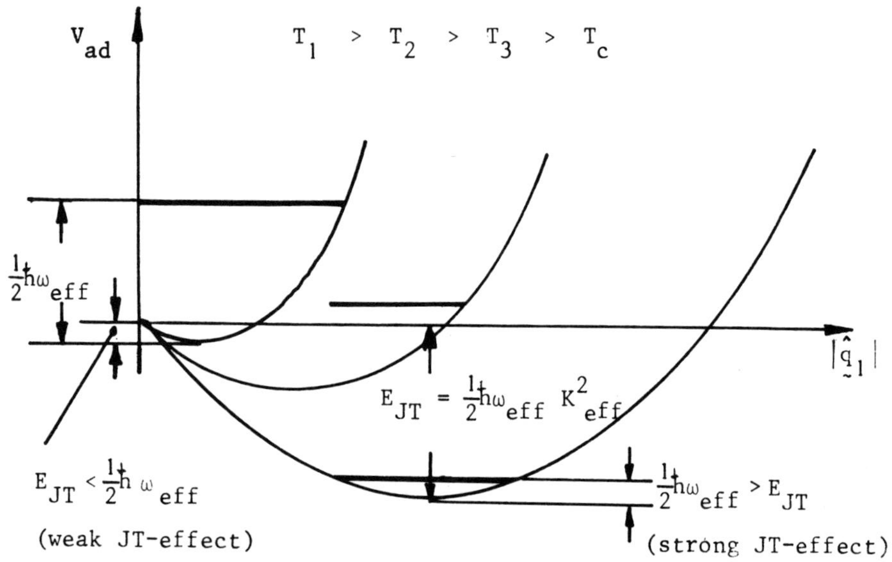

Fig. 1: Critical enhancement of the local JT-effect

oscillation, i.e. the weak JT-effect turns into a strong JT-effect. It turns out that this change occurs for a temperature close to T_c,

$$T_1 - T_c \propto \exp\{-2\hbar\Delta^3/3V^2\} \qquad \Delta^2 = \omega^2_{max} - \omega^2_0$$

Fig. 2 depicts the temperature dependence of the effective JT-coupling constant $K_{eff}(T)$.

The variation from an initially weak JT-effect to a strong JT-effect as the temperature approaches T_c is reflected in a characteristic temperature dependence of the reduction factors p and q of the Ham effect /1/ (Fig. 3). For the weak JT-effect, at high temperatures, p and q are approximately equal to one. As the temperature is lowered the JT-effect increases and p and q decrease. For $T = T_c$, we finally end up with a strong JT effect, and p and q take on the limit values $q = 1/2$ and $p = 0$.

We have furthermore calculated the spatial extension of the JT-induced distortion given by the effective interaction mode $\hat{\tilde{q}}_1$ around the local JT-centre. We find that for $T \neq T_c$, it falls off with an Ornstein-Zernicke law $\langle Q_\theta(\ell)\rangle \propto (1/\ell)\exp(-\ell/\xi)$, where ξ is the correlation length of the lattice distortions associated with the soft optical phonon branch of the host crystal. It is the divergence of ξ at T_c which gives rise to the enhancement of the local JT-effect described above. At $T = T_c$, on the other hand, the spatial dependence is governed by a power law depending on the effective JT-coupling constant $K_{eff}(T=T_c)$. It should be noted that

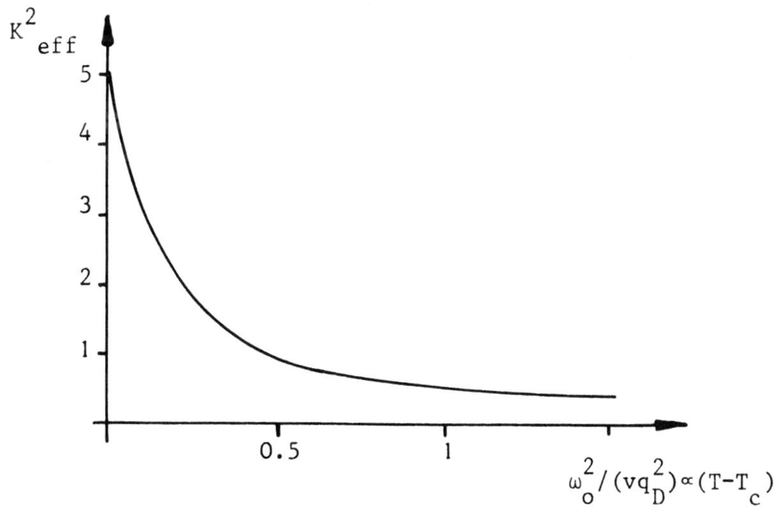

Fig. 2: Effective JT-coupling constant $K_{eff}(T)$

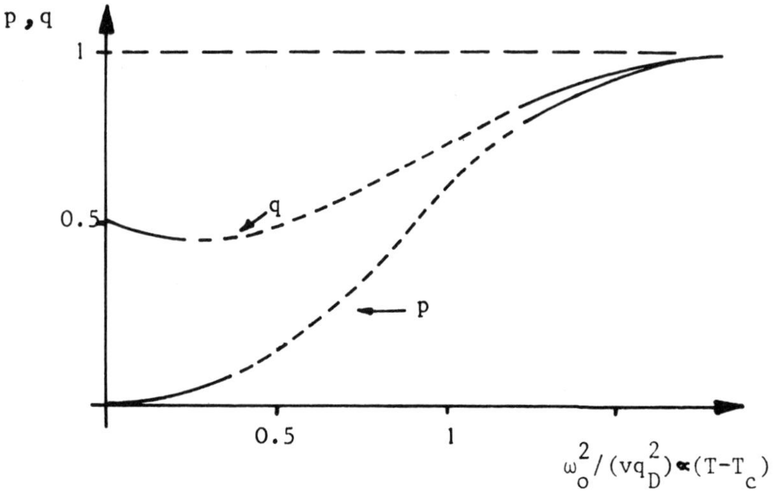

Fig. 3: Temperature variation of the reduction factors p and q.

this is the distortion field associated with the optical phonon coordinates only. Because of the lattice-dynamical coupling to the acoustical phonon coordinates there will occur in addition an elastic strain field falling off with a power law already for $T \neq T_c$. This contribution is, however, not expected to show a critical behaviour in our model of a soft optical phonon.

REFERENCES

/1/ F.S. Ham, in Electron Paramagnetic Resonance, ed. by S. Gschwind (Plenum, New York, 1972)
/2/ M.C.M. O'Brien, 1972, J. Phys. C: Solid State Phys. 5, 2045
/3/ R. Englman, B. Halperin, 1973, J. Phys. C: Solid State Phys. 6, L219
/4/ H. Thomas, in "Structural Phase Transitions and Soft Modes", ed. by E.J. Samuelsen, E.J. Andersen, J. Feder (Universitetsforlaget, Oslo, 1971)
/5/ K.-H. Höck, H. Thomas, to be published
/6/ K.W.H. Stevens, 1969, J. Phys. C: Solid State Phys. 2, 1934.

OPTICAL STUDIES OF JAHN-TELLER TRANSITIONS

R. T. Harley

Clarendon Laboratory, Parks Road, Oxford, England

1. INTRODUCTION

In this chapter we shall consider examples from three areas; firstly isolated 3-d transition metal ions which show localised Jahn-Teller (JT) effects; secondly, rare-earth crystals showing simple cooperative Jahn-Teller (CJT) phase transitions; and finally, examples of more complicated cooperative effects. For each of these systems the low lying electronic and vibrational excitations interact strongly and the coupled modes are of fundamental importance. They can be investigated directly by infra-red absorption and by scattering of light or neutrons and indirectly as splittings in optical absorption and fluorescence spectra. We shall describe results of optical experiments.

2. 3d-TRANSITION METAL IONS

According to the Jahn-Teller theorem (see refs. 1 to 4 and Chapter by H. Thomas in this volume) a symmetrical configuration of a molecule or ionic complex in a solid with orbitally degenerate electronic ground state is unstable because a distortion to lower symmetry must exist capable of producing a linear splitting. The system will therefore distort until reduction in energy of the lowest level (E_{JT}) is just balanced by increased elastic energy, as indicated by Q_0 in Fig. 1a. In principle the distortion will fluctuate thermally between energetically equivalent configurations (e.g. $\pm Q_0$ in Fig. 1a) so that the ground state degeneracy is unchanged. However the fluctuations may be sufficiently slow that a static distortion is observed experimentally.

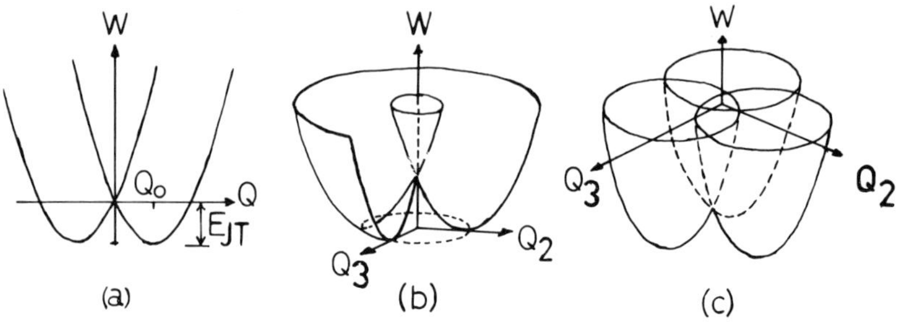

Fig. 1. Electronic energies (W) as a function of distortion coordinate for (a) coupling of one-dimensional distortion (Q) to an electronic doublet (Exβ) and for coupling of two-dimensional distortion (Q_2,Q_3) to (b) a doublet (Exε) and (c) a triplet (Txε).

JT interactions may be classified according to the degeneracy of the electronic state and dimensionality of distortion involved. Figure 1 illustrates schematically three possibilities denoted Exβ, Exε and Exτ, where Roman and Greek letters indicate respectively the electronic and distortional degeneracies. We consider below experiments on 3d ions with Exε and Exτ couplings and in later sections rare-earths coupled to one-dimensional distortions.

The d-orbitals of a 3d-ion in a cubic crystal are split by crystal field and Coulomb effects into singlets (A) which can show no JT effect and into doublets (E) and triplets (T) which can. These orbital states may have spin degeneracy. Strong JT effects occur for E states in octahedral and T states in tetrahedral environments (1) and in general coupling to two-dimensional distortions is predominant.

The low lying excitations of such a complex are determined by the combined effects of JT coupling and spin-orbit interaction; extra splittings may also occur if the crystal field is other than cubic. In general this leads to considerable complication but we shall consider here two simple illustrative examples.

Cu^{2+}:CaO

In CaO, Cu^{2+} occupies an octahedral site and has a ^2E ground state. There is a strong Jahn-Teller coupling to two-dimensional modes of distortion (Exε) and since the spin-orbit interaction cannot split an orbital doublet in first order it can be neglected. If the nuclear motion is harmonic, then the electronic levels as a function of displacements may be represented by the surface in

Fig. 1b where Q_2 and Q_3 represent the two components of ε mode distortion. This is simply the surface formed by revolving the diagram for a one dimensional distortion (Fig. 1a) about the vertical axis. The excitations are found by including the nuclear kinetic energy in the calculation and they correspond to rotational energy levels in (Q_2,Q_3) space, each of which is doubly degenerate (see ref. 1). The nuclear and electronic motions are dynamically coupled to give vibronic excitations (see discussion by H. Thomas in this volume).

Anharmonicity and higher order JT couplings cause local minima at 120° intervals in the circular trough of Fig. 1b corresponding to compression or extension of the complex along the cube axes and lead to splittings of the rotational levels. The results of a calculation for various values of anharmonicity assuming E_{JT} large are shown in Fig. 2a and the Raman spectrum of $CaO:Cu^{2+}$ together with the transition assignments in Fig. 2b (5). Analysis of the spectra gave values of anharmonic barrier height $(2\beta) = 43\ cm^{-1}$, $E_{JT} = 6000\ cm^{-1}$, and effective vibrational frequency $\hbar\omega \sim 350\ cm^{-1}$. Similar measurements for $Ni^{3+}:Al_2O_3$ gave $2\beta = 120\ cm^{-1}$, $E_{JT} = 1100\ cm^{-1}$ and $\hbar\omega \sim 465\ cm^{-1}$ (6).

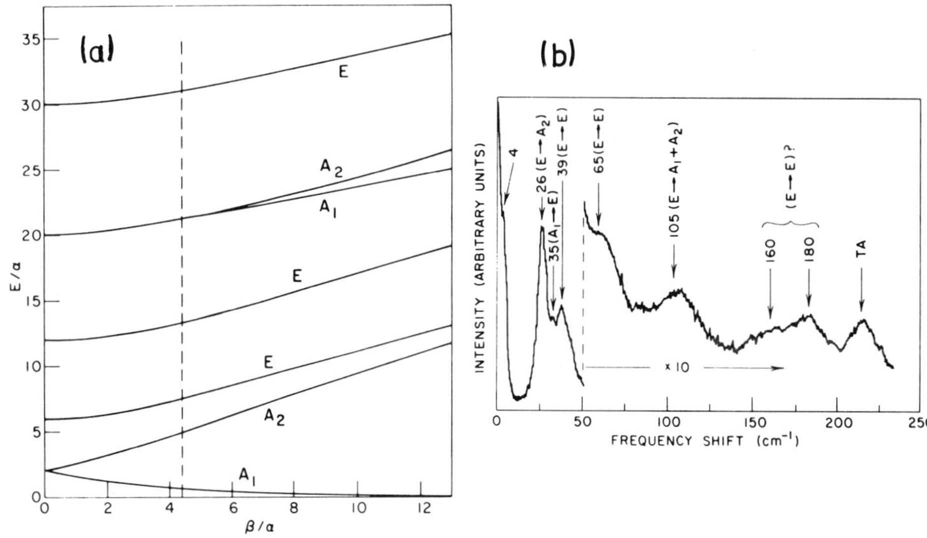

Fig. 2. (a) Calculated quasi-rotational energy levels above the doublet (E) ground state for $(E\times\varepsilon)$ coupling as a function of anharmonicity (β). (b) Raman spectrum of $CaO:Cu^{2+}$ at 4.2 K with lines assigned to transitions in (a); the measured frequencies indicate $\beta \simeq 4.3\alpha$ and $\alpha = 5.2\ cm^{-1}$ (Ref. 5).

If the anharmonicity is large (Fig. 2a) the ground state becomes triply degenerate. This corresponds to the case of very deep potential minima in the trough of Fig. 1b with the complex confined to one of them. For smaller values of anharmonicity a splitting develops due to tunneling between minima. For $CaO:Cu^{2+}$ this tunneling splitting is ~ 4 cm^{-1} (see Fig. 2b) and for $Ni^{3+}:Al_2O_3$ ~ 60 cm^{-1}.

$Ti^{3+}:Al_2O_3$

Our second example is Ti^{3+} in Al_2O_3 which shows $T\times\varepsilon$ coupling. The site symmetry is predominantly octahedral with a small trigonal component and Ti^{3+} has a $^2T_2$ ground state showing weak JT coupling (see ref. 1). The triplet may couple to both two and three-dimensional distortions, but in practice the latter may be neglected. Figure 1(c) shows the splitting of an electronic triplet as a function of two-dimensional distortions (Q_2,Q_3); there are three disconnected minima at 120° intervals corresponding to distortions along the cubic axes. The system will be confined to one of these and its low lying excitations are pure vibrations about the displaced equilibrium configuration which may be represented crudely by a series of harmonic oscillator levels with spacing $\hbar\omega$ (see chapter by H. Thomas). Small perturbations due to spin-orbit coupling and anharmonicity may mix the electronic and vibrational motions and allow tunneling between minima comparable to the case of $Cu^{2+}:CaO$ discussed above. For Ti^{3+} however the JT coupling is weak compared with spin-orbit and trigonal crystal field effects and the Raman spectrum (Fig. 3a) (7) shows little resemblance to that of Cu^{2+} (Fig. 2b). Fig. 3(b) summarises the situation (8); from the right hand side (columns 7-5) it shows the sequence of calculated splittings of the free ion state omitting the JT (vibronic) coupling and from the left (columns 1-4) including it. The observed splittings which have energies δ_1 and δ_2 are a compromise between the splittings calculated for zero JT coupling (column 5) and strong JT coupling (column 3). In the latter case the ground state is a triplet corresponding to the three independent distorted configurations of Fig. 1(c). This reduction of spin-orbit and trigonal field splittings due to the vibronic character of the states favoured by JT coupling is an example of the Ham effect (1,2). The excitations of Ti^{3+} and V^{4+} in Al_2O_3 have also been observed using infra-red techniques (9).

Few experiments of the type illustrated in Figs. 2 and 3 in which the excitations of isolated JT ions are observed directly have been performed so far and there is likely to be increasing interest in them. There is a wide variety of ions and host crystals to investigate. At high concentrations interaction between the

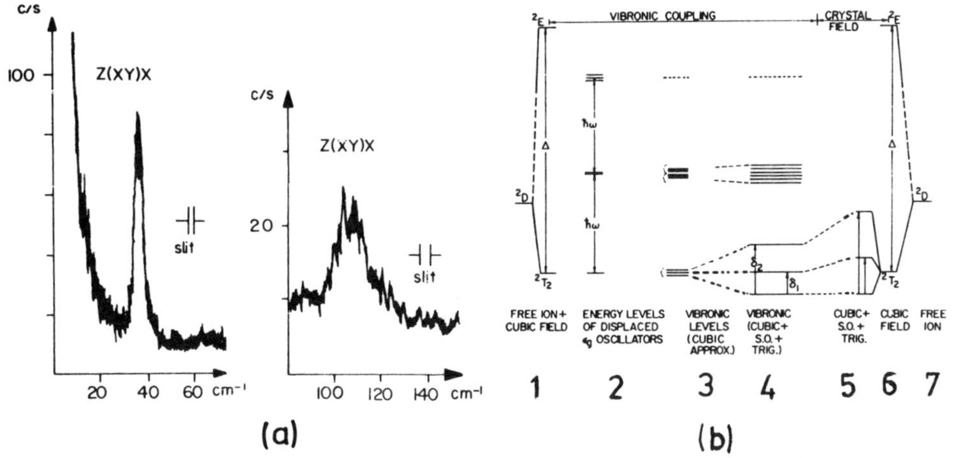

Fig. 3. (a) Raman spectrum of $Ti^{3+}:Al_2O_3$ at 10 K (Ref. 7). (b) Illustration of the origin of the observed transitions δ_1 and δ_2 discussed in text (Ref. 8).

individual JT complexes can occur and measurements as a function of both temperature and concentration will be interesting. Theoretical treatments of cooperative systems of 3d transition metal ions often consider only a ground triplet split by tunneling (see ref. 10 and chapter by H. Thomas in this volume); it is clear from figure 2 that this may not be adequate if the transition temperature is greater than ~ 20 K.

3. COOPERATIVE JAHN-TELLER EFFECTS IN RARE-EARTH CRYSTALS

The second order structural phase transitions in rare-earth vanadates and arsenates ($ReVO_4$ and $ReAsO_4$ where Re is Tm^{3+}, Dy^{3+} or Tb^{3+}) are particularly clear examples of cooperative Jahn-Teller (CJT) effects (4). JT distortions associated with individual rare-earth ions interact because they share ligands and this results in a cooperative instability of the entire crystal. The excitations of the system consist of coupled electronic and phonon modes (which have been called 'vibrons' (4)) and the phase transition is accompanied by strong temperature dependence of these modes. Optical methods of investigating these modes are illustrated in Fig. 4. Raman and infra-red transitions (A) give energies near k = 0. A Brillouin-zone-average of the vibron energy may be obtained from the separation of optical absorption lines due to transitions C and B to a highly excited electronic state, assumed to have no dispersion. This average is expected to be very close

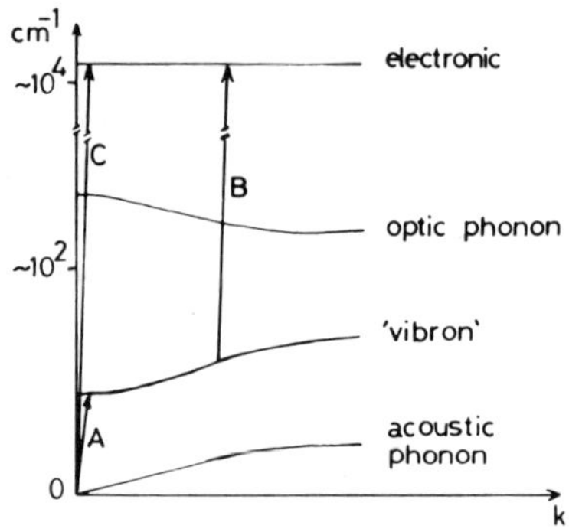

Fig. 4. Schematic diagram of dispersion curves in a CJT system. A is a Raman or infra-red transition; C is an optical transition from the ground state; B is an optical transition from a low-lying 'vibron' mode.

to the uncoupled electronic transition energy (11,4). These crystals are tetragonal (D_{4h}) at high temperatures and below the transition temperature (T_D) orthorhombic (D_{2h}) with two molecules per unit cell in both phases. The distortion is one-dimensional; B_{1g} (B_2 at the rare earth site) for Dy^{3+} or B_{2g} (B_1) for Tb^{3+} and Tm^{3+} (12).

Fig. 5 shows the temperature dependence of the lower electronic states observed as splittings in optical absorption spectra

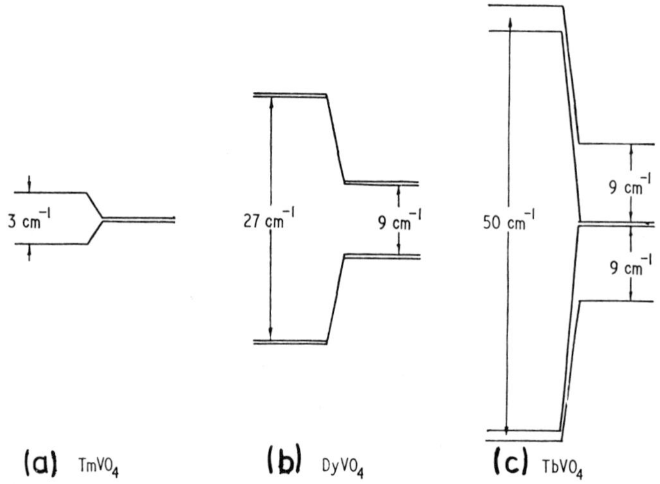

Fig. 5. Low-lying electronic levels of rare-earth ions in $ReVO_4$ crystals above (right) and below T_D (left). The levels for $ReAsO_4$ are similar.

Table 1. Transition Temperatures in $ReVO_4$ and $ReAsO_4$

	Tm		Dy		Tb	
	VO_4	AsO_4	VO_4	AsO_4	VO_4	AsO_4
T_D (measured)	2.14(14)	6.04(15)	14.0(12)	11.2(16)	33.5(17)	25.5(18)
T_D (calc.)*	2.14	6.35	19.1	~17	34.5	28
T_N (measured)	-	-	3.05	2.44	0.51	1.35

*Calculated from electronic level splittings using molecular field theory.

(4); Tm^{3+} and Tb^{3+} are non-Kramers ions whereas Dy^{3+} is a Kramers ion. Molecular field theory may be used to calculate values of T_D from the observed splittings for $T \to 0$ (12). These are compared with experimental values of T_D in Table 1; the correlation of calculated and measured values of T_D shows that electronic energy drives the transitions. However a close look at Table 1 shows that Tm and Tb compounds obey molecular field theory much better than Dy. This and other evidence (13) shows that the dominant site-site interaction is virtually infinite range for Tm and Tb but nearest neighbour only for Dy. (Dy and Tb compounds order antiferromagnetically at very low temperatures (see Table 1).)

Optic phonon modes of B_{1g} or B_{2g} symmetry at $k \simeq 0$ observed by Raman scattering have very little temperature dependence, although the phase transitions involve lattice distortions of these symmetries (12). To a first approximation the JT interaction causes a shift of the origin of vibration without changing the frequency. The process is analogous to that for the isolated complex of Fig. 1a in which the JT effect shifts the origin of the parabolic potential without altering its shape.

The doubly degenerate (E_g) optic phonons of the tetragonal phase show small splittings below T_D which are proportional to the orthorhombic distortion and are due to anharmonic effects (Fig. 6(a)). They can be used to monitor the distortion and hence the order parameter of the transitions; solid curves in Fig. 6(a) are derived from molecular field theory (12). Measurement of optical birefringence is a more precise method for monitoring the distortion very close to T_D. Fig. 6(b) shows results for $TbVO_4$ and

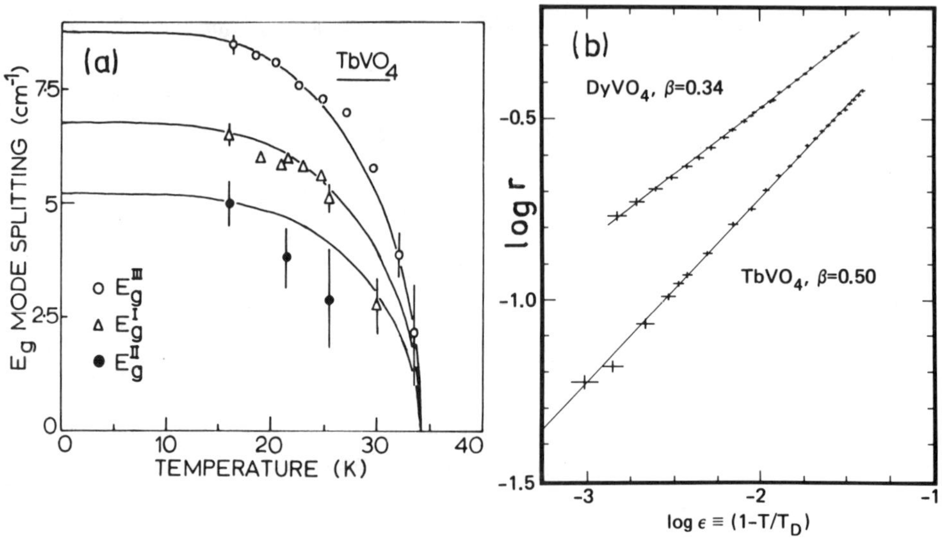

Fig. 6. (a) Measured splitting (points) of E_g optic phonon modes of $TbVO_4$ due to CJT distortion compared with molecular field theory (curves) (Ref. 12). (b) Log-log plot of temperature dependent birefringence (r) of $TbVO_4$ and $DyVO_4$ (Ref. 13).

$DyVO_4$ which have exponents of 0.50 ± .03 and 0.34 ± .03 respectively (13). These values are consistent with long-range interactions in $TbVO_4$ and short-range in $DyVO_4$. The reason for the dominance of short-range interactions in $DyVO_4$ is not known. As discussed by J. K. Kjems in this volume, renormalization group theory predicts (44) a classical value of exponent asymptotically close to T_D so it appears that the measurements (Fig. 6(b)) have not reached this region.

Low lying coupled electronic-phonon modes or vibrons can be observed by both infra-red absorption and Raman scattering. $TbVO_4$ is a good illustrative example; Fig. 7(a) shows typical Raman spectra (12) and Fig. 7(b) the temperature dependence of both Raman and infra-red (17) vibron frequencies. The upper Raman line (open circles) which is that displayed in Fig. 7(a) originates from uncoupled electronic transitions indicated by a circle in the inset. The infra-red line (solid circles) is also thought (17) to be associated with this transition; there is a 17 cm^{-1} Davydov splitting between the Raman and infra-red vibron frequencies because the even parity (Raman) mode is linearly coupled to both optic and acoustic modes near k = 0 whereas the odd parity (infra-red) mode is coupled to optic modes only (17). The solid curves in Fig. 7(b) are the results of calculations based on molecular field and random phase approximations (12).

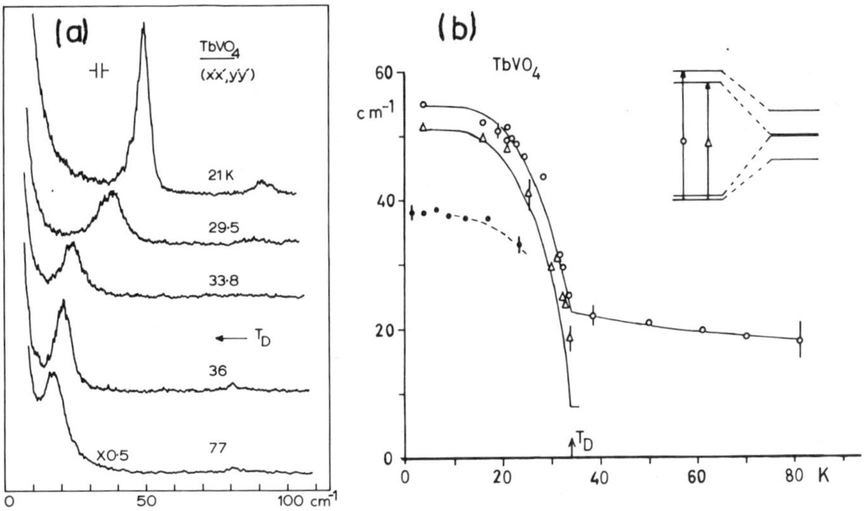

Fig. 7. (a) Raman spectra of $TbVO_4$ at various temperatures. (b) Temperature dependence of Raman (open symbols) and infra-red (solid circles) spectra of $TbVO_4$. Electronic transitions involved are indicated in insert. Solid curves are molecular field and R.P.A. theory (Refs. 12 and 17).

Interaction between the even parity vibrons and acoustic modes near k = 0 causes dramatic changes in the acoustic slopes associated with the transition (19,20) (Fig. 8). For $DyVO_4$ (Fig. 8(a)) the elastic constants at ultrasonic and Brillouin scattering frequencies are the same but there is a marked difference for $TbVO_4$ (Fig. 8(b)). A static elastic constant is the ultimate soft mode ($\frac{1}{2}(C_{11}-C_{12})$ for Dy and C_{66} for Tm and Tb compounds), and measurements at ultrasonic frequencies (~10 MHz) are in the static regime. The difference between ultrasonic and Brillouin measurements in $TbVO_4$ is caused by coupling of the acoustic mode to electronic transitions between the doublet levels of Tb^{3+} (Fig. 5(c)). Above T_D the electronic transition is centred at zero frequency and it will have a width $\Delta\omega$. Measurements made at low frequencies ($\omega \ll \Delta\omega$) see the full acoustic softening at T_D, whereas at higher frequencies ($\omega > \Delta\omega$) there will be only a partial softening caused by coupling to transitions between the singlet levels (Fig. 5(c)) which give rise to the 'vibron' at ~ 20 cm^{-1} (Fig. 7(b)). The frequency of Brillouin measurements is apparently comparable to $\Delta\omega$ because the experimental curve (Fig. 8(b)) lies between the fully softened ultrasonic curve and the value C_I computed for $\omega > \Delta\omega$; a value of 8.4 GHz has been deduced for the 'single ion' value of $\Delta\omega$ and this will be expected to show critical narrowing at T_D due to cooperative

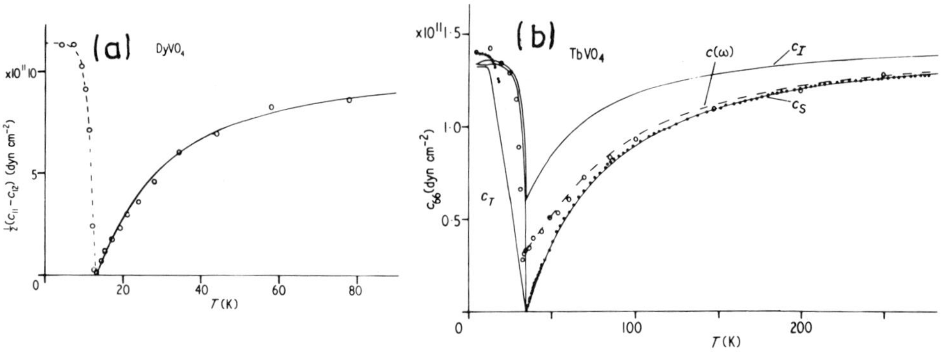

Fig. 8. Elastic constants measured using Brillouin scattering (open circles) and ultrasonics (solid circles) for (a) $DyVO_4$, (b) $TbVO_4$. Curves in (b) are molecular field and R.P.A. theory for $\omega > \Delta\omega$ (C_I), $\omega \sim \Delta\omega$ ($C(\omega)$) and $\omega \ll \Delta\omega$ (C_S) (Ref. 20).

effects (20). A zero frequency mode which shows some of the expected features has been observed by neutron scattering in $TbVO_4$ (ref. 21 and seminar by S. R. P. Smith in this volume) and study of Fig. 5a shows that in Tm compounds there should be a similar zero frequency mode above T_D which would give rise to frequency dependent elastic constants.

Applied magnetic fields may help or inhibit CJT distortions. In $TmVO_4$ and $TmAsO_4$ the ground doublet (Fig. 5a) may be split by a magnetic field parallel to the c-axis or by crystallographic JT distortion; these two perturbations have different symmetry and are mutually exclusive. As a field is applied below T_D the JT distortion is reduced and eventually becomes zero when the magnetic energy ($g\beta H$ where $g \sim 10$) just cancels the energy associated with the CJT distortion (11). The splitting of electronic levels (Fig. 9(a)) is constant at low fields and shows a normal linear increase at high fields (22). Fig. 9(b) shows the behaviour of the JT distortion monitored by splitting of an E_g optic phonon mode. The quenching of the field dependence of the levels at low fields by the JT coupling (Fig. 9(a)) may be regarded as a rather unusual example of the Ham effect (section 2).

Dy and Tb compounds have large g-values in the ab plane which lead to enhancement of JT distortions by applied fields. Figure 10 shows measurements of E_g phonon splittings (proportional to orthorhombic distortion) for $DyVO_4$ up to 13 T; substantial distortions can be induced well above T_D (23). At temperatures above T_D the two Kramers doublets of Dy (Fig. 5(b)) have equal g-values (~10) but as $T \to 0$, due to JT admixture of the wavefunctions below

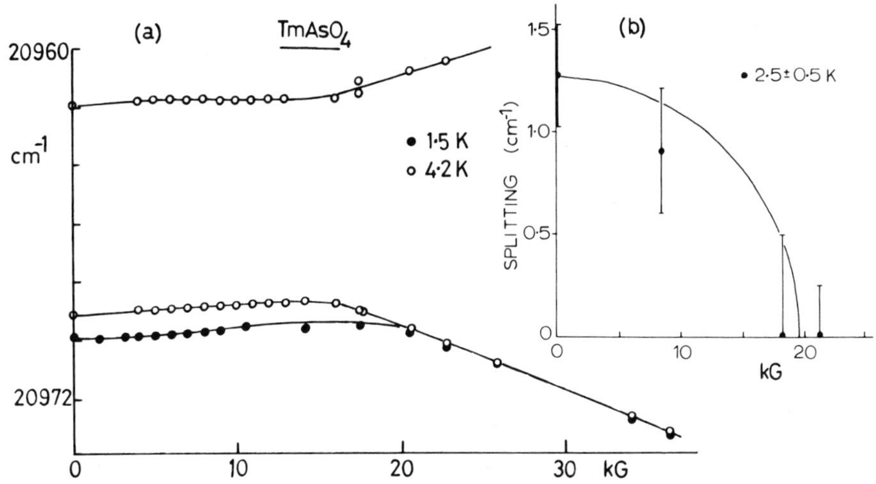

Fig. 9. Magnetic field dependence in $TmAsO_4$ of (a) electronic levels observed by optical absorption (Ref. 22) and (b) splitting of an E_g phonon mode due to CJT distortion observed by Raman scattering (Ref. 11).

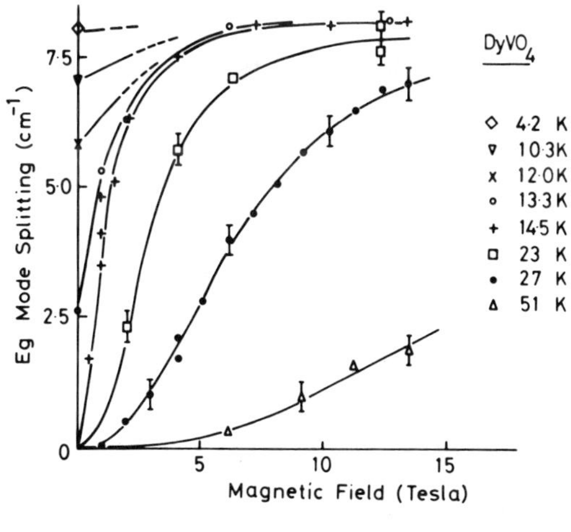

Fig. 10. Magnetic field dependence of E_g phonon splitting in $DyVO_4$ (Ref. 23).

T_D, the g-value of the lower state becomes ~20 and that of the upper state ~0 (12,24). Thus at finite temperatures the system can gain energy, when a field is applied, by undergoing a distortion which increases the g-value of the lower doublet and so decreases the ground state energy. A similar argument can be applied to Tb (25). A mechanism of this type may be responsible for the magnetic field induced splittings observed in optic phonon modes of other rare-earth crystals notably CeF_3 (26). Although no cooperative transition occurs a JT interaction may nevertheless be present.

Interesting phenomena are observed when the JT ions are diluted with inactive ions. Fig. 11 shows measurements of transition temperatures for mixed crystals $Dy_pY_{1-p}VO_4$ and $Tb_pGd_{1-p}VO_4$ compared with molecular field theory in which the molecular field parameter is scaled linearly with concentration of JT active ions (27). This gives a good description of Tb/$GdVO_4$ for which site-site interactions are long range (solid curve Fig. 11b) but not for Dy/YVO_4 (Fig. 11a) which has short-range interactions. The results for Tb/$GdVO_4$ (Fig. 11b) show that over a narrow range of composition near p = .35 the crystal undergoes two transitions. The energy level scheme of Tb^{3+} in the undistorted phase (Fig. 5c) has a singlet lowest with a doublet at ~ 9 cm^{-1} and the strength of the CJT coupling or equivalently Tb concentration (p) in the mixed crystal must exceed a threshold before a distortion will occur at T = 0. However just before this threshold is reached

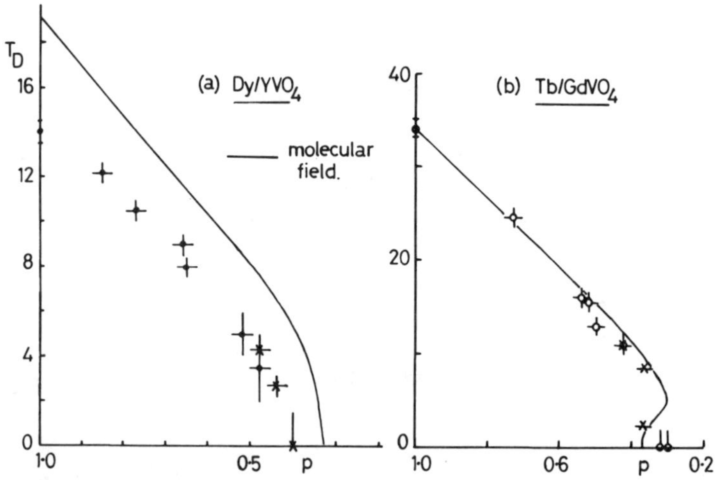

Fig. 11. Dependence of T_D on concentration of JT ions (p) in (a) $Dy_pY_{1-p}VO_4$ and (b) $Tb_pGd_{1-p}VO_4$. Solid curves are scaled molecular field theory (Ref. 27).

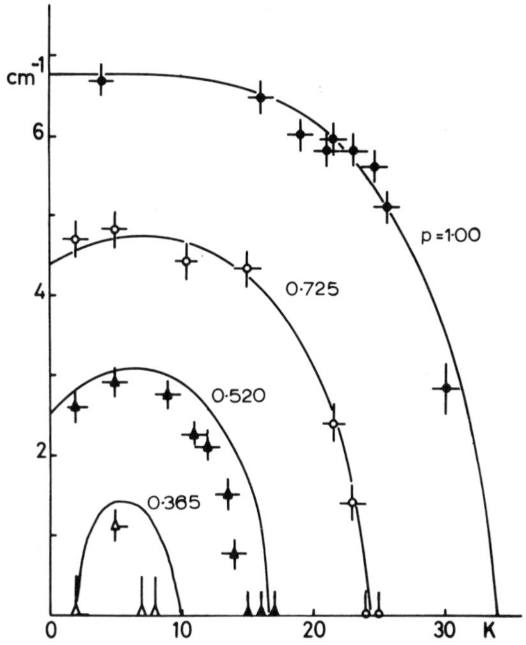

Fig. 12. Variation of E_g phonon splitting (proportional to orthorhombic distortion) in $Tb_pGd_{1-p}VO_4$ (Ref. 27).

distortion is energetically favoured at a non-zero temperature which populates the 9 cm^{-1} doublet; at higher temperatures populations of all the levels increase and the distortion is suppressed. Variation of the distortion with composition and temperature is shown in Figure 12. Measurements of one of the Raman active vibron modes in $Tb/GdVO_4$ crystals are shown in Fig. 13 (cf. Fig. 7b) and compared with calculations in the scaled molecular field and random phase approximations.

4. OPTICAL STUDIES OF COMPLICATED JAHN-TELLER TRANSITIONS

(a) UO_2 has cubic fluorite structure at high temperature and undergoes simultaneous cooperative Jahn-Teller distortion and magnetic ordering at T_N = 30.8 K resulting from the triplet groundstate of U^{4+} (28). Recent experiments (29) have shown that the CJT distortion involves a zone-boundary mode and therefore the existing microscopic theory of the transition (28) must be modified. The atomic displacements below T_N (see Fig. 14) correspond to an X_5^+ optic phonon mode which is observed at 233 cm^{-1} by Raman (30) and neutron scattering (31) and generates a T_{2g} symmetry distortion at the U^{4+} site. This mode shows little temperature dependence.

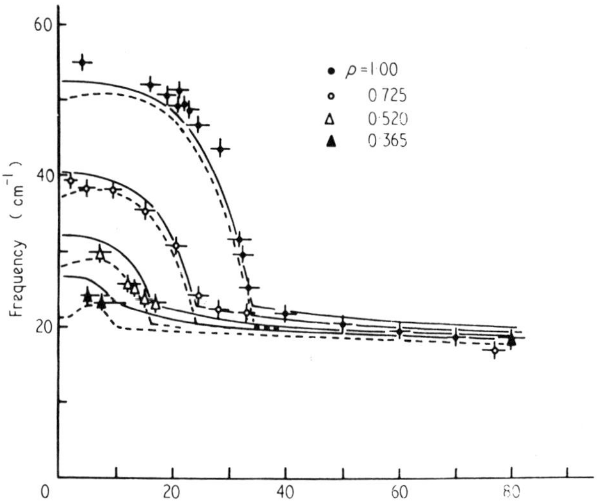

Fig. 13. Measured frequencies of the vibron in $Tb_pGd_{1-p}VO_4$ due to transitions indicated by the circle in the insert to Fig. 7(b). Solid and dashed curves are molecular field and R.P.A. theory for two different choices of parameters (Ref. 27).

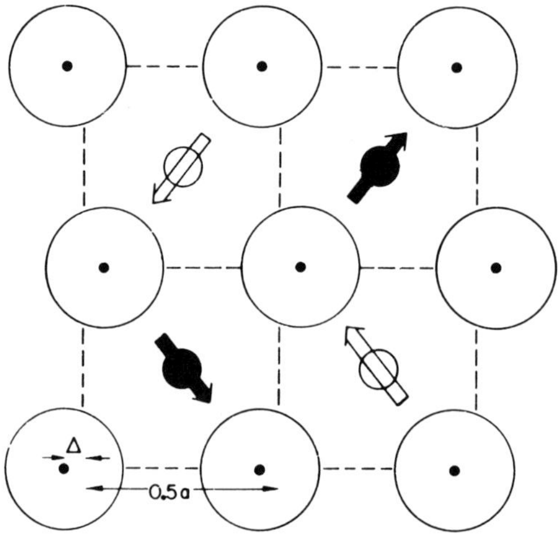

Fig. 14. <001> projection of low temperature structure of UO_2. Oxygen atoms (large circles) are in the plane and uranium atoms a/4 above (small solid circles) and below (small open circles). The values of Δ is 0.014 Å. Arrows indicate a possible arrangement of U^{4+} magnetic moments (Ref. 29).

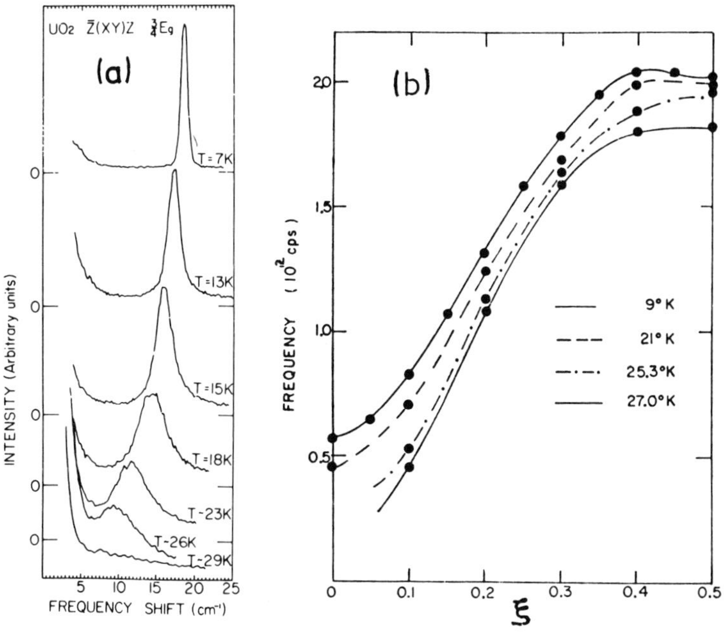

Fig. 15. Temperature dependence of electronic excitations in UO_2 in (a) Raman spectra (Ref. 30), and (b) neutron scattering in the $[\xi,\xi,1-\xi]$ direction (Ref. 32).

Phonon and electronic (spin-wave) excitations have been measured using neutron scattering (31,32), infra-red (28) and Raman scattering (30). The strongest feature in the Raman spectra (Fig. 15a) softens below T_N and corresponding behaviour is observed for a spin-wave at the X-point of the high temperature cell in the neutron spectrum (origin in Fig. 15b). This temperature dependence is compared with that of magnetic Bragg intensity in Fig. 16. The elastic constant C_{44} measured by ultrasonics softens by 24% between 240 K and T_N (34).

The new results for UO_2 suggest that the transition is more nearly second order than was previously believed and that the magnetic order may be of the four-sublattice type. Above T_N the strongest electron-phonon interaction must occur on the zone-boundary but there must also be substantial coupling to acoustic modes at the zone-centre which gives rise to the anomaly in C_{44} at T_N.

(b) $PrAlO_3$ has a distorted perovskite structure and undergoes phase transitions at 210 K and 151 K caused by interaction of Pr^{3+}

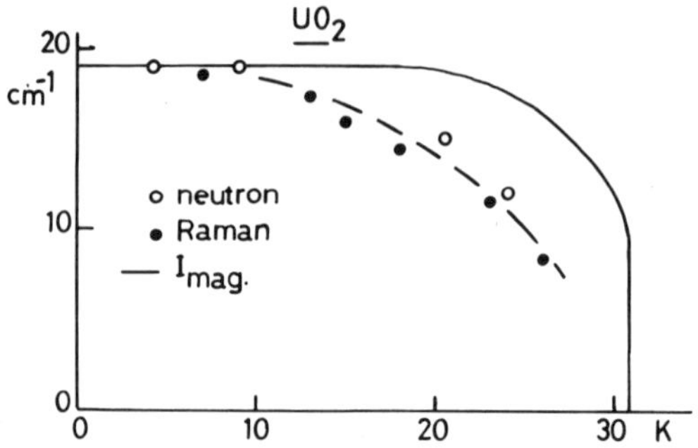

Fig. 16. Temperature dependence of X-point electronic mode of UO_2 observed by Raman and neutron scattering and of the intensity of a magnetic Bragg reflection (Refs. 32,33).

electronic states with small staggered rotations of the AlO_6 units and accompanying lattice strains (35,36). A third transition at 118.5 K (37) probably involves similar interactions (see below). Splittings of the $3H_4$ ground multiplet observed by optical fluorescence can be accounted for quite accurately by calculations (38) (Fig. 17) in which the crystalline electric field at the Pr^{3+} site is modified by including effects of AlO_6 rotations and lattice strains measured directly by neutron diffraction (36) and e.s.r. (39). The resultant AlO_6 rotation (~9°) is almost independent of temperature; above 210 K its axis is along [111], between 210 K and 151 K along [101] and below 151 K it swings continuously towards [001] approaching this direction asymptotically as $T \to 0$.

The temperature dependence of excitations observed by Raman scattering is shown in Fig. 18 (40). Brillouin scattering measurements (37.41) show that the transitions at 151 K and 118.5 K are accompanied by soft acoustic modes which may be linearly coupled to the Raman mode which lies lowest at each temperature.

The splitting of the lowest two Pr^{3+} levels (B_1 and A_1 in Fig. 17) associated with the 151 K transition is analagous to the behaviour of the levels in $DyVO_4$ (Fig. 5b) and the temperature dependences of the Raman excitations (40) (B_1 in Fig. 18) and the elastic constant (41,36) are quite well accounted for by the theory developed for $DyVO_4$ (12, see chapter by K. J. Kjems in this volume). Transitions between the electronic levels (B_1 and A_1 Fig. 17) couple linearly to an optic phonon of B_1 symmetry resulting in two vibron modes (B_1 in Fig. 18). The lower branch is coupled to the acoustic mode which is the soft mode at 151 K. At 118.5 K the lowest Raman mode (Fig. 18) is a pure phonon with eigenvector ϕ_y (staggered rotation of AlO_6 about [010]) (40) which may couple linearly to strain e_{yz} (42). This will cause a renormalisation of

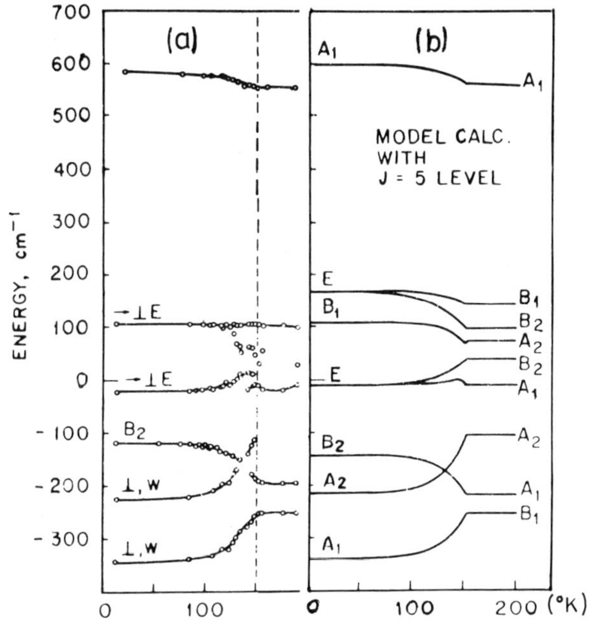

Fig. 17. Electronic energy levels of $^3H_4$ multiplet of Pr^{3+} in $PrAlO_3$ (a) measured using optical fluorescence (Ref. 35) and Raman scattering (Ref. 38), and (b) calculated as described in the text (Ref. 38).

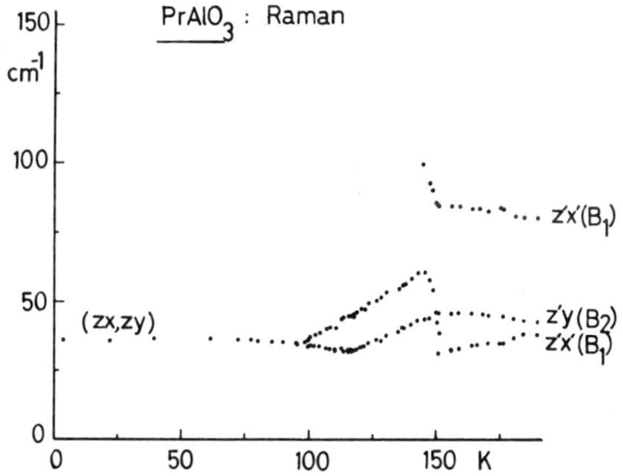

Fig. 18. Frequency of low-lying modes in the Raman spectrum of $PrAlO_3$. x, y and z refer to the cube axes and z' and x' are at 45° to the cube axes in the [010] plane. Modes marked B_1 are coupled electron-phonon modes and B_2 is a pure phonon mode (Ref. 40).

the elastic constant C_{44} given by

$$C_{44} = C_{44}^{c} - \text{Const.}\ \Phi_z^2 \omega^{-2} \tag{1}$$

where ω is the frequency of the lowest Raman mode (Fig. 18) and Φ_z is the static component of AlO_6 rotation along [001]. This expression with the constant chosen to make C_{44} zero at 118.5 K, is compared with the Brillouin measurements in Fig. 19 (43). The very asymmetric shape is caused by the temperature dependence of Φ_z which increases continuously below 151 K (see above). The measured elastic constant is actually $\frac{1}{2}(C_{44}+C_{55})$ whereas the calculation is for C_{44}; a more detailed calculation would probably remove this discrepancy.

5. CONCLUSION

Optical and infra-red techniques are extremely useful for the study of excitations in Jahn-Teller systems. Crystals containing rare-earth ions for which the electron-phonon interaction involves one dimensional distortions and is relatively weak have received most attention and can be described by quite simple theoretical models. Further measurements of the odd-parity vibron modes in rare-earth systems and their contributions to dielectric behaviour (see Seminar by D. R. Taylor in this volume) and of effects of J.T. couplings involving two-dimensional distortions would be interesting. By contrast 3d-ions which have much stronger interactions have not been investigated extensively; the excitations of isolated 3d complexes and effects of cooperative interactions will probably be subjects of increasing interest in the future.

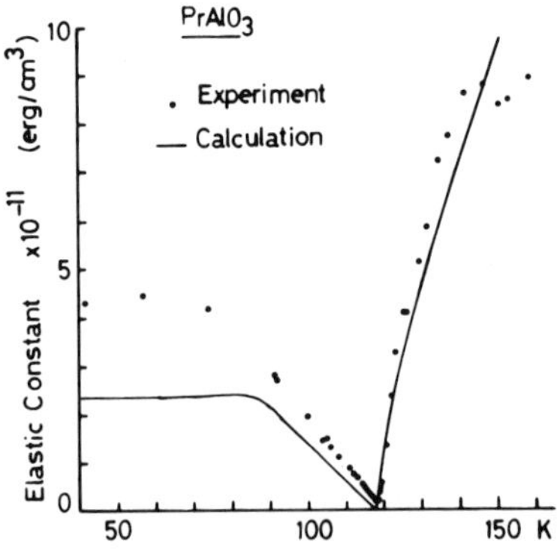

Fig. 19. Brillouin scattering measurements of $\frac{1}{2}(C_{44}+C_{55})$ for $PrAlO_3$ (points, Ref. 37) compared with calculation of C_{44} (curve) using eq. 1 (Ref. 43).

REFERENCES

Reviews: (1) M. D. Sturge, Solid State Physics, 20 91 Ed. F. Seitz, D. Turnbull and H. Ehrenreich (New York, Academic Press, 1967).

(2) F. S. Ham in "Electron Paramagnetic Resonance" p.1 Ed. S. Geschwind (New York, Plenum, 1972).

(3) R. Englman "The Jahn-Teller effect in molecules and crystals" (London, Wiley, 1972).

(4) G. A. Gehring and K. A. Gehring, Rep. Prog. Phys. 38 1 (1975).

(5) S. Guha and L. L. Chase, Phys. Rev. B12 1658 (1975).

(6) L. L. Chase and C. H. Hao, Phys. Rev. B12 5990 (1975).

(7) B. F. Gächter and J. A. Koningstein, Solid State Commun. 14 361 (1974).

(8) R. M. Macfarlane, J. Y. Wong and M. D. Sturge, Phys. Rev. 166 250 (1968).

(9) R. R. Joyce and P. L. Richards, Phys. Rev. 179 375 (1969).

(10) H. Thomas and K. A. Müller, Phys. Rev. Letts. 28 820 (1972).

(11) R. T. Harley, W. Hayes and S. R. P. Smith, J. Phys. C5 1501 (1972).

(12) R. J. Elliott, R. T. Harley, W. Hayes and S. R. P. Smith, Proc. Roy. Soc. A328 217 (1972).

(13) R. T. Harley and R. M. Macfarlane, J. Phys. C8 L530 (1975).

(14) P. J. Becker, M. J. M. Leask and R. N. Tyte, J. Phys. C5 2027 (1972).

(15) J. E. Battison, A. Kasten, M. J. M. Leask, J. B. Lowry and K. J. Maxwell, J. Phys. C9 1345 (1976).

(16) H. G. Kahle, L. Klein, G. Müller-Vogt and H. C. Schopper, Phys. Stat. Sol. (b) 44 619 (1971).

(17) H. B. Ergun, K. A. Gehring and G. A. Gehring, J. Phys. C9 1101 (1976).

(18) L. Klein, W. Wüchner, H. G. Kahle and H. C. Schopper, Phys. Stat. Sol. (b) 48 K139 (1971).

(19) R. L. Melcher and B. A. Scott, Phys. Rev. Letts. 28 607 (1972).

(20) J. R. Sandercock, S. B. Palmer, R. J. Elliott, W. Hayes, S. R. P. Smith and A. P. Young, J. Phys. C5 3126 (1972).

(21) M. T. Hutchings, R. Scherm and S. R. P. Smith, AIP Conference Proceedings 29 372 (1975).

(22) B. W. Mangum, J. N. Lee and H. W. Moos, Phys. Rev. Letts. 27 1517 (1971).

(23) R. T. Harley, C. H. Perry and W. Richter, J. Phys. C (in press).

(24) E. Pytte, Phys. Rev. B8 3954 (1973).

(25) J. W. McPherson and Y. L. Wang, J. Phys. Chem. Sols. 36 493 (1975).

(26) G. Schaak, Solid State Commun. 17 505 (1975).

(27) R. T. Harley, W. Hayes, A. M. Perry, S. R. P. Smith, R. J. Elliott and I. D. Saville, J. Phys. C7 3145 (1974).

(28) S. J. Allen, Phys. Rev. 166 530 (1968) and 167 492 (1968).
(29) J. Faber and G. H. Lander, Phys. Rev. B14 1151 (1976).
(30) P. J. Colwell, L. A. Rahn and C. T. Walker "Light Scattering in Solids" Ed. M. Balkanski, R. C. C. Leite and S. P. S. Porto, p.239 (Paris, Flamarion, 1976).
(31) G. Dolling, R. A. Cowley and A. D. B. Woods, Canadian J. Physics 43 1397 (1965).
(32) R. A. Cowley and G. Dolling, Phys. Rev. 167 464 (1968).
(33) B. C. Frazer, G. Shirane, D. E. Cox and C. E. Olsen, Phys. Rev. 140 A1448 (1965).
(34) O. G. Brandt and C. T. Walker, Phys. Rev. Letts. 18 11 (1967).
(35) R. T. Harley, W. Hayes, A. M. Perry and S. R. P. Smith, J. Phys. C6 2382 (1973).
(36) R. J. Birgeneau, J. K. Kjems, G. Shirane and L. G. Van Uitert, Phys. Rev. B10 2512 (1974).
(37) P. A. Fleury, P. D. Lazay and L. G. Van Uitert, Phys. Rev. Lett. 33 492 (1974).
(38) K. B. Lyons, R. J. Birgeneau, E. I. Blount and L. G. Van Uitert, Phys. Rev. B11 891 (1975).
(39) M. D. Sturge, E. Cohen, L. G. Van Uitert and R. P. van Stapele, Phys. Rev. B11 4768 (1975).
(40) R. T. Harley, W. Hayes, A. M. Perry and S. R. P. Smith, J. Phys. C8 L123 (1975).
(41) P. A. Fleury "Light Scattering in Solids" ed. M. Balkanski, R. C. C. Leite and S. P. S. Porto, p.747 (Paris, Flamarion, 1976).
(42) J. C. Slonczewski and H. Thomas, Phys. Rev. B1 3599 (1970).
(43) R. T. Harley, J. Phys. C (in press).
(44) R. A. Cowley. Phys. Rev. B13 4877 (1976).

ELECTRIC SUSCEPTIBILITY STUDIES OF COOPERATIVE JAHN-TELLER ORDERING IN RARE-EARTH CRYSTALS

D. R. Taylor

Clarendon Laboratory Oxford University, U.K. and

Queen's Univ., Kingston, Canada (permanent address)

The study of dielectric anomalies at structural phase transitions, for many years a standard technique, has only recently been applied to Jahn-Teller (JT) phase transitions. A number of interesting results have already been obtained, some of which will be described below. They include the identification of novel JT behaviour in $PrCl_3$ and some of its isomorphs[1-3], and the observation of interesting dielectric anomalies in previously studied JT systems such as $DyVO_4$[4] and CeES[5] (ES = ethyl sulphate). For convenience only rare-earth insulating JT systems will be discussed. Many of their properties have been reviewed in lectures by Harley, by Kjems, and by Stinchcombe at this conference, and in an article by Gehring and Gehring[6].

Consider first the conditions under which a significant electric dipole moment (EDM) can occur at the site of a rare-earth ion. The point symmetry must of course be non-centric, but if the low-temperature ordering of the crystal is to be affected the ion must also have degenerate or approximately degenerate electronic levels. The latter case does not further restrict the symmetry, but for exact degeneracy the symmetry cannot be too low. For the important case of non-Kramers doublet levels Müller[7] showed that an EDM occurs for only seven point groups: C_3, C_{3v}, C_{3h}, D_3, D_{3h}, D_{2d}, S_4. The existence of an EDM implies a linear splitting of the electronic levels in an applied electric field. In addition it should be clear that ions satisfying the above conditions are JT active, although the converse is not true.

<u>$PrCl_3$.</u> $PrCl_3$ is representative of a number of compounds containing Pr ions in threefold symmetry in which electric dipole splittings were first investigated[8,9]. The ground state is a non-

Kramers doublet transforming as the E' representation of the C_{3h} point group. The symmetric product $[E' \times E'] = A' + E'$. The doublet can therefore be split by a magnetic field H_z parallel to the threefold axis, which transforms like A', and electric field components E_x, E_y which transform like E'. It is convenient to describe these splittings by an effective spin $S = \frac{1}{2}$ Hamiltonian[8,9]:

$$\mathcal{H} = g_z \beta H_z S_z + \gamma(E_x S_x + E_y S_y) \qquad (1)$$

where γ is the "uncorrected" EDM (if E_x, E_y are laboratory fields). Eqn. (1) shows the formal similarity of electric and magnetic dipole moments in these systems, but leaves unanswered questions about the physical origin and typical magnitude of γ. The major contribution to γ can be visualized as follows: E_x or E_y distort the rare earth site by displacing anions and cations in opposite directions. This removes the threefold symmetry and consequently splits the electronic doublet in proportion to the field. The magnitude of γ will depend on the electronic wave functions but is typically comparable to, and sometimes much greater than, the magnetic dipole moment[10].

In $PrCl_3$ the Pr ground doublets will also be split by distortins of E' symmetry. These distortions will be coupled by the lattice and this can be described in terms of an effective spin-spin interaction[6] by transforming the electron-phonon Hamiltonian as described in the lecture by Stinchcombe. Since the distortion mode is an E' doublet the interaction takes the XY form[11]

$$\mathcal{H}_{ij} = J_{ij}(S^i_x S^j_x + S^i_y S^j_y) \qquad (2)$$

Such an interaction has been directly confirmed by EPR measurements[12] on diluted $PrCl_3$, which also established the magnitude $J_{ij} \approx 2\ cm^{-1}$ for nearest neighbours. Each Pr has two nearest neighbours along the threefold axis and it is clear that more distant interactions are small since the low-temperature properties show marked one-dimensional behaviour. For example the specific heat shows a broad maximum which is fit quite well by the XY chain specific heat[1,12], followed by a narrow spike indicating long-range ordering at only 0.4 K. Since E_x and E_y couple to the E' distortions as described above the E' distortions carry an EDM. The JT ordering is therefore expected to be antiferroelectric, and this was confirmed by electric susceptibility measurements[1] as shown in fig. 1. Above the broad maximum due to short-range ordering the data are fitted to the high-temperature series for the 1-D XY susceptibility, yielding $J = 2.0\ cm^{-1}$ in agreement with other experiments. The data also determine the EDM γ ($\gamma = 6.6 \times 10^{-31}$ C.m[1,2]), and it turns out that the nearest-neighbour interaction agrees in magnitude with the point electric dipole-dipole interaction: $J \sim \gamma^2/4\pi\varepsilon_0 R^3$. As yet the symmetry of the low temperature phase has not been determined, but the distortion is expected to be E_{2g}

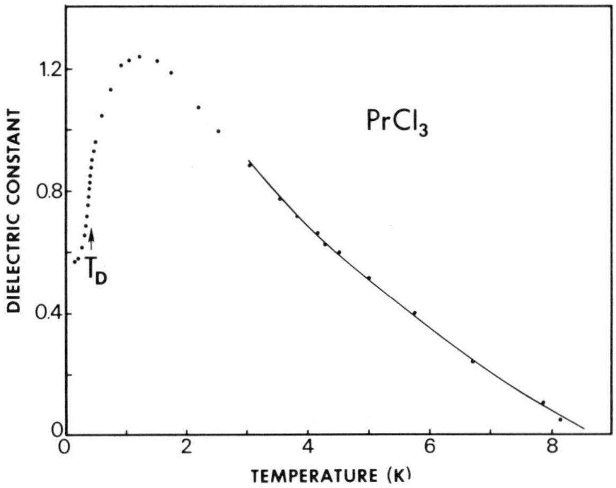

Fig. 1. Electric susceptibility of $PrCl_3$

or E_{1u} with the unit cell doubled along the threefold axis. In summary $PrCl_3$ has some unique properties compared to JT systems previously studied, in particular the quasi 1-D antiferroelectric ordering. On the other hand the low ordering temperature is inconvenient for structural studies. Similar behaviour has been observed in PrES[2], which shows the same 1-D ordering but no long-range order down to ~0.06 K, and in PrBr[3], which has a phase transition at 0.37 K.

<u>$DyVO_4$</u>. One of the most extensively-studied JT systems, $DyVO_4$ differs in many ways from $PrCl_3$. The Dy ions are at D_{2d} sites, and their lowest energy levels are two nearly degenerate Kramers doublets. At the 14 K phase transition a B_{1g} lattice distortion occurs, corresponding to a B_2 distortion at Dy sites. Because the distortion mode is a singlet the effective spin-spin interactions are Ising rather than XY. There is reason to expect a dielectric anomaly in this system because D_{2d} symmetry allows an EDM parallel to the tetragonal axis, and this EDM belongs to the B_2 representation. Hence the B_2 distortion carries an EDM which is therefore an order parameter. Recently the electric susceptibility of $DyVO_4$ was measured[4,13] with results as shown in fig. 2. This curve resembles the typical "parallel" magnetic susceptibility in a 3-D antiferromagnet and was interpreted[4] in terms of improper antiferroelectric ordering; that is the EDM's of the two Dy ions in each unit cell order antiparallel but the unit cell is not doubled. This conclusion is reinforced by the relatively large Dy EDM in this crystal. The high-temperature data of fig. 2 indicate $\gamma \sim 10^{-30}$ Cm[13], which, together with the lattice structure suggests strong antiparallel

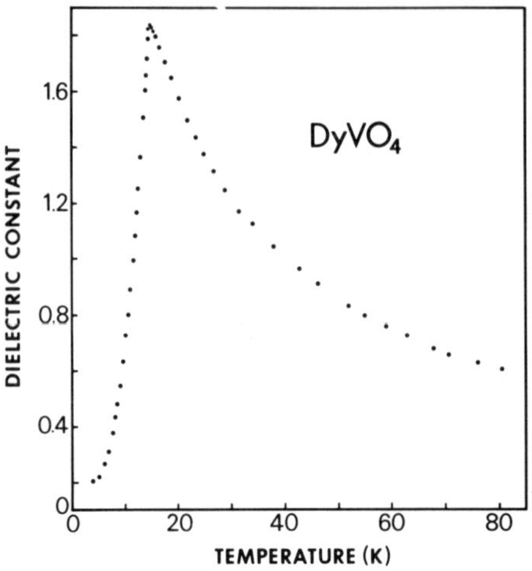

Fig. 2. Electric susceptibility of $DyVO_4$

nearest-neighbour electric dipole interactions. There is considerable experimental evidence, as discussed by Harley and Kjems in their lectures, that the spin-spin interactions in $DyVO_4$ are very short-range, in contrast to other JT systems, such as $TmVO_4$ which undergoes a different distortion and which clearly has long range interactions. It is interesting that in both $DyVO_4$ and $PrCl_3$ where electric dipole ordering occurs the effective spin-spin interactions are short-range.

CeES. The Ce electronic levels in CeES resemble those in $DyVO_4$: two Kramers doublets separated by only a few cm^{-1}. The JT behaviour of CeES is entirely different however because the Ce levels couple most strongly to a uniform (A') lattice distortion. This behaviour[14], originally proposed to explain the specific heat anomaly, is only possible in systems with accidental degeneracy because separation of the levels can be increased, and hence the total energy lowered, by a change in unit cell parameters without lowering the symmetry. The effective spin-spin coupling in this case is that of an Ising interaction in a longitudinal field[5]. No phase transition occurs: the doublet separation and the lattice strain increase monotonically from high- to low-temperature limits[14]. Although in this case no electric dipole ordering occurs, electric susceptibility is a useful probe because an electric field perpendicular to the hexagonal axis couples the two non-Kramers doublets and hence allows their separation to be monitored. Using mean field theory[5] the electric susceptibility is given by

$$\chi_E = (N\gamma^2/2\Delta)\ \tanh(\Delta/2kT) \tag{3}$$

where Δ is the separation between the doublets. Δ varies with temperature according to[14]

$$\Delta - \Delta_\infty = (\Delta_0 - \Delta_\infty)\tanh(\Delta/2kT) \quad (4)$$

where Δ_∞ and Δ_0 are the high- and low-temperature limits. This gives χ_E a "sharper" temperature dependence than if Δ is constant (no JT coupling) just as the specific heat peak is sharper than a simple Schottky peak[14]. Electric susceptibility measurements[5] confirm this interpretation and give reasonable quantitative agreement with the specific heat results[14].

In summary, in JT systems where the crystal symmetry allows electric dipole coupling between degenerate or nearly degenerate crystal field levels the electric susceptibility technique is a very effective probe of JT distortions.

REFERENCES

1. J.P. Harrison, J.P. Hessler, D.R. Taylor, Phys. Rev. B14, 2979 (1976).
2. J.T. Folinsbee, J.P. Harrison, D.B. McColl, D.R. Taylor, J. Phys. C 10, 743 (1977).
3. D.R. Taylor, J.P. Harrison, D.B. McColl, Physica 86-88B, 1164 (1977).
4. H. Unoki, T. Sakudo, Phys. Rev. Letters 38, 137 (1977).
5. D.R. Taylor et al (to be published).
6. G.A. Gehring, K.A. Gehring, Rep. Prog. Phys. 38, 1 (1975).
7. K.A. Müller, Phys. Rev. 171, 350 (1968).
8. F.I.B. Williams, Proc. Phys. Soc. 91, 111 (1967).
9. J.W. Culvahouse, D.P. Schinke, D.L. Foster, Phys. Rev. Letters 18, 117 (1967).
10. J.W. Culvahouse, L. Pfortmiller, D.P. Schinke, J. Appl. Phys. 39, 690 (1968).
11. E. Pytte, Phys. Rev. B8, 3954 (1973).
12. J.W. Culvahouse, L. Pfortmiller, Bull. Am. Phys. Soc. 15, 394 (1970).
13. R.T. Harley, D.R. Taylor (to be published).
14. J.R. Fletcher, F.W. Sheard, Solid St. Commun. 9, 1403 (1971).

NEUTRON SCATTERING STUDIES OF THE COOPERATIVE JAHN-TELLER EFFECT

J.K. Kjems

Research Establishment Risø

DK-4000 Roskilde, Denmark

ABSTRACT

The results of neutron scattering studies of the Jahn-Teller transitions in $PrAlO_3$, $TbVO_4$, $TmVO_4$, $TmAsO_4$, and $PrCu_2$ are summarized. It is found that RPA theory gives an excellent description of both the static and dynamic behaviour of these systems. Within this theory the momentum dependences of the effective electron-phonon coupling parameters have been derived from the measurements on $PrAlO_3$ and $TmVO_4$, and in the latter case an excellent agreement is found with the results of a similar analysis of ultrasonic and Brillouin scattering experiments. The soft phonon response in both $PrAlO_3$ and $TbVO_4$ contains a diverging quasielastic component which only in the case of $TbVO_4$ can be accounted for by the RPA theory adapted by Smith for this problem.

Lectures presented at the NATO Advanced Study Institute, Geilo, Norway, April 1977.

1. INTRODUCTION

The cooperative Jahn-Teller Effect (CJTE) has many fascinating aspects and the efforts that have been devoted to this subject have often had quite different motivations. One may view the CJTE as a basic mechanism for structural phase transitions belonging to the class of pseudo-spin-phonon coupled systems which also induces the hydrogen bonded ferroelectrics[1] and molecular crystal systems[2] where the pseudo-spin relates to the proton positions and the molecular orientations, respectively. In the CJTE-systems the driving force for the phase transition can be readily identified. The free energy of the electronic system with a degenerate or near degenerate ground state is reduced by lowering of the local symmetry around the Jahn-Teller active ions. The electronic energy gain is linear in the distortions whereas the balancing elastic energy has a quadratic dependence so at low enough temperatures the total free energy will be minimized by a finite distortion. The coupling between the lattice deformations and the electronic degrees of freedom can be pictured as an effective quadrupole-quadropole interaction mediated by the phonons[3] in analogy to the usual Rudermann-Kittel exchange interaction in metals where the electrons play the roles of the go-betweens. As in the magnetic case it is a useful first approximation to neglect the microscopic origin of the interactions and to work with model Hamiltonians which then describe the thermodynamics of the phase transitions quite well on the basis of effective ion-ion interaction parameters. Regarded in this manner the simplest Jahn-Teller systems are realizations of the 3-dimensional Ising model with or without a transverse field.

The CJTE systems differ from the usual magnetic systems in the sense that the energy bands of the interaction mediators, the phonons, overlap with the energies of the localized electronic excitations and this gives rise to strong mode mixing[3] which in turn enables accurate measurements of the electron-phonon coupling matrix elements including their q-dependence[4].

A variety of experimental methods have been applied in the study of the CJTE. They range from bulk methods like specific heat, susceptibility and thermal expansion measurements to the use of microscopic probes like x-ray, neutron and light scattering as well as resonance and ultrasonic methods. These notes focus on the results that have been obtained by the use of neutron scattering to characterize both the static and the dynamic properties of systems that display the CJTE. The theoretical background is to a large extent covered in the lectures by Stinchcombe at this school and also in the recent review articles by Gehring and Gehring[5] and by Fulde[6]. Thus the theoretical discussions in this paper will be limited to what is needed in order to comprehend the analysis of the data that are presented.

The systems that so far have been studied in various degrees of detail by neutron scattering all contain rare earth ions and they are the insulators $PrAlO_3$[4] (T_D = 151 K), $TbVO_4$[7] (T_D = 33 K), $TmVO_4$[8] (T_D = 2.1 K), and $TmAsO_4$ (T_D = 6.0 K) and the metallic compound $PrCu_2$ (T_D = 8.0 K). The experimental results for these systems are reviewed with the following organization of these notes. First the properties of the neutron as a spectroscopic tool is summarized followed by a short description of the zoology of the crystal structures and symmetries of the systems involved. Then follows a short theoretical discussion aimed at the fixing of the nomenclature. The experiments are then summarized beginning with the static and critical properties followed by the results for the normal modes and the mode mixing. Finally some conclusions are drawn together with some guesses at what may result from the continued studies.

2. THE NEUTRON PROBE

All the known Jahn-Teller transitions involve zone centre modes and so neutron scattering is only one of many tools that can be applied. The first spectroscopic studies were made using Raman and Brillouin scattering and ultrasonic methods[5]. The power of the neutron probe lies partly in the fact that excitations can be followed throughout the Brillouin zone and partly in the versatility of neutron spectrometry which allows you to study both the structural changes through Bragg diffraction and the excitation spectrum with the same set-up. The neutron couples both to the translational degrees of freedom of the atoms in the crystal and to the electronic degrees of freedom of the Jahn-Teller ions. The deformations of the unit cell that result from a CJT-transition give rise to shifts in the positions of Bragg peaks and the internal displacements of atoms within the unit cell can in principle be detected through intensity changes. An alternative method to determine the internal displacements that result from a CJTE-transition is impurity ESR by which one can monitor the orientation of the electric field gradient tensor at the impurity site. This method has been very successful in the study of the 151 K transition in $PrAlO_3$ as will be discussed later.

The change in the electronic energy levels can in some cases be observed directly via magnetic dipole transitions which couple to the neutron moment. The cross section for such processes is given in the non interacting limit by the "crystal field" formulae[9]

$$\frac{\partial^2\sigma}{\partial\Omega\partial\omega} = N\left(\frac{1.91\ e^2}{2mc^2}\ g_J\right)^2 f^2(\vec{\kappa})\ \frac{k_f}{k_i} \sum_{n,m} \rho_n |\langle n|\vec{J}_\perp|m\rangle|^2 \, e^{-2W}\delta(E_n - E_m + h\omega) \tag{1}$$

where g_J is the Landé factor, f the form factor, $\vec{\kappa}=\vec{k}_i-\vec{k}_f$ the momentum transfer, and ρ_n the occupation probability for the $\langle n|$ electronic state in the J manifold with the energy E_n. The energy transfer is $h\omega = \frac{\hbar^2}{2M_n}(k_i^2-k_f^2)$. M_n is the neutron mass and m the electron mass. $J_\perp$ is the component of the total angular momentum operator <u>perpendicular</u> to the momentum transfer $\vec{\kappa}$. It should be stressed that the polarization information is given in the negative sense that the component <u>parallel</u> to the momentum transfer <u>does not</u> contribute to the scattering. The cross section for crystal field transitions changes if there is noticeable dispersion due to ion-ion interactions and as an example we quote the result for a J^z transition" with a dipole matrix element α and energy splitting 2Δ

$$\frac{\partial^2\sigma}{\partial\Omega\partial\omega} = \left(\frac{1.91\ e^2}{2mc^2}\ g_J\right)^2 \frac{k_f}{k_i}\ f^2(\vec{\kappa})\sin^2\psi\alpha^2\ \frac{2\Delta}{E_{\vec{\kappa}}}\ \rho_0(T)\ e^{-2W} \tag{2}$$

where ψ is the angle between the z-axis and $\vec{\kappa}$. E is the dispersed energy of the mode given by the RPA formulae

$$E_{\vec{\kappa}} = 4\Delta(\Delta - \alpha^2 J(\vec{\kappa})) \qquad (3)$$

where $J(\vec{\kappa})$ is the Fourier transform of the ion-ion interaction. More general expressions for a many level system with interactions have been worked out by Fulde and Peschel[12], Young[13] and by Buyers[14].

The phonon cross section has some characteristic differences compared to the magnetic cross section. The one phonon creation cross section[9] is given by

$$\frac{\partial^2\sigma}{\partial\Omega\partial\omega} = \frac{(2\pi)^3}{2v_o}\frac{k_f}{k_i}\frac{1}{\omega}\left|\sum_j a_j\, e^{+i\vec{\kappa}\cdot\vec{r}_j}\frac{\vec{\kappa}\cdot\vec{u}_j(\vec{\kappa})}{\sqrt{M_j}}\right|^2 \frac{1}{1-e^{-h\omega\beta}}\, e^{-2W} \qquad (4)$$

with $u_j(\vec{\kappa})$ being the eigenvector for the jth atom for the mode in question. It is important to note that in general the phonon scattering increases with the square of the momentum transfer whereas the magnetic cross sections decrease with the form factor. This can be used together with the different temperature dependences to identify the origin of the observed scattering.

As we shall see later the CJTE leads to mixed crystal field-phonon modes and in principle this could give rise to interesting interference effects. However, in most cases only one of the modes has an appreciable cross section and when the modes mix they "share" this scattering power and may both become observable in certain regions of the Brillouin zone.

3. SYMMETRIES AND CRYSTAL FIELDS

One of the keys to the understanding of the CJTE is the way in which the symmetry of the crystalline electric field at the active ion sites determines the electronic levels, the crystal field splittings. For a number of years this has been an active field of research on its own and a large number of materials have been characterized[15]. In cases of high local symmetry only few parameters are needed to define the crystal field Hamiltonian

$$H_{CEF} = \sum_{\substack{n=2,4,6 \\ m=-n,n}} B_n^m \chi_n \mathcal{O}_n^m(J) \quad (5)$$

where B_n^m are the crystal field parameters, χ_n the reduced matrix elements tabulated by Elliot and Stevens[16], and $\mathcal{O}_n^m(J)$ are the angular momentum operator equivalents. The actual number of parameters needed in the different crystallographic phases one for example encounters in a system like $PrAlO_3$ are the following[4]:

O_h,2; D_3,6; C_{2v},9; C_s,15; and D_{2d},5.

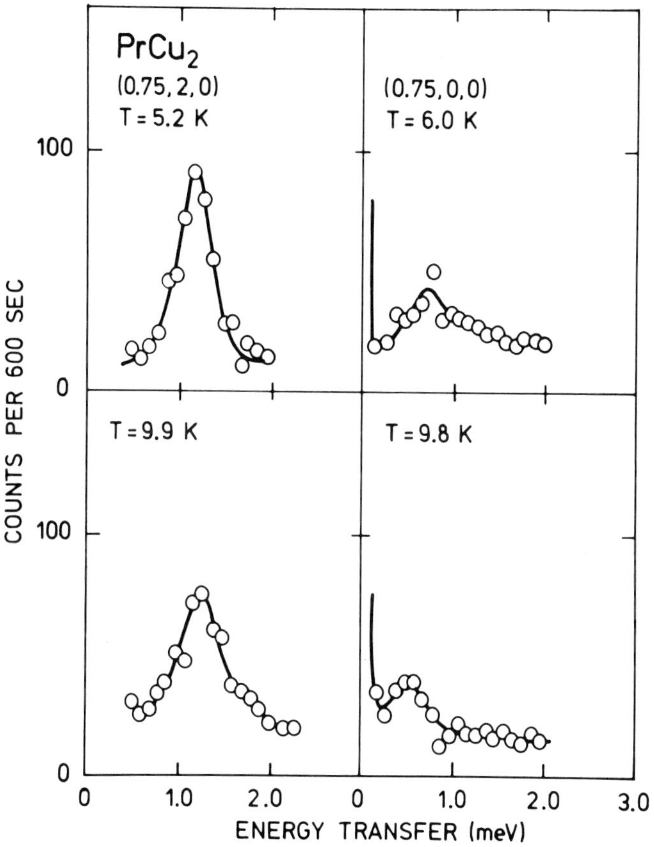

Fig. 1. Left hand side: Constant Q scans at (0.75,2,0) at 5.2 K and 9.9 K in $PrCu_2$.
Right hand side: Constant Q scan at (0.75,0,0) at 5.6 K and 9.8 K in $PrCu_2$.
The Jahn-Teller transition occurs at T_D = 8 K.

Hence it is often necessary to make some simplifying assumptions in order to arrive at a qualitative description as it was done for $PrAlO_3$ by Lyons et al.[17]

As an example we discuss the crystal field level diagram for Pr^{+3} $^3H_4$ in orthorhombic, C_{2v}, symmetry. This pertains both to $PrAlO_3$ and to $PrCu_2$ above T_D. In the low C_{2v} symmetry, which contains two mirrorplanes and a two-fold axis, the ninefold multiplet is split into 9 singlets, namely $3A_1 + 2A_2 + 2B_1 + 2B_2$. As an example of an observed transition Fig. 1 shows some recent inelastic neutron constant Q-scans obtained with a single crystal of $PrCu_2$ which was oriented with the a-b plane in the scattering plane of the spectrometer. The scans refer to the same point (0.75,0,0) in the reduced Brillouin zone with different polarizations. The strong transition at 1.3 meV in Fig. 1.a disappears when $\vec{\kappa}$ is directed along the a* axis, Fig. 1.b. Hence one concludes that this transition is induced by the J^x component of the total angular momentum operator. One then deduces that the two levels connected must be either A_1-B_2 or A_2-B_1. Unfortunately, this piece of information is clearly insufficient to make level assignments, but hopefully with enough data collected one can make some realistic attempts. Fig. 1 also shows that the phase transition at 8 K essentially leaves the strong J^x transition unaffected whereas the much weaker J^y or J^z transitions in Fig. 1.b are shifted appreciably in energy. Tentatively, the present results are interpreted to give a crystal field level scheme for the lowest 4 levels in $PrCu_2$ as indicated in Fig. 2 together with the much more firmly established results for the other CJTE-systems.

In $PrAlO_3$ a combination of fluorescence, neutron and Raman spectroscopy paired with model calculations have resulted in a detailed understanding of the full level scheme (see figure 17 in the proceeding paper by Harley). At T_D = 151 K only the B_1 transition between the lowest two that couples to the lattice and drives the phase transitions.

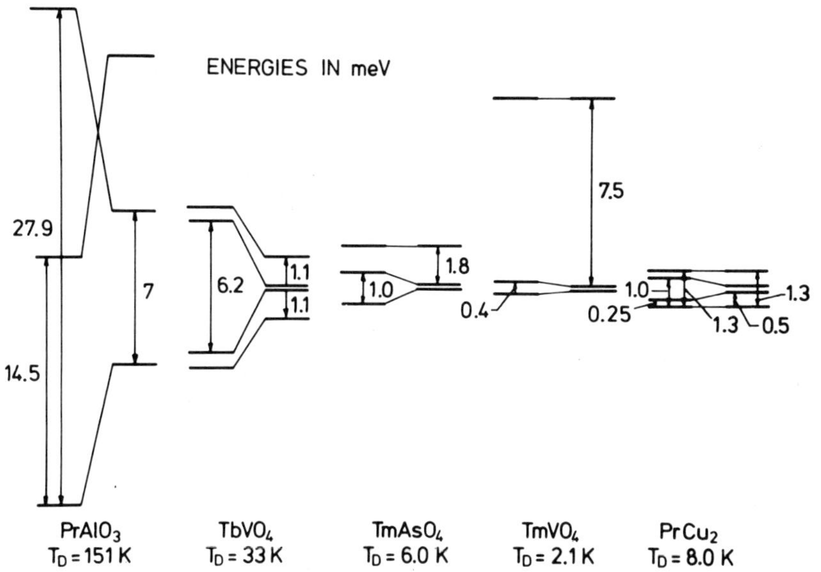

Fig. 2. Schematic representation of the change in the lowest electronic energy levels for the CJTE systems that so far have been investigated by neutron scattering.

4. THEORY

The theory developed by Elliott et al.[1] for a two level system with a splitting, 2Δ, is briefly summarized. The interaction with the lattice is expressed as a linear coupling between electric quadrupole operators, often in a pseudospin representation, and the normal modes of the lattice of compatible symmetry.

$$
\begin{aligned}
H = & \sum_{n,\vec{k},p} \xi_p(\vec{k}) e^{i\vec{k}\cdot\vec{R}_n} S_n^z (a_{p,\vec{k}} + a^+_{p,\vec{k}}) \\
+ & \sum_{\vec{k},p} \hbar\,\omega_p(\vec{k})(a^+_{p,\vec{k}} a_{p,\vec{k}} + \tfrac{1}{2}) \qquad (6) \\
+ & \Delta \sum_n S_n^x
\end{aligned}
$$

where S_n^z and S_n^x are the pseudospin operators representing two electronic states at the energies $\pm\Delta$, $a^+_{p,\vec{k}}$ is the operator for a phonon of wavevector $\vec{k}$, band p and frequency $\omega_p(\vec{k})$ and $\xi_p(\vec{k})$ denotes the corresponding Jahn-Teller coupling. The detailed solutions of this Hamiltonian are discussed in the paper by Stinchcombe. The first step is a displaced oscillator transformation which gives

$$H = \sum_{\vec{k}} \hbar\, \omega_p(\vec{k})(\gamma^{+}_{p,\vec{k}}\gamma_{p,\vec{k}} + \tfrac{1}{2}) - \sum_{\vec{k}} J(k)\, \vec{S}_{\vec{k}}\vec{S}_{-\vec{k}} + \Delta N^{\frac{1}{2}}\, S^x_0$$

where

$$\gamma_{p,\vec{k}} = a_{p,\vec{k}} + \frac{\xi_p(\vec{k})}{\hbar\, \omega_p(\vec{k})} S^z_{\vec{k}} \tag{7}$$

and

$$K_p(\vec{k}) = \frac{|\xi_p(\vec{k})|^2}{\hbar\, \omega_p(\vec{k})}, \quad J(\vec{k}) = \sum_p K_p(\vec{k}) - N^{-1} \sum_{p,\vec{k}} K_p(\vec{k})$$

This shows that the linear Jahn-Teller coupling gives rise to an effective quadrupole-quadrupole interaction, $J(\vec{k})$, which here includes the static strain contributions. As a first approximation for the static properties one ignores the non-commutation of the displaced oscillator and the transverse field parts of Eq. 6 and simply treats the Ising model in a transverse field via molecular field theory. The essential results are the following[1,5]

(a) T_D is defined by

$$\frac{\Delta}{J(0)} = \tanh\left(\frac{\Delta}{kT_D}\right) \tag{8}$$

(b) Below T_D the order parameter $\langle S^z \rangle$ is given by

$$\langle S^z \rangle = \frac{J(0)\, \langle S^z \rangle}{W} \tanh\left(\frac{W}{kT}\right) \tag{9}$$

where

$$W^2 = J^2(0)\langle S^z \rangle + \Delta^2 \quad \text{and} \quad \langle S^x \rangle = \frac{\Delta}{W} \tanh\left(\frac{W}{kT}\right)$$

(c) If the elastic strain energy is taken as $\frac{1}{4} N V_o C e^2$ then the equilibrium strain is

$$\langle e \rangle = 4\, K_a(0)\langle S^z \rangle / V_o C \tag{10}$$

and similarly for the optic mode displacements

$$\langle Q_0 \rangle = - K_0(0)\langle S^z \rangle / M\omega^2(0) \qquad (11)$$

The results for the dynamics are quoted later under the discussion of the mixed modes.

5. STATIC AND CRITICAL PROPERTIES

The temperature dependence of the order parameter below T_D has been thoroughly studied in $PrAlO_3$. The static strain has been determined from the twinning patterns that evolves due to domain formation[4] below T_D. The optic phonon order parameter[10] has been deduced from measurements of the electric field gradient tensor around a Gd^{+3} impurity as sensed by ESR and the electronic order parameter, as manifested in the splitting, 2W, has been measured by fluorescence[10,17,20] relative to the $^3P_0$ state of the Pr^{+3} ion. Fig. 3 summarizes all these results on normalized scales together with the simplest mean field calculation[4]. The internal consistency between these different measurements of the temperature dependence of $\langle S^z \rangle$ offers strong support to the basic

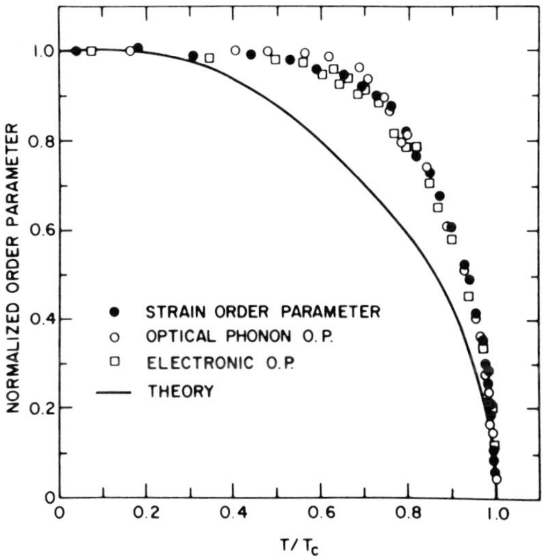

Fig. 3. The experimentally determined order parameters in $PrAlO_3$ displayed on normalized scales. The full line illustrates the results of the simplest mean field calculation for a two level system.

physical assumptions in the theory of the CJTE. The discrepancy between the mean field calculation and the experimental results for $PrAlO_3$ is probably due to oversimplifications in the description of this rather complicated system. For the simpler $TmVO_4$ system Segmüller et al.[21] find complete agreement between the mean field calculation of the strain order parameter and the results of an X-ray experiment.

The critical exponent, β, for the order parameter has been determined for $PrAlO_3$[4], $TbVO_4$[22], $TmAsO_4$ and $TmVO_4$[21] and in all cases one finds $\beta \doteq 0.50$. The only CJTE system that does not conform to this picture is $DyVO_4$[22] where Harley found $\beta = 0.34$ using birefringence as a probe. The fact that all but one CJTE-systems display the classical exponent $\beta = 0.5$ has often been ascribed to the long range nature of the strain mediated interactions. However, as pointed out by Cowley[23] and by Als-Nielsen and Birgeneau[24] this may be too naive. More rigorous arguments can be made on basis of renormalization group theory paired with the Ginzburg[25] criterion and the concept of marginal dimensionality, d*. In order to obtain a self-consistent picture within mean field theory Ginzburg stated that below the transition temperature the fluctuations in the order parameter, $\delta\sigma$, averaged over a suitable region, Ω, should be small compared to the averaged order parameter, σ_Ω.

$$(\delta\sigma)^2_\Omega << \sigma^2_\Omega$$

The suitable region is the one spanned by the correlation length, ξ, and it is the real space counterpart of the volume in reciprocal space determined by the half value of the critical susceptibility, κ_g. The key point is that the volume of the correlated regions in real space may diverge more rapidly than ξ^d in cases where the interactions are anisotropic.

$$\Omega_\xi \propto \xi^{d+m} \qquad m>0$$

The marginal dimensionality is defined by

$$d^* = (\gamma'+2\beta)/\nu'-m$$

$$d^* = (1+2\cdot\tfrac{1}{2})/\tfrac{1}{2}-m = 4-m \quad \text{(mean field exponents)}$$

where γ', β, and ν' are the usual critical indices for susceptibility, order parameter, and correlation

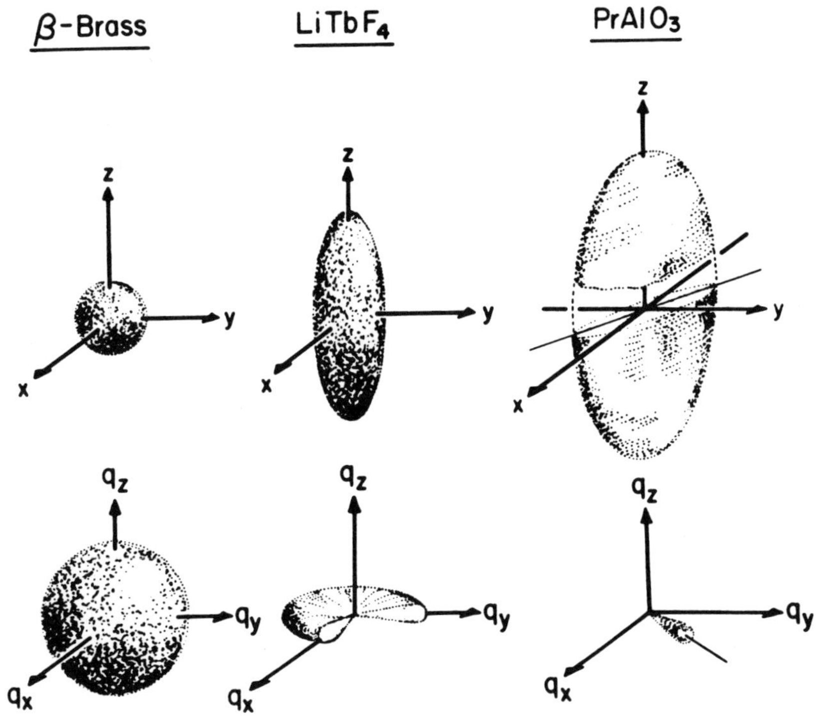

Fig. 4. Examples[24] of the shapes of the half height contours of the susceptibility in reciprocal space and the corresponding correlated regions in real space for 3 model systems with marginal dimensionalities d* = 4 (β-brass), d* = 3 ($LiTbF_4$) and d* = 2 ($PrAlO_3$ and other Jahn-Teller systems).

length, respectively. The Ginzburg criterion can then be expressed as follows. The molecular field theory is self-consistent if $d > d^*$ and not for $d < d^*$. For $d = d^*$ there are logarithmic corrections to the mean field results. Examples of the shape of κ_q and the corresponding correlated regions in real space are shown in Fig. 4 for 3 model systems. The first is β-brass a $d = 3$ Ising system with isotropic interactions. Here $m = 0$ and $d^* = 4$ so mean field theory fails. The second example is $LiTbF_4$, a dipolar coupled Ising ferromagnet where $m = 1$ $d^* = d = 3$, and experiments have shown very beautifully that the exact renormalization group theory applies. The third example shown in Fig. 4 is $PrAlO_3$ which represents the class of transition of interest in the present context. Here the fluctuations are confined to the lines in reciprocal space along which the acoustic anomalies occur. In real space the correlated volumes have "pan-cake" shape, and their volume diverges as ξ^{d+2} i.e. m = 2. ξ measures the thickness of the "pan-cake". Hence d* = 2 and mean field theory is expected to be exact. Indeed one finds that the all Jahn-Teller system discussed here displays classical exponent with the notable exception of $DyVO_4$. The non-classical result for $DyVO_4$ is quite puzzling, and we intend to check it using diffraction methods.

6. NORMAL AND MIXED MODES

The coupling of the electronic mode often called the vibron or the quadrupole exciton to the lattice modes have been observed directly in $PrAlO_3$[4], $TbVO_4$[7] and in $TmVO_4$[8]. The RPA solution for the unperturbed quadrupole mode is given by Eq. 5 and in the simplest picture the frequency of this mode at $\vec{k}$ = 0 tends to zero as T_D is approached. However, the two first terms in the Hamiltonian, Eq. 9, do not commute and this gives rise to mode mixing and anticrossing between the lattice modes and the electronic mode. Ultimately it is the acoustic mode that becomes the soft mode at the transition. This was first demonstrated experimentally for $DyVO_4$[26] by Melcher and Scott who also studied $TmVO_4$[27]. The acoustic anomalies in $TbVO_4$[28] have been studied by Sandercock et al. using ultrasonics and Brillouin scattering and the same probes have been applied to $PrCu_2$[19] and $PrAlO_3$[29], respectively. The first neutron scattering study of the acoustic dispersion in a CJTE-system was the measurement of the Σ_3-branch in $PrAlO_3$[4] which is shown in Fig. 5.

The RPA solution for the mixed mode frequencies, $\omega(\vec{k})$, was derived by Elliott et al. and for > T_D the result is contained in the equation

$$\hbar\omega(\vec{k})(\omega^2(\vec{k})-\omega_v^2(\vec{k})) = \sum_p \frac{4\Delta\langle S^x\rangle\omega^3(\vec{k})K_p(\vec{k})}{\omega^2(\vec{k})-\omega_p^2(\vec{k})} \qquad (12)$$

where ω_v and ω_p refer to the unperturbed vibron and phonon frequencies, respectively. In the limit of small $|\vec{k}|$ the frequency of the mode with predominantly acoustic character is given by

$$\omega^2(\vec{k}) = \omega_a^2(\vec{k}) \frac{\Delta - J(\vec{k})\langle S^x \rangle}{\Delta - (J(\vec{k}) - K_a(\vec{k}))\langle S^x \rangle} \tag{13}$$

This expression has been applied to $PrAlO_3$ at various temperatures and the results are shown as the broken lines labelled 2, 3 and 4 in Fig. 5 where the lines marked 1 and 210 K served to establish the parameter values.

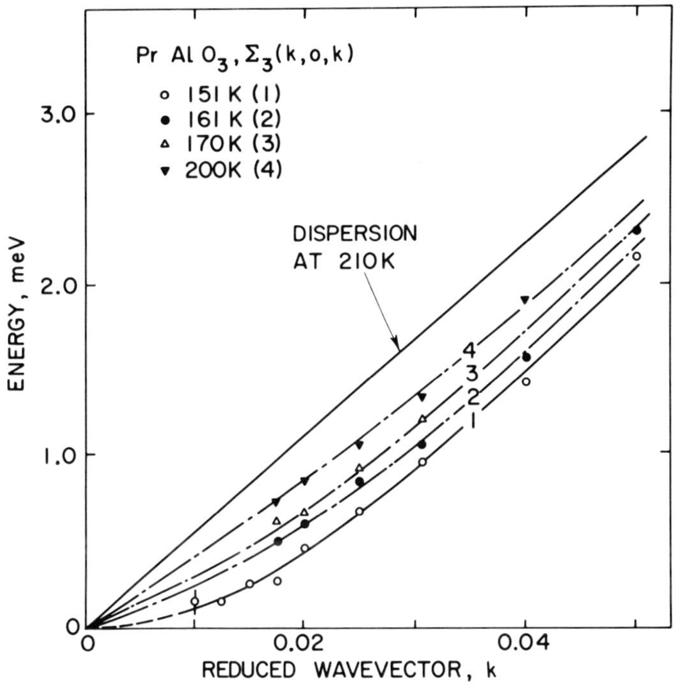

Fig. 5. The observed dispersion of the B_1 acoustic mode in the orthorhombic phase of $PrAlO_3$ at different temperatures. There is an abrupt structural change to a rhombohedral structure at 205 K and the CJTE transition is at 151 K. The unit of the reduced wavevector is $2\pi/3.76$ $Å^{-1}$.

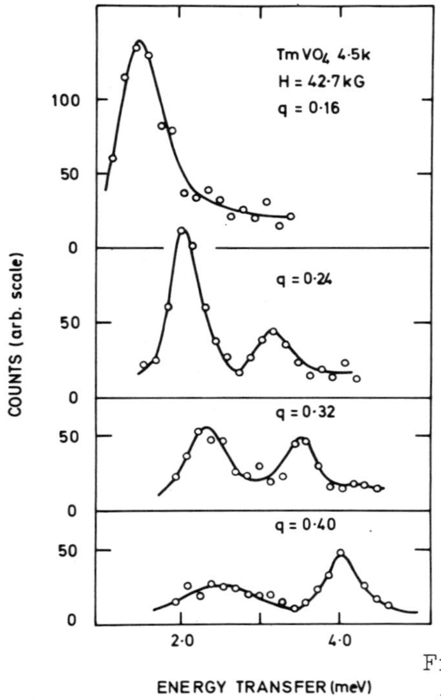

Figure 6.

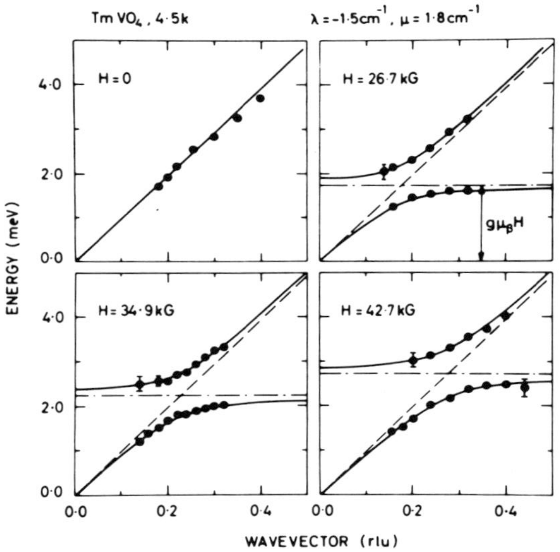

Figure 7.

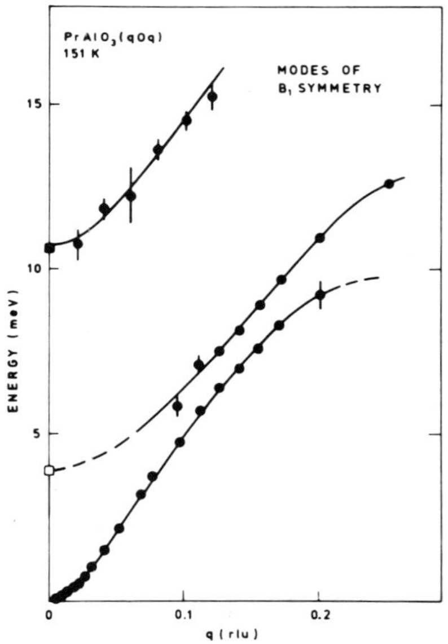

Figure 8.

Fig. 6. Observed spectra for the B_1 acoustic phonon in $TmVO_4$ propagation along (100) and polarized along (010). The zone boundary is at q = 1.0. The phonon anticrosses the Zeeman split doublet ground state at $2D = g\mu_B H = 2.7$ meV.

Fig. 7. The mixed mode dispersion in $TmVO_4$ at 4.5 K for external field along c with the strengths indicated on the figure. The full circles correspond to the peak positions in the inelastic scans. The full lines correspond to calculations with $g = 10.6$, $C_{66} = 1.90\times10^{11}$ erg/cm^2, $\rho = 6.80$ g/cm^3, $K_a(0) = 1.8$ cm^{-1} and $J(0) = +0.3$ cm^{-1}. The broken lines illustrate the unperturbed phonon dispersion and the Zeeman splitting, respectively.

Fig. 8. Dispersion of the mixed modes of B_1 symmetry in $PrAlO_3$ at $T = T_D = 151$ K.

The anticrossing of the acoustic mode and the quadrupole exciton in $TmVO_4$[8] was observed with an applied field along the tetragonal axis, c^*. This resulted in a finite splitting $2\Delta = g\mu_B H$ of the ground doublet. Fig. 6 shows the typical shifts in the intensities from the lower to the upper branch one observes with increasing wavevector. Fig. 7 shows the observed dispersion curves together with the calculated ones based on Eq. 15 using the parameters derived from the acoustic anomaly. The corresponding results for $PrAlO_3$ showing all three modes involved are shown in Fig. 8.

In the measurements on $TbVO_4$ Hutchings[7] et al. found a strong feature in their spectra at zero energy transfer, which diverged as $T \rightarrow T_D$ and $|\vec{q}| \rightarrow 0$. They ascribed this scattering to transitions within the excited doublet (see Fig. 2) and the following formulae was derived for the wavevector dependence of the central peak intensity

$$I(k) \alpha \frac{2K_a(\vec{k})g_2}{\beta\omega_a(\vec{k})(1-J(\vec{k})g_1)(1-J(\vec{k})(g_1+g_2))} \qquad (14)$$

Here they used a Green's function formalism akin to the one used by Buyers in his description of singlet ground state dynamics.

$$g_1 = \frac{\tanh(\beta\Delta)}{\Delta} \cdot \frac{\cosh\beta\Delta}{1+\cosh\beta\Delta}$$

and

$$g_2 = \frac{\beta}{1+\cosh\beta\Delta}$$

are single ion susceptibilities arising from the assumed equal strength transitions between the singlets at 0 and 2D and within the doublet at D.

Both the anticrossing dispersion curves and the central peak scattering can in principle be used to derive experimentally determined effective interaction parameters $K_a(\vec{k})$ and Fig. 9 shows the results of such analysis for $TmVO_4$. This $|q|$-independent result, which is consistent with the parameter that was derived in the ultrasonic experiment[27], confirms that the coupling to the acoustic mode is proportional to the strain produced by the phonon mode as given by the Debye-model.

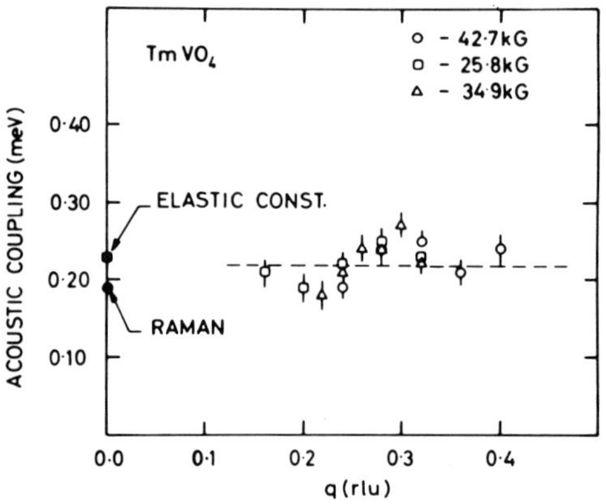

Fig. 9. Wavevector dependence of the CJTE-coupling parameters for $TmVO_4$ as determined from the experimental data shown in Fig. 7 and the use of Eq. 15.

7. DISCUSSION

The overall agreement between the simple mean field and RPA theories and the experimental observations including the temperature dependences is remarkably good. For instance one finds for the prototype

CJTE-system $TmVO_4$ that only two, now well known, parameters are needed to describe all the details of the field and temperature dependences of such varied quantities as the order parameter[20,21] $\Delta(T,H)$, the elastic constant[27], $C_{66}(H,T)$, the susceptibility[30], $\chi(T)$, the specific heat[30] $C_v(T)$ and the mixed-mode frequencies[8] $\omega(\vec{k},H,T)$ in the field and temperature ranges where the ground doublet can be regarded as isolated. This is probably a consequence of the limited role played by the fluctuations in these systems which also makes Landau theory correct for critical exponents. However, there are manifestations of some fluctuations, namely the central peak scattering as observed in $TbVO_4$[8] and less pronounced in $PrAlO_3$[4]. The theoretical description by Hutchings et al. does not apply to $PrAlO_3$ since there are no degenerate electronic levels above T_D. Surprisingly, one does not find similar central peak scattering near T_D in neither $TmVO_4$ not in $TmAsO_4$. Based on Eq. 17 one would expect to observe scattering with intensity $\sim 1/k$ as $T \to T_D$ in $TmAsO_4$. The actual observed scattering falls off much more rapidly and it is confined to the q_y direction within the instrumental resolution of $q_x = \pm 0.01$ rlu. This could indicate that the scattering in $TmAsO_4$ originates in changes in the mosaic distribution as T_D is approached. The paper presented by Møllenbach at this ASI lends support to this interpretation.

Another aspect of the analysis of the CJTE-systems is the understanding of the interaction parameters and their underlying microscopic mechanisms. Dohm[31] has pointed out that second order strain effects may be important for these systems and Fulde[6] has suggested that in the metallic systems the interaction with the conduction electrons also gives rise to quadrupole couplings. In this context a system like $PrCu_2$ may hold some essential clues to be unravelled by skilled searchers although the complexity of the level scheme may prevent any definite conclusions.

Further detailed studies of the soft phonon response in these systems may also be warranted since they could shed light upon the related magnetic problem of ordering in singlet ground state systems where a similar insufficient understanding persists for the relative roles of soft mode and central peak contributions to the generalized susceptibility. As an amusing final point we note that the CJTE-systems that order magnetically presumably are the simplest examples of singlet ground state systems so they may hold the key to the solutions of this problem in more than one sense.

ACKNOWLEDGMENTS

The studies which form the basis for this article were carried out in collaboration with R.J. Birgeneau, W. Hayes, G. Shirane, and H. Ott. The author thanks J. Als-Nielsen and R.J. Birgeneau for the use of Fig. 4 prior to its publication.

REFERENCES

1. R.J. Elliott in "Structural Phase Transitions and Soft Modes", Edited by E.J. Samuelsen, E. Andersen and J. Feder. Universitetsforlaget, Oslo 1971.

2. Y. Yamada, Y. Noda, J.D. Axe and G. Shirane, Phys. Rev. B9, 4429 (1974).

3. R.J. Elliott, R.T. Harley, W. Hayes, and S.R.P. Smith, Proc. R. Soc. A328, 217 (1972).

4. R.J. Birgeneau, J.K. Kjems, G. Shirane, and L.G. Van Uitert, Phys. Rev. B10, 2512 (1974).

5. G.A. Gehring and K.A. Gehring, Rep. Prog. Phys. 38, 1 (1975).

6. Peter Fulde in "Handbook on the Physics and Chemistry of the Rare Earths", Editors K.A. Gschneidner and Le Roy Eyring, to be published.

7. M.T. Hutchings, R. Scherm, and S.R.P. Smith, AIP Conference Proceeding No. 29, p. 372 (1975).

8. J.K. Kjems, W. Hayes, and S.H. Smith, Phys. Rev. Letters 35, 1089 (1975).

9. See for example W. Marshall and S.W. Lovesey "Theory of Thermal Neutron Scattering", Clarendon, Oxford (1971).

10. M.D. Sturge, E. Cohen, L.G. Van Uitert and R.P. van Stapele, Phys. Rev. B11, 4768 (1975).

11. R.J. Birgeneau, AIP Conf. Proc. 10, 1664 (1973).

12. P. Fulde and I. Peschel, Z. Phys. 241, 82 (1971) and Adv. Phys. 21, 1 (1972).

13. A.P. Young, J. Phys. C: Solid State Phys. 8, 3158 (1975).

14. W.J.L. Buyers, AIP Conf. Proc. 24, 27 (1975).

15. For recent reviews see Proceedings of the Second International Conference on "Crystal Field Effects in Metals and Alloys". Edited by A. Furrer, Plenum Press, New York (1977).

16. R.J. Elliott and K.W.H. Stevens, Proc. R. Soc. A218, 553 (1953).

17. K.B. Lyons, R.J. Birgeneau, E.I. Blount, and L.G. van Uitert, Phys. Rev. B11, 891 (1975).

18. M. Wun and N.E. Philips, Phys. Letters 50A, 195 (1974).

19. K. Andres et al., AIP Conf. Proc. 34, 222 (1976).

20. R.T. Harley, W. Hayes, A.M. Perry, and S.R.P. Smith, J. Phys. C6, 2382 (1973).

21. A. Segmüller, R.L. Melcher, H. Kind, Solid St. Comm. (1974).

22. R.T. Harley and R.M. Macfarlane, J. Phys. C8, L53? (1975).

23. R.A. Cowley, Phys. Rev. B13, 4877 (1976).

24. J. Als-Nielsen and R.J. Birgeneau, to be published in Am. Journ. of Phys.

25. V.L. Ginzburg, Soviet Phys. Sol. Stat. 2, 1824 (1960).

26. R.L. Melcher and B.A. Scott, Phys. Rev. Lett. 28, 607 (1972).

27. R.L. Melcher, E. Pytte, and B.A. Scott, Phys. Rev. Letters 31, 307 (1973).

28. J.R. Sandercock et al., J. Phys. C5, 3126 (1972).

29. P.A. Fleury, P.D. Lazay, and L.G. Van Uitert, Phys. Rev. Letters 33, 492 (1974).

30. A.H. Cooke, S.J. Swithenby, and M.R. Wells, Solid State Comm. 10, 265 (1972).

31. Volker Dohm, Z. Phys. B23, 153 (1976).

GAMMA-RAY DIFFRACTION STUDIES OF THE MOSAIC DISTRIBUTION IN $TmAsO_4$ NEAR THE COOPERATIVE JAHN-TELLER TRANSITION AT 6 K

K. Møllenbach and J.K. Kjems

Risø National Laboratory
DK-4000 Roskilde, Denmark

S. H. Smith

Clarendon Laboratory
Oxford University, England

ABSTRACT

The structural cooperative Jahn-Teller phase transformation in $TmAsO_4$ has been studied using gamma-ray diffraction. Precise measurements of the critical exponent, β, confirms the earlier reported classical value $\beta = 0.50 \pm .03$. Precurser effects in the mosaic structure of the sample are observed as T_D is approached from above.

The static and dynamic properties of the cooperative Jahn-Teller effect have been well characterized by the application of a variety of experimental methods to the rare earth insulators[1,2]. Recently a new high resolution gamma-diffraction technique has been developed[3] which allows for very accurate determination of the mosaic distribution of single crystals. In structural phase transitions where strain is an order-parameter one often finds twinning patterns which can be directly related to the magnitude of the strain and hence an accurate measurement of these patterns near T_D can yield the critical behaviour of the order-parameter. Here we report on a study of the cooperative phase-transition in $TmAsO_4$ at T_D = 6 K using this technique. We find that the order-parameter has a classical index β = 0.50 and maybe more interestingly that there are clear precurser effects observable in the mosaic distribution as T_D is approached from above.

The experiment was done using a gamma-ray diffractometer similar to the one described by Schneider[3]. 412 keV gamma-radiation from a 125 Ci Au-foil is collimated in a single slit beam-defining system. The lower limit to the angular width of the primary beam is 15 sec of arc. The instrument takes advantages of the typical small-angle scattering resolution function[4]: A relatively poor resolution along the momentum transfer vector, $\underline{\kappa}$, and an extreme narrow resolution perpendicular to $\underline{\kappa}$. The high energy of the gamma-rays makes the

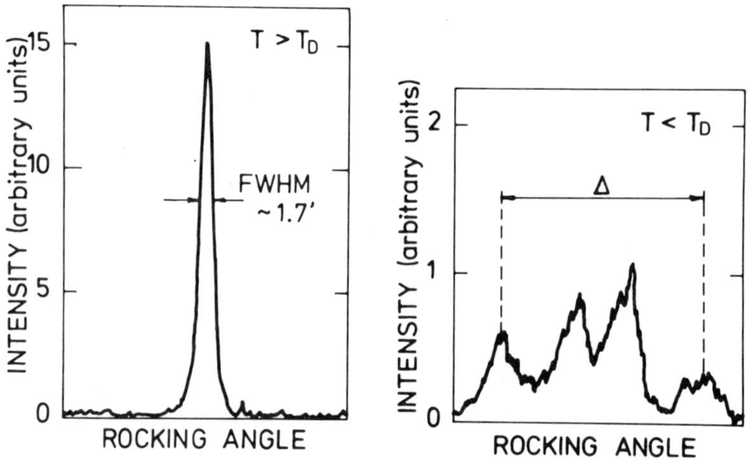

Fig. 1.
Examples of mosaic distributions of the (200) reflection in $TmAsO_4$ above and below T_D. In the right hand side figure T = 5.92 and Δ = .26°.

scattering probability per unit path length in the crystal small giving a negligible secondary extinction and a very small primary extinction. These characteristics make this instrument well suited for a study of the Jahn-Teller phase-transition. The $TmAsO_4$ sample in this experiment had the size 1.5 × 1.5 × 10 mm^3, and was grown in the Clarendon Laboratory.

The phase-transition in $TmAsO_4$ is driven by the coupling of the electronic E-doublet of the tetragonal phase to the B_{2g} strain. As the orthorhombic distortion proceeds a twinning pattern evolves which gives rise to a splitting of the mosaic distribution of the (200)-Bragg reflection. The splitting is proportional to the B_{2g} strain[5]. Examples of mosaic distributions are shown in fig. 1. In the right-hand-side picture the method of measurement of the twinning angle is indicated. The ideal twinning pattern consists of three equally spaced peaks. The outer ones originate from domains with (020) twinning planes and the centre peak from (200) twinning planes. In our measurements the centre peak was also split, probably due to external strains. In figure 2 we show the twinning angle as a function of temperature. These data together with neutron measurements show that for $10^{-1} > 1-T/T_D > 5 \cdot 10^{-4}$ the twinning angle follows the simple power law

$$\Delta \sim (1-T/T_D)^{0.50\pm0.03} , \quad T_D = 6.130 \text{ K}.$$

Changes in the crystal mosaic structure can be observed above T_D both on cooling and heating of the sample. One finds both a broadening of the rocking-curve and a decrease in the peak intensity on approaching T_D from above. This is shown in the right-hand-side of figure 2. This phenomenon is consistent with neutron scattering observations. The present measurements cannot distinguish between the possible origins of the observed changes in the mosaic structure above T_D. As T_D is approached the elastic constant C_{44} tends to zero and if the crystal has internal stresses it may relax in the observed fashion. In a more speculative interpretation one could relate the relaxation to the fluctuations of the quadrupole moments although at present we have no evidence for any dynamical effects.

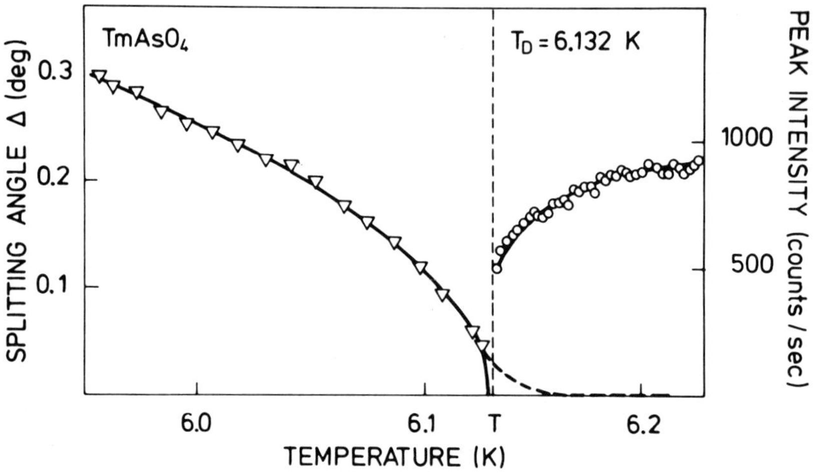

Fig. 2.
Left-hand-side: The splitting angle of the (200) mosaic distribution as a function of temperature (∇). Right-hand-side: The peak intensity of the (200) reflection above T_D (O). The dashed line at the bottom indicates the precurser changes in the observed width.

REFERENCES

1. Kjems, J.K.: Lecture at this NATO ASI.
2. Harley, R.T.: Lecture at this NATO ASI.
3. Schneider, J., J. Appl. Cryst. 7, 541 (1974).
4. can be deduced from:
   Bjerrum Møller, H. and Nielsen, M., Acta Cryst. A25, 547 (1969).
5. Birgeneau, R.J., Kjems, J.K., Shirane, G., and Van Uitert, L.G., Phys. Rev. B10, 2512 (1974).

THE CENTRAL PEAK IN $TbVO_4$

S.R.P. Smith* and M.T. Hutchings†

*Department of Physics, University of Essex
Colchester, Essex, U.K.
†Materials Physics Division
Harwell Didcot, Oxon., U.K.

A central peak (i.e. a low frequency anomaly) has been observed in neutron scattering experiments on $TbVO_4$ above the Jahn-Teller phase transition at T_D = 33 K (1). The measurements provide what is probably at present the most clearly understood example of a central peak occurring in a structural phase transition, though the frequency width has still to be measured directly.

The lowest levels of Tb^{3+} consist of two singlets separated by a splitting 2ε = .49 THz, with a degenerate doublet midway between them (2). The mode of distortion at the phase transition (B_{2g} at $\underline{q}$ = 0) has roughly equal matrix elements between the singlet levels and within the levels of the doublet. Thus, one expects an anomalous response near zero frequency in measurements which involve a coupling to the electronic doublet, and because this coupling is directly involved in the phase transition, one also expects to observe critical effects near T_D in such measurements. The central peak in the neutron scattering measurements is therefore an illustration of the anomalous low frequency response of the doublet, and this condition is similarly illustrated by elastic constant measurements of C_{66} (3), which show that at ultrasonic (10 MHz) frequencies, the elastic constant falls to zero at T_D, but remains finite in the Brillouin scattering ($\sim$ 10 GHz) measurements (2).

Fig. 1(a) shows how the central peak grows as T falls towards T_D, and also shows the softening of the acoustic phonon along [010] and transversely polarized along [100]; more detailed scans of the central peak are given in Fig. 1(b). The width of the peak is determined by the instrumental resolution. The variation with temperature of the central peak intensity vs. q_y is shown in

Fig. 2(a) and at $T = T_D$ in Fig. 2(b). The measurements are restricted by Bragg contamination at low q_y, but demonstrate the type of divergence expected in the intensity as $T \to T_D$ and $\underline{q} \to 0$. The peak is strongly localised along the [010] axis, with a width perpendicular to the axis of $\Delta q_x \sim .015$ r.l.u. at $T = T_D$, $\underline{Q} = (0, 1.95, 0)$.

The measurements have been analysed using the theoretical techniques of linear response function theory, as discussed by Stinchcombe (4). In the mean field approximation, one can write the single site susceptibility for Tb^{3+} (for $T > T_D$) as

$$\Pi_s(\omega) = 4\varepsilon^2 g_1/(4\varepsilon^2 - \omega^2) + \Gamma_2 g_2/(\Gamma_2 - i\omega)$$

where g_1 and g_2 are the static susceptibilities of the singlets and the doublet respectively, and Γ_2 is the frequency width of the doublet response ($\sim$ 10 GHz). From this (using, for example, eqs. (38) and (46) of Stinchcombe's paper), one can write the coupled acoustic phonon response near the central peak ($\omega << \omega_q, \varepsilon$) as

$$G_{aa}(\underline{q},\omega) = \langle\langle c_q^+ + c_{-q}; c_q^+ + c_{-q}\rangle\rangle = \frac{2}{\omega_q}\,\frac{\Gamma_2[1-\mathcal{J}(\underline{q})(g_1+g_2)] - i\omega[1-\mathcal{J}(\underline{q})g_1]}{\Gamma_2[1-J(\underline{q})(g_1+g_2)] - i\omega[1-J(\underline{q})g_1]}$$

where $J(\underline{q})$ is the total Jahn-Teller coupling, and $\mathcal{J}(q) = J(\underline{q}) - K_a(\underline{q})$, where $K_a(\underline{q})$ is the coupling to the acoustic phonon branch. This result shows that the central peak has a width

$$\Gamma(\underline{q}) = \Gamma_2\,[1 - J(\underline{q})(g_1 + g_2)]/[1 - J(\underline{q})g_1]$$

which tends to zero as $q \to 0$ and $T \to T_D$, when $1 - J(0)(g_1 + g_2) = 0$. Unfortunately, the critical narrowing of the linewidth cannot be resolved because of the frequency resolution of the spectrometers (.012 THz for IN2 and .022 THz for IN3 at ILL, Grenoble), but the validity of the analysis has been checked by convoluting the theoretical expressions for the central peak response with the instrumental response, as shown by the solid curves in Fig.1(b), assuming that J(q) has the form $J(\underline{q}) \sim J(0)(1 - \alpha q^2)$ for $\underline{q}$ along [010]. Further comparisons are given in ref. (1). The agreement between theory and experiment is satisfactory at each temperature; the details of the temperature dependence are qualitatively correct, though there are difficulties in obtaining quantitative agreement which may be partly due to the problems associated with the convolution procedure.

One concludes from this work that the mechanism for and description of the central peak in $TbVO_4$ are satisfactorily understood. One can expect this type of central peak in any phase transition in which the mode of distortion (the "soft mode") couples to a

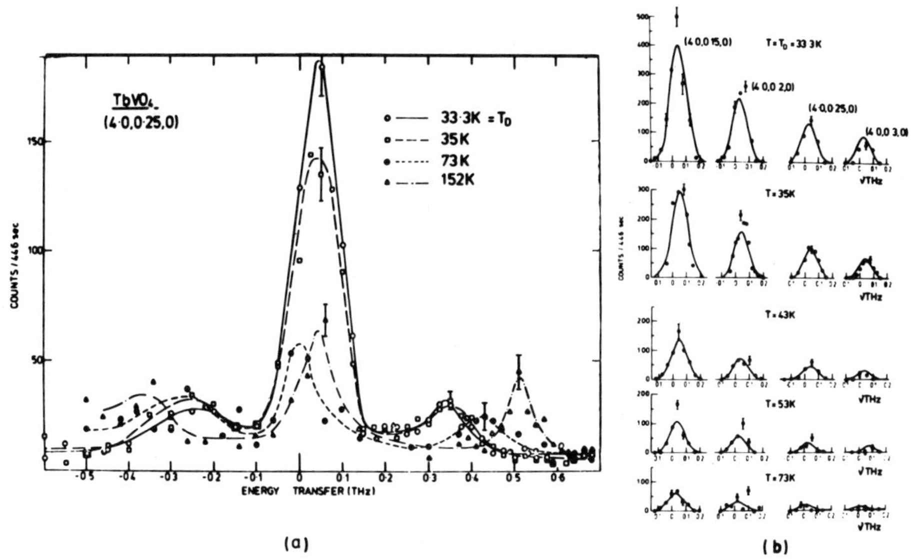

Fig. 1. (a) Neutron groups observed on IN3 scanning energy transfer at $\underline{Q}$ = (4.0, 0.25, 0).

(b) Central peak scans (with background subtracted) at four values of $\underline{Q}$ and five temperatures. The solid lines are the results of convoluting the theoretical response with the instrumental resolution function (IN3).

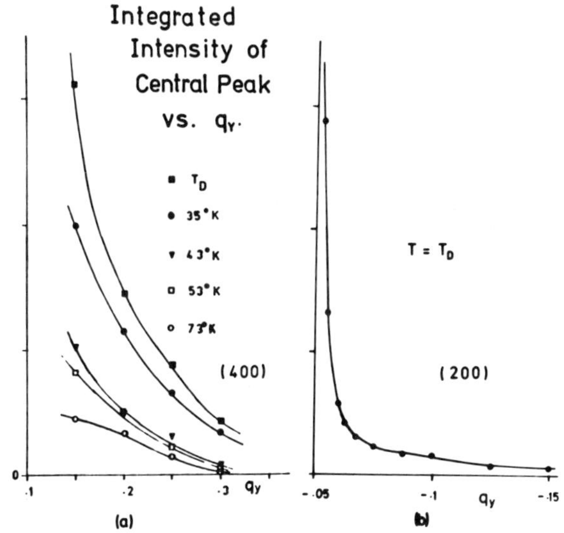

Fig. 2. (a) Central peak intensity vs. wavevector (4.0, q_y, 0) at various temperatures (IN3).

(b) Central peak intensity vs. wavevector (2.0, q_y, 0) at $T = T_D$ at higher resolution (IN2).

degenerate electronic doublet (as is often illustrated, of course, in magnetic phase transitions). It is also significant that the localisation of the central peak response along <010> axes indicates that the effective exchange coupling J_{ij} is long range in {010} planes, which is consistent with the success of classical mean field theory in describing the behaviour of $TbVO_4$ (2).

References

1. M.T. Hutchings, R. Scherm and S.R.P. Smith, "Magnetism and Magnetic Materials - 1975", ed. J.J. Becker, G.H.Lauder and J.J. Rhyne (A.I.P. Conf. Proc. No. 29; 1975) pp.372-8.

2. See the article by R.T. Harley.

3. J.R. Sandercock, S.B. Palmer, R.J. Elliott, W. Hayes, S.R.P. Smith and A.P. Young, J.Phys.C. 5, 3126-46 (1972).

4. See the article by R.B. Stinchcombe.

THE NATURE OF THE EIGENFUNCTIONS IN A STRONGLY COUPLED JAHN-TELLER PROBLEM

Mary Jane Shultz

DEAP/Pierce Hall, Harvard University

Cambridge, Massachusetts 02138 U.S.A.

I. INTRODUCTION

In the lectures of Stinchcombe, Harley, Thomas, and Kjems we have heard about the role of the cooperative Jahn-Teller effect in driving a phase transition. These lectures have dealt primarily with doubly degenerate electronic states coupled to either nondegenerate or double degenerate vibrational modes. (For an introduction to the Jahn-Teller effect, see the lecture of Professor Thomas in this volume.) In this seminar we are interested in the nature of the wavefunctions for a single Jahn-Teller complex which exhibits strong coupling between a triply degenerate electronic state and a triply degenerate vibrational mode ($T \times \tau$). The nature of these wavefunctions is of interest because, as Professor Thomas said, "The excitations of the single Jahn-Teller complexes give rise to bands of collective modes, and an analysis of the collective modes requires a thorough understanding of the excitations of a single complex."

The purpose of this seminar is to give a physical understanding of what the eigenfunctions and distorted configurations look like, rather than to present a detailed calculation. Those readers who are interested in the details are referred to a recent paper by Shultz and Silbey[1]. The organization of this seminar is as follows. The next section describes the physical setting for the $T \times \tau$ Jahn-Teller problem; the third section describes the Hamiltonian and details the effect of the electron-phonon coupling on the electronic state; the fourth section describes the spectral calculation; and the last section presents the summary and conclusions.

II. THE PHYSICAL SETTING

We are interested in calculating the absorption spectrum for an S→P transition in octahedral symmetry. An example of this transition occurs in the alkali halide phosphors. These are simple cubic crystals like NaCl with a few of the Na^+ ions replaced by heavy metal impurities such as In^+ or Tl^+. Since the electron associated with these impurities is fairly well localized on the impurity center, we will employ the quasimolecular approximation. That is, we will assume that the electronic state is only affected by motion of the impurity and its six nearest halide neighbors. Due to the localization mentioned above, this is a fairly good approximation.

Since the P excited state is triply degenerate in octahedral symmetry, it will have a linear Jahn-Teller interaction with the vibrational modes of the ligands giving structure to the absorption spectrum. From group theory it is known[2] that the triply degenerate excited state will couple to the α_{1g}, ε_g, and τ_{2g} vibrational modes. However, since coupling to the α_{1g} and ε_g modes is trivial, we will treat coupling to the τ_{2g} vibrational modes only. These modes are shown in Fig.1a-c. Note particularly the combination mode $Q_4 + Q_5 + Q_6$. This combination corresponds to motion of the six halide ligands toward one of the cubic diagonals as shown in Fig.1d. In this configuration one expects that the electronic state $p_x + p_y + p_z$ will have a lower energy than the two orthogonal states, because an electron in this orbital will spend more time near the three halide ligands. In fact, it is a balance between this electronic energy lowering and the halide-halide repulsion which determines the minimum point of the adiabatic potential surface.

The details of the adiabatic potential surface are in fact fairly complicated[3,4]. However, the important features are the four minimum points in configuration space which are displaced from the origin in the directions I=(1,1,1), II=(-1,-1,1), III=(-1,1,-1) and IV=(1,-1,-1). These configurations correspond to distortions of the octahedron along one of the four cubic diagonals and are the four trigonal distortions referred to by Professor Thomas in his lecture. We shall refer to this surface again later when generating approximate eigenfunctions for the strong coupling case. First, however, we will present the Hamiltonian for the excited P state and indicate how the electron-phonon coupling affects the electronic state.

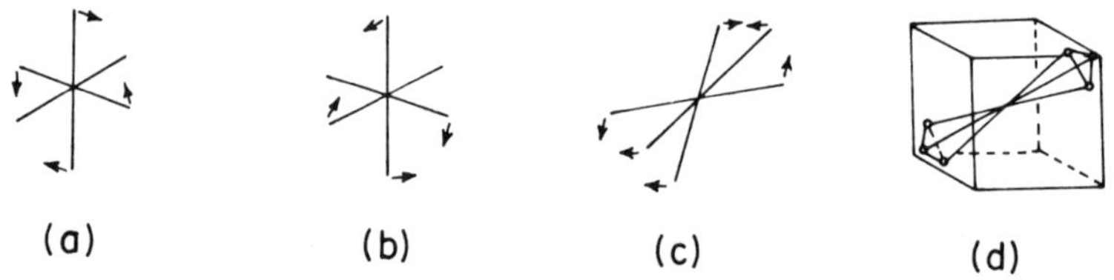

Fig. 1a-c: The τ_{2g} Normal Modes. (a) Q_4 (b) Q_5 (c) $Q_4 + Q_5 + Q_6$.
Fig. 1d: Distorted Configuration for Large Finite k.

III. THE HAMILTONIAN

The Hamiltonian for the T×τ Jahn-Teller interaction has been derived using group theory by earlier authors[3] and is given in second quantized notation as $H = H_o + V$ where

$$H_o = (b_4^\dagger b_4 + b_5^\dagger b_5 + b_6^\dagger b_6 + 3/2)\underset{\approx}{1} \tag{1}$$

$$V = (k\omega/\sqrt{2})[(b_4 + b_4^\dagger)\underset{\approx}{\tau}_1 + (b_5 + b_5^\dagger)\underset{\approx}{\tau}_2 + (b_6 + b_6^\dagger)\underset{\approx}{\tau}_3]$$

where

$$\underset{\approx}{1} = \begin{bmatrix}1&0&0\\0&1&0\\0&0&1\end{bmatrix}, \quad \underset{\approx}{\tau}_1 = \begin{bmatrix}0&0&0\\0&0&1\\0&1&0\end{bmatrix}, \quad \underset{\approx}{\tau}_2 = \begin{bmatrix}0&0&1\\0&0&0\\1&0&0\end{bmatrix}, \quad \underset{\approx}{\tau}_3 = \begin{bmatrix}0&1&0\\1&0&0\\0&0&0\end{bmatrix}.$$

In this notation, the operator $b_j(b_j^\dagger)$ destroys (creates) a quantum of vibrational excitation in mode Q_j, k is a unitless coupling constant which represents the relative energy of interaction between the electronic and vibrational modes, and the matrices are written with respect to the electronic orbitals p_x, p_y, p_z. Since the matrices $\underset{\approx}{\tau}_1$, $\underset{\approx}{\tau}_2$, $\underset{\approx}{\tau}_3$ do not commute, this Hamiltonian entails a non-trivial coupling between the electronic and vibrational modes.

To appreciate the effect of this electron phonon coupling on the electronic state, it is useful to think in the terms of scattering theory. That is, if we put an electron in the p_x orbital at time zero and let the interaction V act repeatedly, what happens to the electron? The answer is shown schematically below

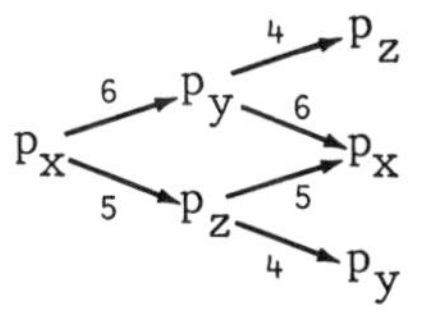

(2)

When V acts a single time, the electron is scattered from the p_x orbital into the p_y and p_z orbitals with a coefficient that depends on the value of the Q_5 and Q_6 vibrational modes. Similarly, when V acts again, the electron is scattered out of the orbital that it was in and into the remaining two. This coupling makes the problem difficult to solve when k is large, because then the eigenfunctions of H_o are not very close to eigenfunctions of H.

In the next section we will develop a canonical transformation which will generate a basis of approximate eigenfunctions. We will then use these eigenfunctions to calculate the absorption spectrum.

IV. ABSORPTION SPECTRUM

The absorption spectrum for a dipole allowed transition can be written in terms of the electronic autocorrelation function as [5]

$$I(\Omega) = 2\ \mathrm{Re} \int_0^\infty dt\ \exp(i\Omega t) \sum_{i=1}^{3} \langle C_i(t) C_i^\dagger \rangle \tag{3}$$

where $C_i(C_i^\dagger)$ destroys (creates) an electron in orbital i. Thus, we need to know the time evolution of the electronic operator C_i, or equivalently the time evolution of the electronic state. But remember that the time evolution is given by

$$C_i(t) = \exp(iHt)\ C_i\ \exp(-iHt). \tag{4}$$

Thus, if the electronic orbitals were eigenfunctions of H, we would be finished. However, the electronic orbitals are not eigenfunctions of H. In fact, for the strongly coupled case, they are not even close. (This is a direct result of the breakdown of the Born Oppenheimer Approximation.) The idea for calculating the absorption spectrum, then, is to find a set of wavefunctions such that $H\Psi \simeq E\Psi$. These wavefunctions are then related to the electronic wavefunctions and the spectrum is calculated.

The approximate wavefunctions are generated by looking at the adiabatic potential surface. The important feature of the potential surface is that it has four minimum points which are displaced from the origin by an amount proportional to k and which are separated by barriers of height proportional to k^2. Thus, for infinite k, the molecule cannot get from one configuration to another. Therefore, we first solve for the wavefunction in each well independently.

We calculate the separate well wavefunctions in two steps. Firstly, we displace the origin of the vibrational coordinant system with a canonical transformation which is analogous to a phonon shift operator. Secondly, we make a transformation which accounts for the difference in potential felt by an electron localized between the halides and one localized in the perpendicular directions. The resulting first order wavefunctions localized in well I are

$$\psi_I^{inm\ell} = \exp(S)(\alpha_1)^n(\alpha_2)^m(\alpha_3)^\ell \bar{C}_i^\dagger|0\rangle/(n!m!\ell!)^{\frac{1}{2}} \tag{5}$$

where

$$S = -k\{(2/3)^{\frac{1}{2}}(\alpha_3 - \alpha_3^\dagger)\underset{\approx}{1} + (1/6)^{\frac{1}{2}}(\bar{C}_2^\dagger\bar{C}_2 + \bar{C}_3^\dagger\bar{C}_3)(\alpha_3 - \alpha_3^\dagger)$$
$$-(2)^{\frac{1}{2}}(\bar{C}_2^\dagger\bar{C}_2 - \bar{C}_3^\dagger\bar{C}_3)(\alpha_1 - \alpha_1^\dagger)\}$$

$$\bar{C}_1 = (3)^{-\frac{1}{2}}(C_1 + C_2 + C_3),\quad \bar{C}_2 = (2)^{-\frac{1}{2}}(C_1 - C_3),$$

$$\bar{C}_3 = (6)^{-\frac{1}{2}}(C_1 - 2C_2 + C_3),\quad \alpha_1 = (6)^{-\frac{1}{2}}(-b_4 + 2b_5 - b_6),$$

$$\alpha_2 = (2)^{-\frac{1}{2}}(-b_4 + b_5),\quad \alpha_3 = (3)^{-\frac{1}{2}}(b_4 + b_5 + b_6).$$

Here $|0\rangle$ is the electron-phonon vacuum state. Ignoring the exp(S) term, this wavefunction is merely a Born Oppenheimer product of an electronic state times a vibrational state. The exp(S) factor is the phonon dressing of the electronic state and is a direct result of the breakdown of the Born Oppenheimer Approximation.

Wavefunctions localized in the other wells are similarly calculated, and we have the following important results for the lowest energy set of wavefunctions. Firstly, the interwell overlap vanishes for large k

$$\lim_{k\to\infty} \langle \psi_J^{1nm\ell} | \psi_K^{1'n'm'\ell'} \rangle = 0 \qquad J \neq K \tag{6}$$

Secondly, the interwell interaction energy decreases with k

$$\langle \psi_J^{1nm\ell} | H | \psi_K^{1'n'm'\ell'} \rangle / \langle \psi_J^{1nm\ell} | \psi_K^{1'n'm'\ell'} \rangle \propto k^2 \exp(-8k^2/9) \qquad J \neq K \tag{7}$$

Finally, the second order wavefunctions are nearly eigenfunctions for large k

$$\lim_{k\to\infty} \langle \psi_J^{1nm\ell} | H - E_{exact} | \psi_J^{1nm\ell} \rangle / \langle \psi_J^{1nm\ell} | \psi_J^{1nm\ell} \rangle \ll \omega \ . \tag{8}$$

where E_{exact} is the known asymptotic energy. These results indicate that in the limit of strong electron phonon coupling, our separated well model is a good one.

The final step is to recover the full octahedral symmetry of the original problem by forming appropriate (un-normalized) functions which transform as irreducible representations of the octahedral group by combining states of the four wells. For example, for the lowest A_2 and T_1 states, we have

$$\Phi_{T_1} = \frac{1}{2} \begin{Bmatrix} \psi_I + \psi_{II} - \psi_{III} - \psi_{IV} \\ \psi_I - \psi_{II} + \psi_{III} - \psi_{IV} \\ \psi_I - \psi_{II} - \psi_{III} + \psi_{IV} \end{Bmatrix}$$

$$\Phi_{A_2} = \frac{1}{2} (\psi_I + \psi_{II} + \psi_{III} + \psi_{IV}) \tag{9}$$

Using these wavefunctions, we calculate the matrix elements of the total Hamiltonian to second order and find a singlet triplet splitting of [6]

$$E_{A_2} - E_{T_1} = (272/243)\, k^2\, \omega\, \exp(-8k^2/9) \tag{10}$$

Note that due to the octahedral symmetry of the original problem, the lowest state remains a triplet independent of k. The above result (valid asymptotically for $k\to\infty$) is in good agreement with the result $0.8\, k^2 \omega \exp(-0.8k^2)$ given by Caner and Englman [7] as a reasonable fit of their numerical data for $k \simeq 2.5$.

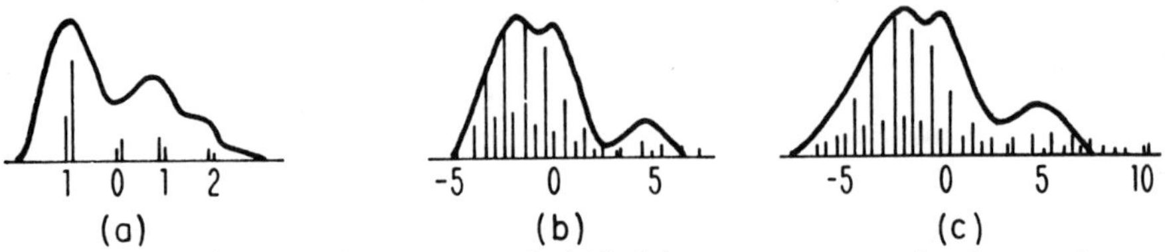

Fig. 2: Absorption Spectra. Solid lines are numerical results of Englman et al.[*8*], bar spectra are this calculation. Energy is in units of ω. (a) k=1, $k_BT=0$; (b) k=2.3, $k_BT=0$; (c) k=2.3, $k_BT=0.5$.

Finally, we use the above wavefunctions to calculate the absorbtion spectrum and find the spectra shown in Fig.2. (For details of the spectral calculation, see[*1*].) Fig.2 compares our spectra with the exact numerical spectra of Englman et al.[*8*] and we find good agreement even for fairly small k. Further, the analytical formula has the advantage of being valid for larger values of k.

V. SUMMARY AND CONCLUSIONS

In this work we have used a canonical transformation technique to generate approximate eigenfunctions for a strong coupling problem. This transformation can be viewed as replacing the description of the motion of the strongly coupled individual electrons and nuclei by a description in terms of weakly interacting collective modes. After transformation, the interaction parameter is $k^2 \exp(-k^2)$; therefore, even for large k, the perturbation is small.

Using the transformed Hamiltonian, we have derived analytic expressions for the asymptotic eigenvalues and for the spectral density. In both cases we find good agreement with earlier numerical results [*7*,*8*].

REFERENCES

1. Shultz, M.J. and Silbey, R. (1976). *J. Chem. Phys.* 65, 4382.
2. Toyozawa, Y. and Inoue, M. (1966). *J. Phys. Soc. Japan* 21, 1663.
3. Van Vleck, J.H. (1939). *J. Chem. Phys.* 7, 72.
4. Opik, U. and Pryce, M.H.L. (1957). *Proc. Roy. Soc. London* 238A, 425.
5. See eg. Gordon, R.G. (1967). In *Advances in Magnetic Resonance*, Vol. 3 (ed. Waugh, J.S.).
6. Shultz, M.J. and Silbey, R. (1974). *J. Phys. C* 7, L325.
7. Caner, M. and Englman, R. (1966). *J. Chem. Phys.* 44, 4054.
8. Englman, R.; Caner, M.; and Toaff, S. (1970). *J. Phys. Soc. Japan* 29, 306.

COOPERATIVE PSEUDO JAHN TELLER MODEL OF THE SEQUENCE OF FERRO-ELASTIC TRANSITIONS IN BARIUM SODIUM NIOBATE

D. Paquet

Centre National d'Etudes des Télécommunications

196 rue de Paris, 92220 Bagneux (France)

Barium sodium niobate ($Ba_2NaNb_5O_{15}$ or BSN) has recently been shown (1) to posses four phases, whose labelling, temperature stability range and most probable space symmetries are listed in Table I. In this paper we will point attention only on the three lower temperature polar phases. The room temperature one is a ferroelastic orthorhombic mm2 phase sandwiched between two tetragonal phases of same point symmetry 4mm which, moreover, are expected to belong to the same space symmetry.

Fig. 1 shows the temperature behaviour of the birefringence Δn_{ab} and the spontaneous strain e_6 obtained from the work of Schneck et al. (1). The two phase transitions display first order (hysteresis) and diffused characters. The diffuseness has been attributed (2) to the local disorder of the cations in the structure.

The high temperature phase transition has been studied by Brillouin scattering (3). The difference between elastic constants (C_{11}-C_{22}) associated to the onset of the spontaneous strain e_6 does not go monotonically to zero on heating the crystal to the transition temperature (Fig. 2). Tolédano (4) has shown that this anomalous behaviour is due to the fact that the phase transition is of the improper type : the order parameter is not the spontaneous strain, but a physical quantity transforming as an irreducible representation of the paraelastic space group situated at the Z point of the Brillouin zone boundary of the P4bm space group. The phase transition is thus associated to a breakdown of the translational symmetry corresponding to a doubling of the unit cell along the polar c-axis. This prediction has been checked using X-ray diffraction (5) : the intensity of the superstructure line vanishes

Phase labelling	Temperature range	Space group
I	T > 858 K	P4/mbm
II	858 K > T > 573 K	P4bm
III	573 K > T > 105 K	$Ccm2_1$
IV	T < 105 K	P4bm

Table I - Different stable phases of BSN

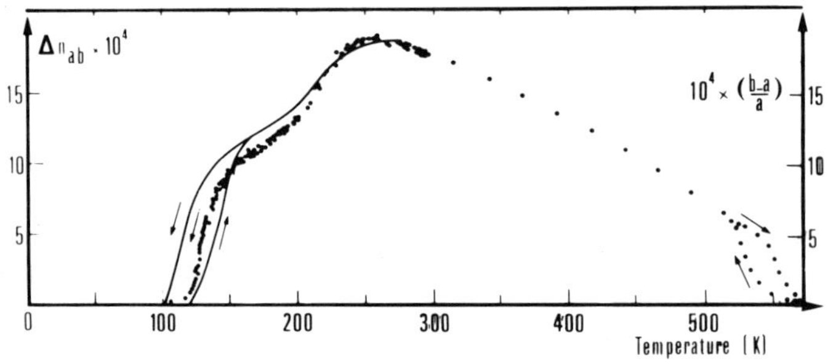

Fig. 1 - Temperature dependence of the birefringence Δn_{ab} in the (001) plane (dotted line) and of the spontaneous strain (b-a)/a (solid line). (From Ref. 1).

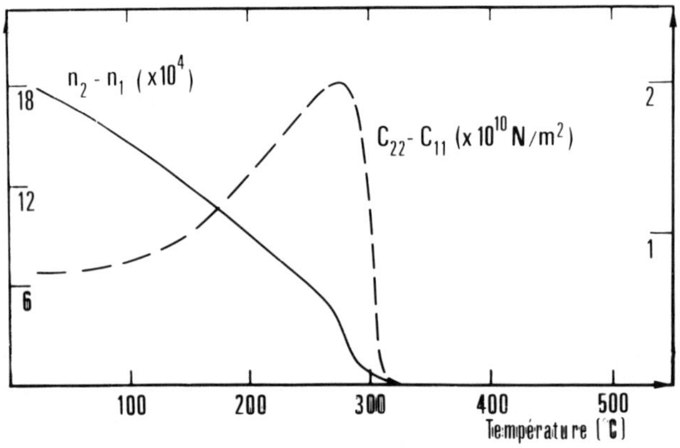

Fig. 2 - Compared variations of the optical and elastic anisotropies in the (001) plane. Both should decrease in a proper ferroelastic. They show opposite variations in the improper ferroelastic transition of BSN . (From Ref. 2).

at 573 K, and exhibits also a spread of the transition temperature.

The low temperature ferroelastic phase transition has been recently discovered by Schneck et al.(1). Birefringence and pyroelectric measurements show that below 105 K the crystal recovers a tetragonal symmetry and remains polar. X-ray print indicates that it belongs to the 4/mmm Laue class. Its point symmetry is thus 4mm, as confirmed by nonlinear optics experiments. Assuming that Landau theory can be used to describe this phase transition, the space group of the low temperature phase must be a supergroup of the ferroelastic phase one. Every supergroup of $Ccm2_1$ corresponding to the 4mm point symmetry is associated to an increase of the translational symmetry, and among those satisfying the Landau-Lifschitz criteria, the only one compatible with a continuous change in the positions of the various atoms is P4bm. The two ferroelastic phase transitions thus appear as completely symmetrical, the two extreme phases (II and IV) being identical.

This sequence of two inverse ferroelastic phase transitions is puzzling because, while entropy is expected to be an increasing function of temperature, the "configurational" entropy decreases with increasing temperature at the low temperature phase transition. Hence in phase IV, an "internal" degree of freedom, different from atomic vibrations, should order to compensate the excess of configurational entropy. We assume that this internal degree of freedom is of electronic nature. A careful analysis of the tungsten bronze structure of BSN (6) (Fig. 3) actually shows that the onset of the spontaneous strain corresponds to tilts of the oxygen octahedra surrounding the niobium atoms. Though all the octahedra in the tetragonal unit cell are non equivalent, they all undergo the same distortion (a Γ_4^+ rotation and a Γ_5^+ shear), and the cooperative onset of these octahedral distortions builds up a bidimensional crystal normal mode possessing the same symmetry properties as the abstract order parameter deduced from the phenomenological theory. We thus believe that the ferroelastic transitions are driven by the cooperative tilt of the oxygen octahedra induced by a Jahn-Teller interaction between the octahedral vibrations and some d-electrons localized on the niobium ions (7).

If one accepts that BSN is a pure ionic compound, then one finds that the d states of the Nb^{5+} ion are empty. However several experimental and theoretical data show that, in oxygen octahedra ferroelectrics, the effective charge of the transition metal ion is substantially reduced from its formal value. Infrared absorption in $LiNbO_3$ (8), where the oxygen octahedra have exactly the same dimensions as in BSN, leads to an effective charge of +3.2. Some $SrTiO_3$(9) band calculations assume highly correlated electrons in the same octahedral unit. We thus built up an effective hamiltonian for the

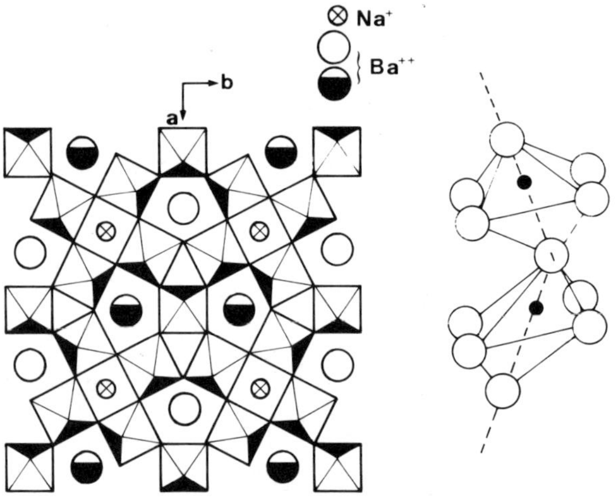

Fig. 3 - Schematic representation of the distorted tungsten bronze structure of BSN at room temperature. Left : half unit cell. In the consecutive half cell along |001|, the tilt of the octahedra and the positions os the shaded barium atoms are inverted. Right : two consecutive octahedra along the |001| axis. The c-parameter corresponds to two octahedra heights.

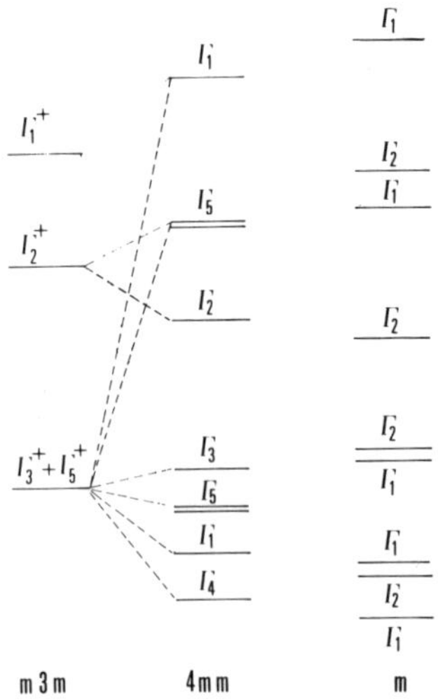

Fig. 4 - Splitting of the electronic energy levels of the Nb^{3+} ion in phases I (site symmetry m3m), II and IV (site symmetry 4mm) and III (site symmetry m).

octahedral units whose levels exhibit the same symmetry properties as those of two correlated d electrons in the crystal field at the niobium site. We found a nine-dimension manyfold, as sketched in Fig. 4, described by the local hamiltonian :

$$H = \zeta\, h_{so} + d\, h_p + a\,(\sigma_1 q_1 + \sigma_2 q_2)$$

where h_{so}, h_p, σ_1, σ_2 are electronic operators corresponding respectively to the spin orbit interaction, the tetragonal part of the crystal field induced by the spontaneous polarization of the crystal, and the coupling to the bidimensional (q_1, q_2) octahedral distortion. ζ, d and a are constants measuring the strength of these interactions. It must be outlined that, in the paraelastic phases, where the Nb site symmetry is 4mm, the ground state is a singlet, due to the tetragonal components of the crystal field induced by the spontaneous polarization.

The hamiltonian of the whole crystal becomes :

$$H = \sum_{j=1,2} \left\{ \frac{1}{2M} P_j(\vec{k})\, P_j^*(\vec{k}) + \frac{M\omega^2}{2} Q_j(\vec{k})\, Q_j^*(\vec{k}) \right\}$$

$$+ \frac{1}{2} V\, C_{66}\, e_6^2$$

$$+ g\, e_6 \left\{ Q_1(\vec{k})\, Q_1^*(\vec{k}) - Q_2(\vec{k})\, Q_2^*(\vec{k}) \right\}$$

$$+ \sum_{\ell,\alpha} \left\{ \zeta\, h_{so}(\ell,\alpha) + d\, h_p(\ell,\alpha) + a \sum_{j=1,2} \sigma_j(\ell,\alpha)\, q_j(\ell,\alpha) \right\}$$

The first term corresponds to the energy of the bidimensional crystal normal mode, the second one to the elastic energy associated with the spontaneous strain e_6, the third one to the coupling between the spontaneous strain and the normal mode, the last one to the sum over each individual octahedral unit (ℓ,α) of the local hamiltonians. This total hamiltonian, though more complicated, is very similar to the one discussed by Stinchcomb in his lecture. Using the same technics (displaced operators, neglect of the self energy and of the non commutation of the σ_j with h_{so} and h_p, mean field approximation), we obtain the single site hamiltonian :

$$h_m = \zeta\, h_{so} + d\, h_p - b_o \langle \sigma \rangle \sigma$$

with $b_o = n_o a / M\omega^2$, n_o being the number of Jahn Teller active ions

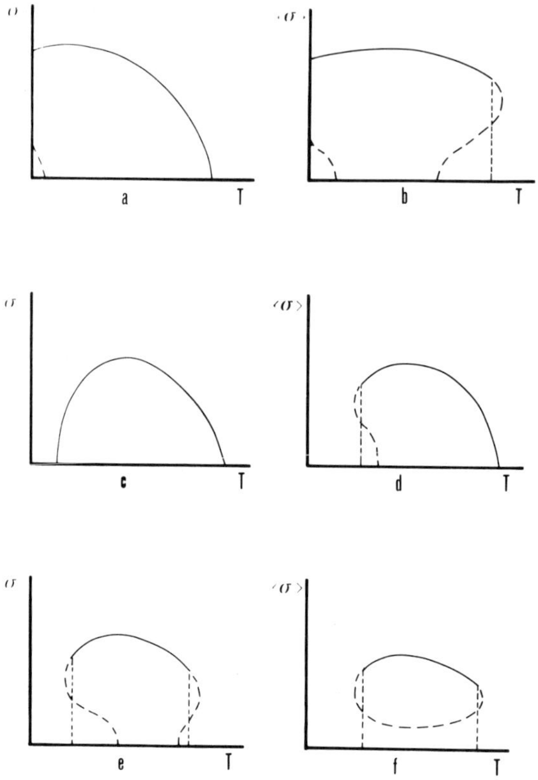

Fig. 5 - Different behaviours of the order parameter $< \sigma >$ versus temperature, depending on the relative values of ζ, d, b_o. The experimental behaviour of BSN corresponds to case e.

in the tetragonal unit cell. The individual octahedral distortions and the spontaneous strain are then given by :

$$q(\ell,\alpha) = -\frac{b_o}{a} < \sigma > \cos(\vec{k}.\vec{h}_\ell)$$

$$e_6 = \frac{3gb_o}{C_{66}\ v\ M\omega^2} < \sigma >^2$$

The mean field self consistency equation has been solved in a computer, and a phase diagram has been drawn. Six different behaviours have been found (Fig. 5) depending on the relative values of ζ , d, b_o : zero, one, or two phase transitions, of first or second order type. The three coefficients have been fitted, with physical meaningful values, to describe the two first order ferroelastic phase transitions occuring in BSN.

The model predicts that an electric field applied along the polar axis would change the strength of the tetragonal component of the crystal field (d) and thus change the transition temperatures (Fig. 6). A similar effect, due to the change of the b_0 coefficient by substitution of the Nb atoms by non Jahn-Teller active ions, must occur (Fig. 7). These experiments will be performed to test the model.

As a conclusion we shall emphasize that the occurence of the low temperature phase transition strongly depends on the polarity of the crystal : the tetragonal crystal field splits the five fold degenerate cubic ground state ; the latter becomes a singlet. The electron-vibration coupling thus occurs through pseudo Jahn-Teller effect. At very low temperature the ground state alone is occupied which does not induce any distorsion. However, as this state is a singlet, the electronic entropy vanishes, and the system "stores" the order. At higher temperatures, higher states are occupied, and the cooperative onset of the octahedral distortions occurs through pseudo Jahn-Teller coupling. Then at much higher temperatures, the increase of the electronic entropy overcomes the decrease in energy, which leads to a minimum of the free energy for zero distortion. This process is quite similar to the one discussed by R.T. Harley for $Tb_pGd_{1-p}VO_4$ in his lecture.

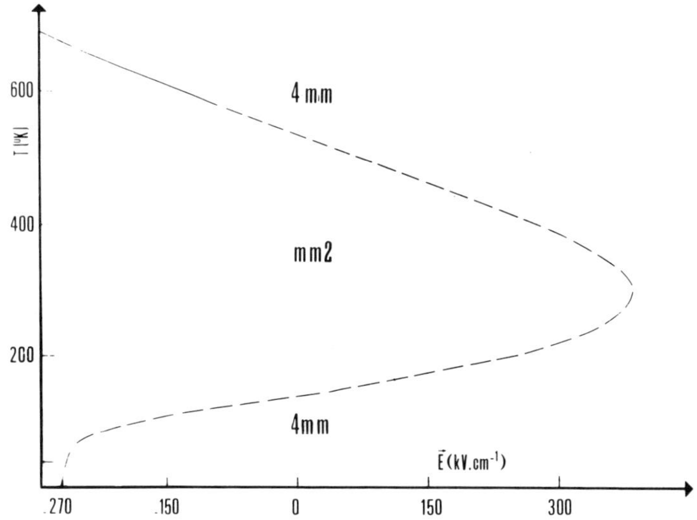

Fig. 6 - Ferroelastic transition temperatures vs applied electric field along |001|. Dashed line : first order phase transition ; solid line : second order phase transition.

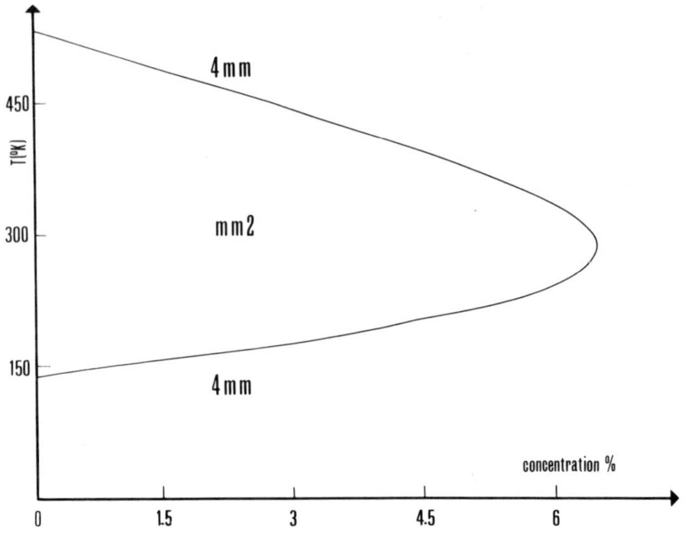

Fig. 7 - Ferroelastic transition temperatures vs concentration of Jahn-Teller inactive impurities substituted to the niobium atoms.

References

1. J. Schneck, J. Primot, R. Von der Mühl, and J. Ravez, Sol. State Commun. 21, 57 (1977)
2. J.C. Tolédano and J. Schneck, Sol. State Commun. 16, 1101 (1975)
3. J.C. Tolédano, M. Busch and J. Schneck, Ferroelectrics 13, 328 (1976)
4. J.C. Tolédano, Phys. Rev. B12, 943 (1975)
5. J. Burgeat and J.C. Tolédano, Sol. State Commun., 20, 281 (1976
6. P.B. Jamieson, S.C. Abrahams, and J.L. Bernstein, J. Chem. Phys. 50, 4352 (1969)
7. D. Paquet, J. Chem. Phys., 66, 886 (1977)
8. F. Gervais, Sol. State Commun. 18, 191 (1976)
9. T.F. Soules, E.J. Kelly, D.M. Vaught, J.W. Richardson, Phys. Rev. B6, 1519 (1972)

SINGLE ION AND COOPERATIVE JAHN-TELLER EFFECT FOR A NEARLY DEGENERATE E DOUBLET

L.F. Feiner

Philips Research Laboratories

Eindhoven, The Netherlands

ABSTRACT

Single ion behaviour in $Fe:CoCr_2S_4$ and cooperative behaviour in $FeCr_2S_4$ of Fe^{2+} are treated within a model where this Jahn-Teller (JT) ion interacts with local displacements and which takes full account of the spin-orbit splitting of the electronic ground doublet. Calculated static properties are compared with experiment. The discussion includes qualitative effects due to the splitting of the ground doublet and the problem as to what extent the strengths of spin-orbit coupling, single ion JT coupling, and interaction between JT complexes can be estimated separately.

1. PHYSICAL SYSTEM AND MODEL

Experimental information is available on two very similar compounds, one ($Fe:CoCr_2S_4$) where Fe^{2+} acts as an isolated Jahn Teller (JT) impurity, and the other one ($FeCr_2S_4$) where the Fe^{2+} ions interact and take part in a cooperative JT effect [1]. I wish to investigate whether it is possible to describe both single ion and cooperative behaviour with one coherent model.

Both materials have the normal spinel structure with Fe^{2+} at the purely cubic tetrahedral site, at the center of a tetrahedron of S^{2-} ions. We notice that each S^{2-} is ligand to at most one Fe^{2+}, and that in $FeCr_2S_4$ the Fe^{2+} ions build up a diamond lattice, which consists of two interlacing Bravais lattices. Below 175 K the compounds are magnetically ordered and each Fe^{2+} experiences a magnetic exchange field H_{ex} from the Cr^{3+} ions, which is directed along one of the cubic axes. Consequently [2], the cubic crystal field splits the ^5D free ion groundstate of Fe^{2+} into a higher $^5T_2$ and a lower ^5E multiplet, separated by about 3500 K, while the spin degeneracy is removed by H_{ex}, which splits the magnetic sub-levels by about 300 K. Second order spin-orbit interaction between E and T_2 states finally separates the components of the electronic grounddoublet by 2 δ_{LS}, which is about 50 K.

The experimental information consists of (i) ^{57}Fe Mössbauer measurements, which, when analyzed on the basis of the usual spin Hamiltonian, yield the magnetic hyperfine field H_{hf} and the components of the electric field gradient (EFG) V_{xx}, V_{yy} and V_{zz} at the Fe nucleus [1], (ii) specific heat measurements on $FeCr_2S_4$ [3].

To describe the local static properties of the Fe^{2+} ion at low temperatures I assume (i) that it is sufficient to take into account only the JT active nearly degenerate E grounddoublet, and (ii) that the (cubic invariant) interaction of the JT ion is only with the E-type vibrations of the ion's own S^{2-} tetrahedron. This gives the following Hamiltonian for a single Fe^{2+} ion in $Fe:CoCr_2S_4$:

$$\mathcal{H}_{single} = \hbar\omega \left[\tfrac{1}{2}(p_\theta^2 + p_\varepsilon^2 + q_\theta^2 + q_\varepsilon^2) + k\,(q_\theta\,\sigma_\theta + q_\varepsilon\,\sigma_\varepsilon) - \delta\,\sigma_\theta \right], \qquad (1)$$

where σ_θ and σ_ε are the usual pseudo spin operators, i.e. Pauli matrices $-\sigma^z$ and σ^x acting on the electronic states E_θ and E_ε [4], $\hbar\omega$ is the vibrational energy quantum, and k and $\delta = \delta_{LS}/\hbar\omega$ are dimensionless parameters characterizing the strength of the JT coupling and the spin-orbit splitting, respectively. In the present case of E states in tetrahedral coordination where the JT coupling is only weak or intermediate, it is important to include the spin-orbit splitting right from the beginning.

To treat the interacting Fe^{2+} ions in $FeCr_2S_4$ I assume that the $[Fe^{2+}S^{2-}{}_4]$ complexes (i) are described each by the same Hamiltonian as in the single ion case, and (ii) are now coupled by lattice dynamical forces between the displacements of S^{2-} ions in complexes that are nearest neighbour to each other. In order to have cubic invariance (apart from the spin-orbit term) the total Hamiltonian is then

$$\mathcal{H} = \sum_i (\mathcal{H}_{single})_i + \tfrac{1}{2}\hbar\omega \sum_{n.n.} \lambda\,(q_{\theta i}\, q_{\theta j} + q_{\varepsilon i}\, q_{\varepsilon j}), \qquad (2)$$

where λ is a dimensionless force constant characterizing the strength of the interaction between complexes.

Crystal field theory [5] yields $H_{hf} = A - B\langle\sigma_\theta\rangle$, $V_{zz} = C\langle\sigma_\theta\rangle$, $(V_{xx} - V_{yy})/\sqrt{3} = C\langle\sigma_\varepsilon\rangle$, where A, B and C are constants. Therefore the task is to calculate the thermal averages of the electronic pseudo spin operators.

2. CALCULATIONS, RESULTS AND DISCUSSION

For the single ion case I have performed numerical diagonalizations of Hamiltonian (1) using the familiar basis of electronic-vibrational product states [4]. For $k \lessgtr 1.5$ and $\delta \lessgtr 0.3$ inclusion of states with up to 7 vibrational quanta was sufficient to obtain an accuracy better than 0.5% in the thermal averages.

The low temperature behaviour can be understood qualitatively by looking at the lowest two levels. Consider the JT interaction first. It changes the exactly degenerate electronic states E_θ and E_ε into degenerate vibronic states Ψ_θ and Ψ_ε, thereby reducing the magnitude of the pseudo spin operators by the Ham factor q. Secondly, the spin-orbit splitting splits the states but also changes their wavefunctions by mixing the vibronic state Ψ_θ (Ψ_ε) with higher lying states of A_1 (A_2) symmetry. It thereby shifts the values of σ_θ in the two states by the same amount, b. For $k_BT/\hbar\omega \lesssim 0.3$ we find, with an accuracy of a few percent,

$$\langle\sigma_\theta\rangle = b + q \tanh\left(\Delta/k_BT\right), \qquad (3)$$

where $\Delta/\hbar\omega \simeq q\cdot\delta$ is half the splitting, and $b \simeq b_1(k)\cdot\delta$, while $\langle\sigma_\varepsilon\rangle = 0$ by symmetry. The result (3) is, due to the constant b, qualitatively different from the result of static crystal field theory. Quantitatively, b can become about one third of q for $k \lesssim 1$ and $\delta \lesssim 0.2$.

For the case of interacting ions I have performed a molecular field calculation assuming (i) that the coupling is antiferrodistortive, i.e. $\lambda > 0$, and (ii) that ordering takes place in two sublattices. For a JT complex, say in sublattice 1, feeling the average displacements of its neighbours in sublattice 2, one gets a molecular field Hamiltonian

$$\begin{aligned}\mathcal{H}_1^{mf} = \hbar\omega\Big[&\tfrac{1}{2}\left(p_{\theta 1}^2 + p_{\varepsilon 1}^2 + \hat{q}_{\theta 1}^2 + \hat{q}_{\varepsilon 1}^2\right)\\ &+ k\left(\hat{q}_{\theta 1}\sigma_{\theta 1} + \hat{q}_{\varepsilon 1}\sigma_{\varepsilon 1}\right) - \delta\,\sigma_{\theta 1}\\ &+ g\left(\langle\sigma_{\theta 2}\rangle\sigma_{\theta 1} + \langle\sigma_{\varepsilon 2}\rangle\sigma_{\varepsilon 1}\right)\\ &- z\lambda g\left(\langle\sigma_{\theta 1}\rangle\sigma_{\theta 1} + \langle\sigma_{\varepsilon 1}\rangle\sigma_{\varepsilon 1}\right),\end{aligned} \qquad (4)$$

where $g = z\lambda k^2/(1-(z\lambda)^2)$, z (=4) is the number of nearest neighbours, and the carets indicate displaced coordinates. The fourth term in Eq. (4) represents antiferrodistortive coupling of the XY type between the pseudo spins on the two sublattices, the fifth term additional ferrodistortive intrasublattice coupling, while the spin-orbit splitting acts as an external field in the XY plane. In contrast with the usual treatment [6] involving displaced phonons and retaining only the resulting coupling between the pseudo spins, I have taken the JT interaction into account, in the same way as in the single ion case. This inclusion of local dynamics makes the details of the temperature dependence different from those of the analogous magnetic system. The qualitative features, though, are the same. At high temperatures, we have a paradistortive phase where the averaged sublattice pseudo spin vectors $\langle\vec{\sigma}\rangle_{1,2}$ are equal and along the θ direction preferred by the field δ. Upon lowering the temperature $\langle\sigma_\theta\rangle$ increases until at a temperature T_D it reaches the value $\delta/2g$. Here a second order phase transition occurs after which $\langle\sigma_\theta\rangle$ remains constant, while ε components of $\langle\vec{\sigma}\rangle_i$ now develop, of equal magnitude but opposite direction on the two sublattices.

The accuracy of the experimental data is not sufficient to determine the parameters uniquely. A typical example of a fit is given in Fig. 1 for the electric field gradient and in Fig. 2 for the hyperfine field. Here, first the model parameters k, δ and $\hbar\omega$ and the constant C have been chosen such as to reproduce the single ion EFG data, and then the parameter $z\lambda$ has been determined from the value of

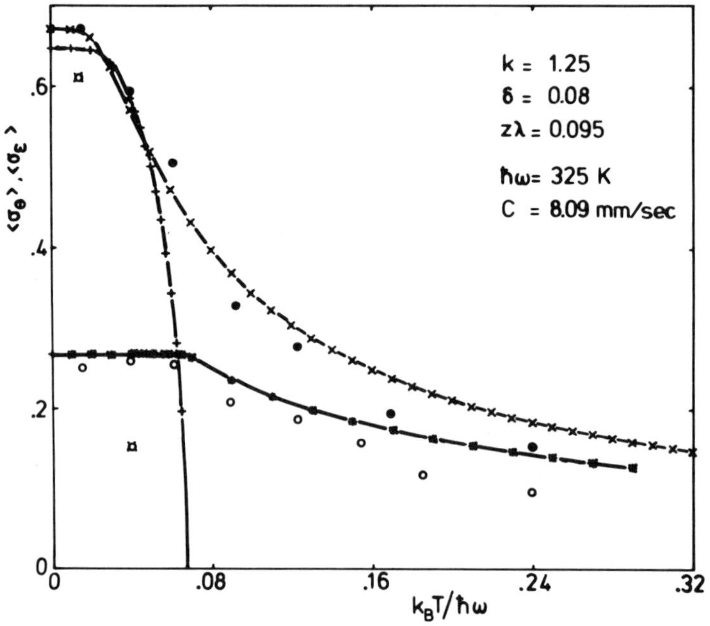

Fig. 1: Calculated (curves) thermal averages of the pseudo spin operators and measured (dots, circles) electric field gradient as a function of temperature. Single ion: $\langle\sigma_\theta\rangle\equiv$ —x—, ● ($Fe:CoCr_2S_4$); interacting ions: $\langle\sigma_\theta\rangle\equiv$ —✱—, ○ ($FeCr_2S_4$); $\langle\sigma_\varepsilon\rangle\equiv$ —+— , ¤ ($FeCr_2S_4$).

V_{zz} in $FeCr_2S_4$ at T = 0. The constants A and B have been fixed from the T = 0 values of H_{hf} in the two compounds. The data allow the range $0.75 \lesssim k \lesssim 1.25$, $0.06 \lesssim \delta \lesssim 0.10$. The corresponding value for $\hbar\omega$ varies between 225 K and 450 K, but δ_{LS} is always between 22 K and 28 K. Both results can be compared with experiments for similar compounds: from optical absorption measurements [7] $\delta_{LS} \simeq 28$ K, from an analysis of Raman and infrared data [8] $\hbar\omega$ is estimated to be of the order of 300 K. In addition, the values obtained for A (258 kG), B (280 kG) and C (8.09 mm/s) are in satisfactory agreement with crystal field calculations [5] without the need to invoke strong covalency.

The calculated temperature dependence of $\langle\sigma_\theta\rangle$ and $\langle\sigma_\varepsilon\rangle$ shows qualitative agreement with the EFG data on $FeCr_2S_4$. The lower values and weaker temperature dependence of V_{zz} in $FeCr_2S_4$ compared with $Fe{:}CoCr_2S_4$ are explained as consequences of the cooperative effect. The calculated value of $V_{xx} - V_{yy}$ at 4 K agrees quantitatively. In fact the model explains, why at low T the EFG has axial symmetry in $Fe{:}CoCr_2S_4$ and not in $FeCr_2S_4$. It also explains that at low T the hyperfine field decreases strongly in $Fe{:}CoCr_2S_4$ (where the agreement is quantitative), but not in $FeCr_2S_4$.

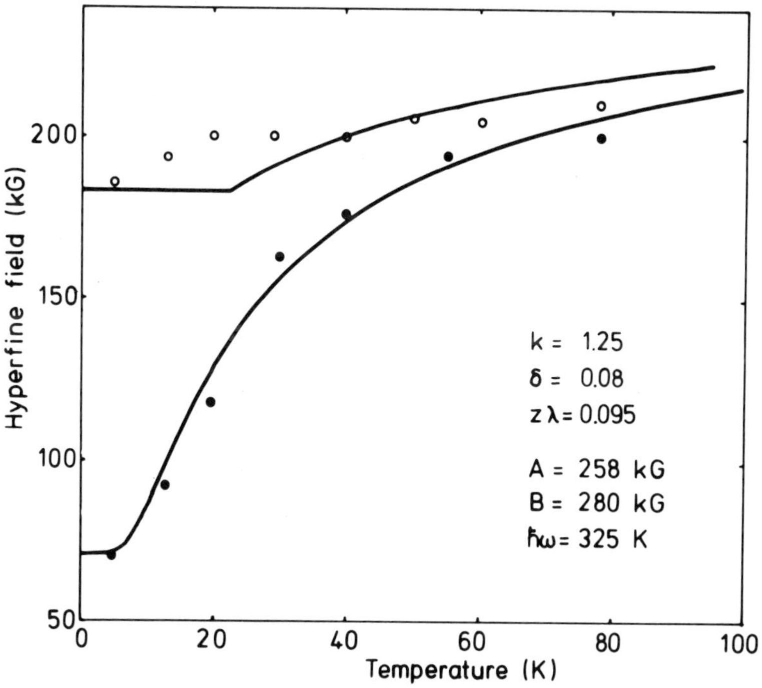

Fig. 2: Calculated (curves) and measured (dots, circles) hyperfine field as a function of temperature; ● ≡ $Fe{:}CoCr_2S_4$, ○ ≡ $FeCr_2S_4$.

On the other hand, the calculated transition temperature is much higher than the experimental value 9.25 K, obtained in specific heat measurements [3]. For the parameters used in the figures one calculates 22 K, and this is the main cause for the quantitative disagreement in the EFG above the transition. The discrepancy is partly due to the use of the molecular field approximation in a situation where the interaction is assumed to be of short range. But even multiplication with a correction factor appropriate for an XY model with four neighbours yields 13.5 K.

3. CONCLUSIONS

The local dynamics associated with a purely electronic splitting in a JT system can have a qualitative effect even on static properties. If one tries to get quantitative information about the parameters describing such a system, it may be dangerous to neglect this dynamics.

The single JT ion system Fe:$CoCr_2S_4$ and the interacting JT ions system $FeCr_2S_4$ can be reasonably well described within one model with quite satisfactory values of the parameters. However, as to the determination of these parameters separately, the spin-orbit splitting is well determined, whereas the other parameters have a rather large range of possible values.

REFERENCES

1. A.M. van Diepen and R.P. van Stapele, Solid State Comm. 13, 1651 (1973).
2. B. Hoekstra, R.P. van Stapele, and A.B. Voermans, Phys. Rev. B6, 2762 (1972).
3. F.K. Lotgering, A.M. van Diepen and J.F. Olijhoek, Solid State Comm. 17, 1149 (1975).
4. R. Englman, 'The Jahn-Teller Effect in Molecules and Crystals' (Wiley, London, 1972).
5. A.M. van Diepen and R.P. van Stapele, Phys. Rev. B5, 2462 (1972).
6. G.A. Gehring and K.A. Gehring, Rep. Progr. Phys. 38, 1 (1975).
7. S. Wittekoek, R.P. van Stapele, and A.W.J. Wijma, Phys. Rev. B7, 1667 (1973).
8. P. Brüesch and F. d'Ambrogio, Phys. Stat. Sol. (b) 50, 513 (1972).

STUDY OF THE MOTT TRANSITION IN n.TYPE CdS BY SPIN FLIP RAMAN SCATTERING AND FARADAY ROTATION

R. Romestain

Laboratoire de Spectrométrie Physique

Boîte Postale 53, 38041 Grenoble-Cédex

I. Introduction

The importance of correlations between electrons in a metal has been pointed out 40 years ago. Though band calculations which treat the interaction of electrons through a self consistent potential provides a good description of most atoms and solids, they are unable to explain why a theoretical crystalline array of monovalent atoms would either be a good conductor or an insulator depending on the distance d between atoms. First treatment of this effect was proposed in 1949 by N.F. Mott[1]. He showed that one gets an insulator until the density of electrons is such that they screen the Coulomb interaction ; there should be a critical density when the screening is large enough so that electrons are no more bound to the positive ions, this in turn would improve the effectiveness of the screening. One thus expects a quite sharp transition which can be estimated at a density N such that

$$N^{1/3} a_o = 0.2$$

where a_o is the Bohr radius of localized electrons. Another approach was put forward by Hubbard[2]. He introduces a Hamiltonian which takes into account the repulsion U between two electrons on the same site, and the transfer energy T which represents the overlap between orbitals on different sites and thus the ability of electrons to hop from one site to the next. The larger T/U, the easier one forms a conduction band. A crystalline array with adjustable interatomic spacing has not been studied yet. Nature provides us with solids which do demonstrate a metal-non metal transition but they also show a simultaneous structural transition due to electron-vibration interaction ; they are too much coupled to differentiate the effects and causes of each. On the other hand there exists a class of materials which undergo a purely electronic transition : the doped semiconductors. The radius of electron orbital is so large (around 20 Å) that the transition occurs at a very small density (around 10^{17} cm^{-3}), one does not expect the lattice to be sensitive to a concentration of impurity of 10^{-5}. A lot of work has been done on Si doped with P donors. The transition happens where expected, and it has been studied by a variety of experiments (see ref. 3 and following papers). There is a conduction which does not require any activation energy, so that even at low temperature Si:P has most of the properties of a metal.

However it still does not correspond to the theoretical model : donors are introduced in a random way and distance between sites can only be determined statiscally. Near critical concentration electron may hop between close-by sites but may find regions where the distance is too large to jump and thus conduction might not occur through a macroscopic crystal. Coexistence of two phases (a conducting and a non conducting one) in the same crystal is controversial. Magnetic susceptibility measurements[4] can be interpreted as showing such a coexistence but it was demonstrated[5] that taking correlations into proper account, could lead to similar results.

The experiments we are going to describe shed some light on this problem. CdS has the simplifying feature that being a direct gap semiconductor wave functions of electrons are simple to describe when Si conduction band minimum is on the edge of Brillouin zone.

We will first review the basic properties of CdS before going into the detail of the calculation of spin flip Raman Scattering (SFRS). Emphasis will be given on SFRS linewidth since it gives a direct estimate of the diffusion of electrons in the vicinity of the Mott transition. Results obtained by Faraday rotation will also be presented.

II. Basic properties of Cadmium Sulfide

CdS is a hexagonal crystal (wurtzite structure). Optical properties of "pure" CdS have been extensively studied[6] so that a great deal is known about the detailed structure of the conduction and valence band together with the excitonic levels.

Band structure

The CdS crystal has some covalent character but for simplicity let us assume that it is ionic so that orbitals for electron can be thought of being localized either on cation Cd^{++} or anion S^{--}. Then the highest occupied levels are the 3 p electrons of the S^{--} ion. The next level is a 5 s orbital on the Cd^{++}. From these levels one

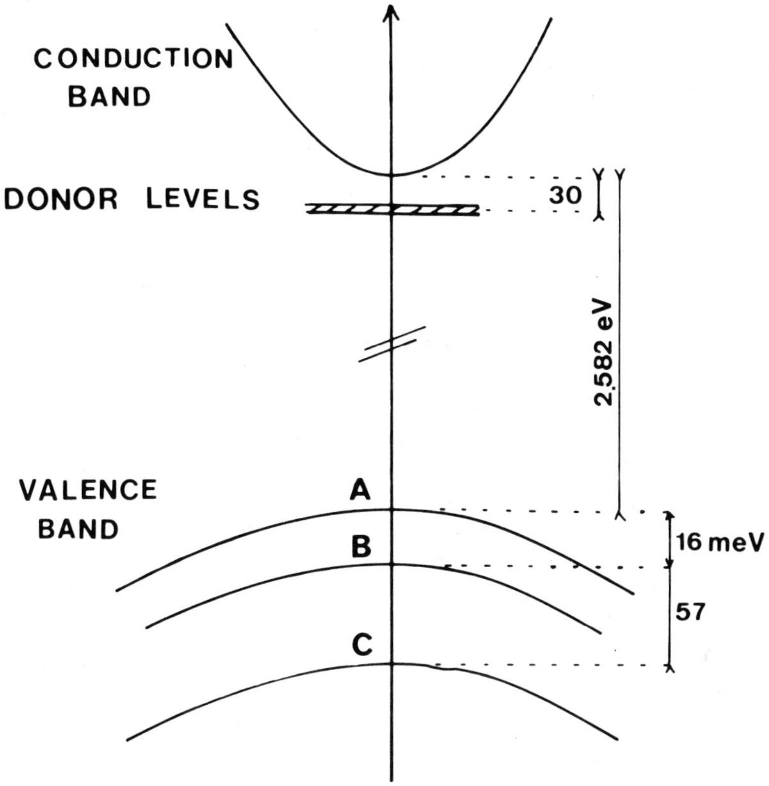

Fig. 1 : Band structure of Cadmium Sulfide

builds up bands shown in fig. 1 the 3 p level is actually split by the combined action of spin orbit interaction leaving a J = 3/2 highest and J = 1/2 lowest and the crystal field of hexagonal symmetry : this leaves 3 doublets Γ_9 (or A) Γ_7 (or B) Γ_7 (or C) respectively 16 and 57 meV from each other. The A valence band is thus made up of electrons whose orbital momentum and spin are parallel to the C-axis of the crystal. The maximum of each band is obtained at the center of the Brillouin zone i.e. $\vec{k} = 0$. The effective mass is roughly equal to m_0, mass of the electron.

The conduction band, made up from the 5 s orbital has its minimum at $\vec{k} = 0$. It lies at $E_g = 2.58$ eV above the A valence band maximum. The effective mass m^* is much lighter, almost isotropic with $m^* = 0.19$ m. Detailed calculations show that there is a mixing of valence band and conduction band (this is often carried out through the so called k.p approximation), this results in a orbital magnetic moment which appears as a modification of the g factor of the conduction electron from its g = 2 expected value. In small gap materials this can be huge (g = 50 for In Sb). For CdS the g-shift is much smaller and it was found that $g_{//} = 1.79$ while $g_{\perp} = 1.77$.

Impurity levels

If an impurity enters the crystal substitutionnally, it should have the same electric charge that the ion it substitutes for. However it usually has a tendency to keep a closed shells electronic structure. For instance when In substitutes for Cd^{++}, it first looses 3 electrons to become In^{+++}. This would leave the equivalent of one positive charge which will attract a conduction electron if any is available. If the dielectric constant of the lattice is large the Coulomb force will be very attenuated and the lowest bound state will actually be very loosely bound (30 meV for CdS). The radius is $a_0 = 28$ Å, so that the electron spend so little time on the impurity itself that it is almost insensitive to its chemical nature : the difference, the central cell correction, between different donors accounts for no more than a few percent in the binding energy. Radius a_0 and binding energy E_D are estimated directly from the values of the effective mass m^* and the dielectric constant ε, in the same way as for the hydrogen atom, so that the ground state is similar to a 1 s hydrogenic wavefunction and

$$a_0 = \varepsilon \frac{\hbar^2}{m^* e^2} \qquad E_D = \frac{m^*}{\varepsilon^2} \frac{e^4}{2h^2}$$

Free exciton

Excitation of an electron from the top of valence band to the bottom of conduction band results in a very strong absorption of light as soon as $\hbar\omega > E_g$: this is the fundamental absorption. But the electron can be attracted to the hole left in the valence band forming bound states with hydrogenic behaviour. The binding energy being 28 meV, recombination of electron and hole will generally lead to the emission of light at E_g - 28 meV ; similarly there is absorption at this frequency.

Bound exciton

The free exciton is neutral, still it can be attracted to a neutral donor in the same way two hydrogen atoms link together to form a hydrogen molecule. The corresponding excited state is roughly 8 meV below the free exciton (2.547 eV) : this can be considered as one excited state of the neutral donor. Depending on the valence band A or B which the hole belongs to, there can be two such excited levels corresponding to emission lines called I_2 and I_{2B} respectively. Excitation consists in taking one electron from a valence band, putting it next to the electron of the donor, so that they will have spins antiparallel.The symmetry properties of the complex so formed are thus primarily the ones displayed by the valence band the hole belongs to. For Spin Flip Raman Scattering, these excited states will be the virtual states which will allow matrix elements of electric dipole operator to be taken. Their value has been evaluated either from direct absorption measurement[7] correlated to concentration of donors or from the lifetime of the exciton[8] : it happens to be extremely large since it corresponds to an oscillator strength f = 10 or 6. It is to be remembered that the sum of all oscillator strengths of a n electron system is equal to n. The huge value of f is connected to the fact that during optical excitation not only the electron promoted from the valence band but also the closed shells electrons which are responsible for the dielectric constant, are involved.

III. Spin Flip Scattering

Spin Flip Raman Scattering was detected by Thomas and Hopfield in CdS[7] in a dilute sample. First linewidth studies were performed by Scott, Damen and Fleury[9]. The ground state of the donor has a spin equal to 1/2 and upon application of a magnetic field H it splits into two sublevels which we will call $|a\rangle$ and $|b\rangle$ separated by a Zeeman energy

$$\hbar\omega_Z = g\mu_B H.$$

Since g = 1.8, a 10 kG field yields a 25 GHz splitting.

If a spin is in the $|a\rangle$ state it can scatter a incident photon at frequency ω_L and wavevector $\vec{k}_L$; a photon is emitted at frequency $\omega_L - \omega_Z$ while the spin goes to state $|b\rangle$. Two lines are indeed detected Stokes at $\omega_L - \omega_Z$ and Anti Stokes at $\omega_L + \omega_Z$. They are proportionnal respectively to the population of each state, their ratio is the Boltzmann factor exp $\hbar\omega_Z/kT$ which is typically 1.5 at 2 K and

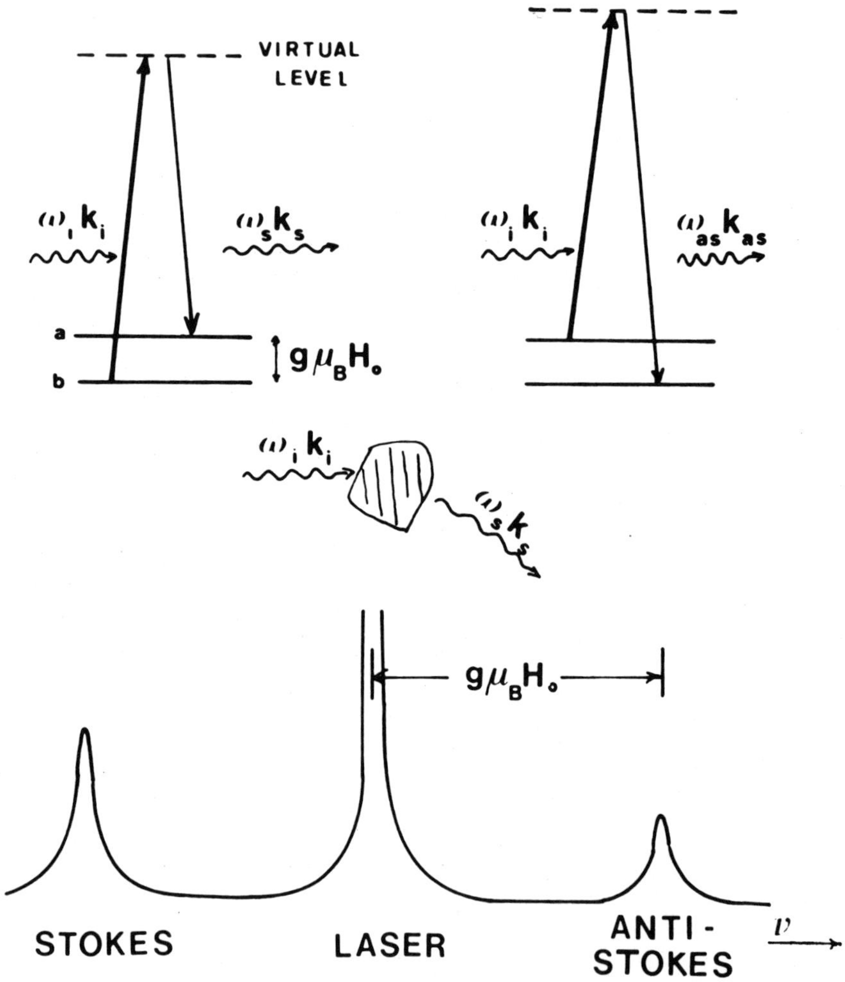

Fig. 2 : Schematic of spin flip Raman Scattering

10 kGauss.

For the experiment the CdS crystal is cooled at low temperature and laser light is focussed on to it. Scattered light at an angle θ from the incident beam is collected by a lens, passed through a Fabry Perot, and then frequency analyzed with the help of a small pinhole. Optimum resolution can be 30 MHz. Signal to noise ratio is improved, when needed, by photon counting and multichannel analyser. The index of the crystal being very large $n \# 3$, the viewing angle corresponds closely to the perpendicular to the exit face, so that bevels had to be cut in the crystals to allow large variations of θ.

To determine the magnitude of the scattering we will use a semi classical treatment[10]. The electric field of the incident beam $\vec{E}_L \cos \omega_L t$ admixes some excited states $|n\rangle$ in the ground state wavefunctions $|a\rangle$ and $|b\rangle$ with energy $h\omega_a$ and $h\omega_b$, which become :

$$(1) \quad |\psi_a\rangle = e^{-i\omega_a t+\phi}\{|a\rangle - \frac{1}{2}\sum_n |n\rangle \frac{\langle n|\vec{E}_L.e\vec{r}|a\rangle}{E_n - E_a - \hbar\omega_L} e^{-i\omega_L t}\}$$

where ϕ is a random phase constant.

Though no electric dipole could appear between states $|a\rangle$ and $|b\rangle$ because they have the same parity, the modified ψ_a and ψ_b display matrix elements of $e\vec{r}$, for instance

$$(2) \quad \langle\psi_a|e\vec{r}|\psi_b\rangle = \sum_n \frac{\langle a|\vec{E}_L.e\vec{r}|n\rangle\langle n|e\vec{r}|b\rangle}{E_n - E_a - h\omega_L} e^{-i(\omega_L \; \omega_b+\omega_a)\, t \,+\, i\phi}$$

$$+ \text{ a similar term with } e^{+i(\omega_L+\omega_b-\omega_a)t \,+\, i\phi}$$

In the classical treatment of spontaneous emission of radiation one also calculates a dipole matrix element between an excited and a ground state : the existence of such an oscillating dipole gives rise to emission of light at its frequency during the transition between the two states. The same applies here but light will be emitted at frequency $\omega_L + \omega_{ba}$ while the system goes from state $|a\rangle$ to state $|b\rangle$. The random phase factor ϕ represents the fact that oscillating dipoles are incoherent, thus one will add emitted power by each spin instead of emitted electric field. One can then define a cross section for spontaneous Raman Scattering so that intensity of scattered radiation per unit solid angle in a direction Ω is given as a function of incident intensity I_L and length l of the region of the crystal under study :

$$I_S = N.l.\frac{d\sigma}{d\Omega} I_L$$

where N is the concentration of scattering centers, that is neutral donors.

One can get a simplifying view of the problem by introducing a so-called Raman dipole D^2 which operates in the 2-dimensional manifold of the unperturbed states $a(t) = |a\rangle\, e^{-i\omega_a t - i\phi}$ and $b(t) = |b\rangle\, e^{-i\omega_b t}$ such that for instance

$$(3)\quad \langle\psi_b(t)|e\vec{r}|\psi_a(t)\rangle = \langle b(t)|\vec{D}^{(2)}|a(t)\rangle$$

In most general fashion the components of any such operator may be expressed in terms of Pauli $\vec{\sigma}$ matrices for a spin S = 1/2

$$(4)\quad D_k^{(2)} = \sum_{ij} \alpha_{ijk}\, E_{Li}\, \sigma_j\, e^{-i\omega_L t} + \beta_{ik}\, E_{Li}\, \tilde{I}\, e^{-i\omega_L t} + C.C.$$

For cubic symmetry this expression simplifies greatly, giving

$$(5)\quad \vec{D}^{(2)} = \vec{\sigma} \times \vec{E}_L\, (\alpha\, e^{-i\omega_L t} + C.C.) + (\beta\, \vec{E}_L\, \tilde{I}\, e^{-i\omega_L t} + C.C.)$$

where

$$(6)\quad \alpha = \sum_n \frac{i}{2} \frac{\langle a|ex|n\rangle\langle n|ez|b\rangle}{E_n - E_a - \hbar\omega_L}$$

In equ. (5), β represents the usual polarizability which gives rise to the index of the crystal as well as Rayleigh scattering. Radiation of the dipole D can be calculated classically and yield a Raman electric field $\vec{E}_R$ polarized parallel to $\vec{D}$, the intensity being given by

$$(7)\quad \frac{d\sigma}{d\Omega} = 4\, |\alpha|^2\, \frac{\omega^4}{c^4}$$

Though expression (5) is valid only for cubic symmetry, and is a little more complicated for hexagonal symmetry as for CdS, it gives the salient features observed in this case. Due to the vector product $\vec{\sigma} \times \vec{E}_L$ Raman Scattering will be polarized perpendicular to the laser polarisation ; a typical configuration would then be for instance $\vec{E}_L \,//\, \vec{u}_z \,//\, \vec{H}$, $\vec{E}_R \,//\, \vec{u}_x \perp \vec{H}$ so that σ_y will have matrix element between $|a\rangle$ and $|b\rangle$. At this point nothing has been said upon the wavevector of light, though obviously it cannot be parallel to the electric field.

An important point to stress is the energy denominator : if the frequency of the laser is such that $\hbar\omega_L$ almost matches the difference in energy between the ground state and one excited state $|n\rangle$ the value of $d\sigma/d\Omega$ is enhanced. This happens to be case when the 4880 Å line of an Ar ion laser is used since the excited state corresponding to the I_2 line lies at 4867 Å. The cross section is so large that real forward scattering has in some cases been detected though the exciting beam actually goes through the detection system. The so-called Rayleigh scattering is often smaller than the Raman Scattering. A typical values of $d\sigma/d\Omega$ is 2.10^{-18} cm^2. Now we are in position to understand how the light can flip a spin though there is no direct interaction between an electric field and a spin.

This occurs because the excited state $|n\rangle$ carries the orbital momentum of the p.like valence band for which there is a strong spin orbit interaction. The state $|n\rangle$ is a superposition of spin up and spin down wave functions and thus can be connected to both $|a\rangle$ and $|b\rangle$ through electric dipole, giving a non-zero value of α. If all spin orbit states $|n\rangle$ were degenerate (no spin orbit interaction) all the contributions of different $|n\rangle$ would cancel out.

IV. Study of the SFRS linewidth

If the donor electron is localized, then the SFRS linewidth should simply reflect the EPR linewidth, be it due to inhomogeneous broadening (strains or local fields) or homogeneous broadening (relaxation time T_2 or T_1). This has been verified for a sample with concentration 10^{+17} where the linewidth of 0.003 cm^{-1} is independent of scattering angle and also equal to the measured EPR linewidth.

On the other hand if the electron moves with velocity $\vec{v}$, then the Raman line will have an additional Doppler shift : it experiences a laser frequency Doppler shifted by $\vec{k}_L.\vec{v}$ and radiates at an angle θ a Raman frequency which will be shifted by $-\vec{k}_S.\vec{v}$, so that $\Delta\omega_{Dop} = \vec{q}.\vec{v}$ if $\vec{q}$ is the scattering vector $\vec{q} = \vec{k}_L - \vec{k}_S$. For electron velocities of $v \sim 10^7$ cm s^{-1} corresponding to Fermi velocities for the concentrations used $\Delta\nu_{Dop} \sim 10^{12}$ Hz. On the other hand, the electrons undergo collisions at a rate τ_c approaching 10^{14} s^{-1} so that Doppler shift is collisionally narrowed giving rise to a diffusionnal line whose effective linewidth is given by

$$(9) \quad \Delta\nu_{eff} = 4\pi(\Delta\nu_{Dop})^2 \tau_c$$

A uniform repartition of the velocities on the Fermi surface yields an angular average

$$(10) \quad (\Delta\nu_{Dop})^2 = \frac{V_F^2}{3}(2 \sin \frac{\theta}{2})^2 k^2 \qquad \text{and}$$

$$(11) \quad \Delta\nu_{eff} = \frac{16\pi}{3} \frac{V_F^2}{\lambda^2} \text{Sin}^2 \frac{\theta}{2} \tau_c$$

Alternate description of the SFRS linewidth[11]

We have derived a formula giving what we called Raman dipole $\vec{D}^2$, this dipole interacts with the radiation field at its frequency giving rise to emission as we have seen. This interaction can be expressed by the Hamiltonian

$$(12) \quad \mathcal{H}_{SF} = \vec{D}^{(2)} \vec{E}_R e^{+i\omega_R t} = (\vec{\sigma} \times \vec{E}_L).\vec{E}_R e^{+i\omega_R t} (\alpha\, e^{-i\omega_L t} + C.C.)$$

if we introduce the spatial dependence of the radiation field one will obtain terms of the form :

$$(13)\quad \mathcal{H}_{SF} = \vec{\sigma}\,(\vec{E}_L \times \vec{E}_R)\; e^{-i(\omega_L-\omega_R) + i\,\vec{q}\,\vec{r}} = \vec{\sigma}\,\vec{H}^+$$

where again $\vec{q} = \vec{k}_L - \vec{k}_R$.

The effect of the radiation will then be the same as a magnetic field interacting with the spins. When the field is constant the effect of the interaction depends on the static susceptibility χ_o, for an oscillating magnetic field $H_1 \cos \omega_1 t$ like the one used for magnetic resonance one would need define a susceptibility at $\omega_1 \neq 0$. Here we introduce a susceptibility $\chi(\omega,q)$ as a response to a field with time and space dependence $H^+ = H\, e^{-i(\ t+\vec{q}\,\vec{r})}$.

Still the q value corresponding to the wavelength of light 0.5 μ is large compared to the average inter electron spacing or to the Bohr radius of a donor. Under such conditions one expects that the spin transverse magnetization M^+ can be described by a modified Bloch equation of the form :

$$(14)\quad \frac{dM^+}{dt} + i\,\omega_s\, M^+ + \frac{M^+ - \chi_o H^+}{T_2} - D_s\, \nabla^2 M^+ = \frac{i\mu g}{\hbar} M_o\, H^+$$

where $\omega_s = \omega_a - \omega_b$ is the precession induced by the static field while $\mu g/\hbar\, H^+$ is the one induced by the oscillating H^+, T_2 is a transverse spin-relaxation time. D_s is the spatial spin diffusion constant.

If H^+ is a plane wave

$$H^+(r,t) = H_o^+\, e^{-i(qr-\omega t)},$$

eq (14) has the solution

$$(15)\quad M^+ = \frac{(-\omega_s + \frac{i}{T_2})\,\chi_o\, H^+}{(\omega-\omega_s) + i(\frac{1}{T_2} + D_s\, q^2)}$$

from which it follows that

$$(16)\quad \chi^+(q,\omega) = \frac{(-\omega_s + \frac{i}{T_2})\,\chi_o}{(\omega-\omega_s) + i(\frac{1}{T_2} + D_s q^2)}$$

For magnetic resonance it is well known that imaginary part of susceptibility gives rise to absorption, that is real transitions. For the same reasons Im χ^+ will describe the actual dependence of spin flip scattering; the linewidth follows :

$$(17)\quad \gamma = \frac{1}{T_2} + D_s q^2$$

We must emphasize the fact that D_s introduced phenomenologically in eq (14) can represent either a real motion of electrons carrying their spin (particle diffusion D) or pure spin diffusion as would occur in case of localized electrons interacting via dipole dipole coupling. In that sense eq (14) is more general than eq (11) but the pure spin diffusion introduced appears difficult to evaluate in a random system.

Mott transition in CdS

We have so far demonstrated the usefulness of SFRS in that it gives direct information upon the motion of the particles scattering the light. One major advantage is its selectivity : at the Stokes frequency only donors can participate, that would not be true for conventional Rayleigh scattering. We will see later that Faraday rotation has also this advantage and we will make use of it. Before going into the results obtained by SFRS let us review the main experiments showing unambiguously the Mott transition in CdS.

The occurence of Mott transition is displayed very clearly by conductivity measurements taken at Helium temperature on samples of different concentrations[12] : this is seen in figure (3) where over a range extending between $N = 10^{18}$ to $N = 10^{17}$ cm^{-3} the conductivity drops by more than 4 orders of magnitude. Above $N = 9\ 10^{17}$ cm^{-3} the number of carriers as determined by Hall effect[13] is almost insensitive to temperature (see fig. 4). In this regime the electrons are definetely moving in a not completely filled band like they would in a metal. Below $N = 5\ 10^{17}$ cm^{-3} on the other hand the number of carriers reflects a thermal activation from bound states, behaving like semiconductors though the dependence in temperature is more complicated.

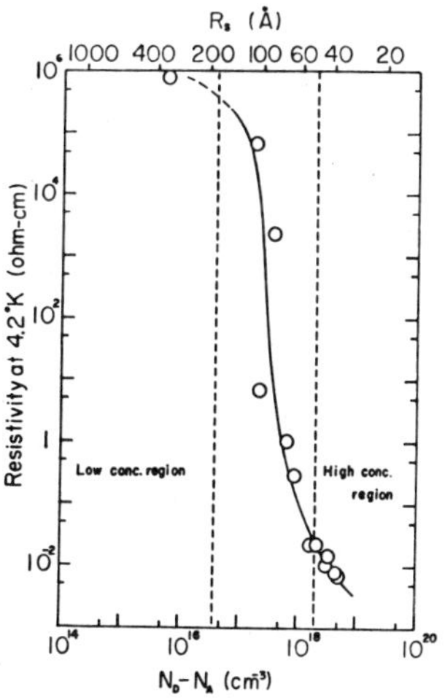

Fig. 3 : Resistivity of CdS as a function of concentration of neutral donors at 4.2 K (from ref. 12)

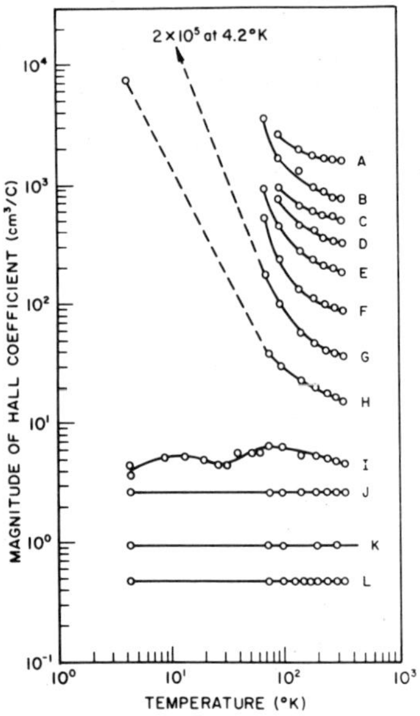

Fig. 4a

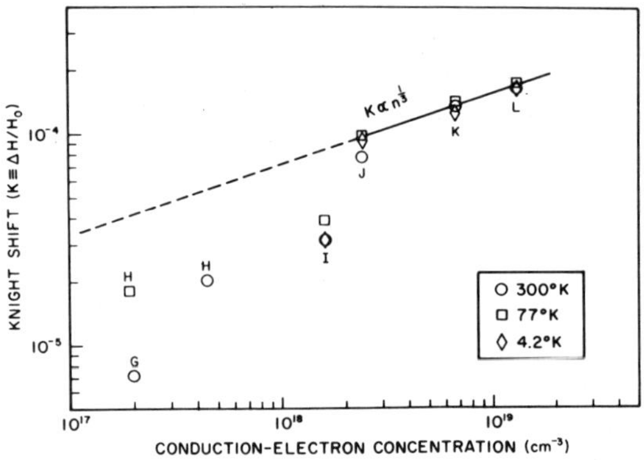

Fig. 4b

Fig. : 4a Hall coefficients of different CdS samples. The labelling of samples is the same as in fig. 4b. Concentration is $2\ 10^{17}$ for sample G ; H = $4.5\ 10^{17}$, I = $1.6\ 10^{18}$, J = $2.5\ 10^{18}$ (from ref. 13)

Experiments performed on the NMR of the ^{113}Cd isotope[13] reveals more precise feature of the transition on the metallic side. The Knight Shift (fig. 4b) has an expected $N^{1/3}$ dependence on concentration when $N > 2\ 10^{18}$. Then the sample has all the caracteristics of a metallic sample : the impurity band formed around $9\ 10^{17}$ cm^{-3} merges with the conduction band.

Analysis of experimental results

The diffusional scattering is already clearly present at a concentration $N = 2.4\ 10^{17}\ cm^{-3}$ which lies at the onset of Mott transition and still has a semiconductor behaviour. The SFRS linewidth is shown in fig. 5 for forward scattering $\theta = 0(\vec{q} = 0)$ and $\theta = 90°$. At $\theta = 90°$, $\Delta\nu = 300$ MHz whereas for $\theta = 0$, $\Delta\nu = 60$ MHz including an instrumental linewidth of 30 MHz. The q^2 dependence of $\Delta\nu$ is illustrated in fig. 6 for a more metallic sample with $N = 1.4\ 10^{18}$ for scattering angles close to the forward direction.

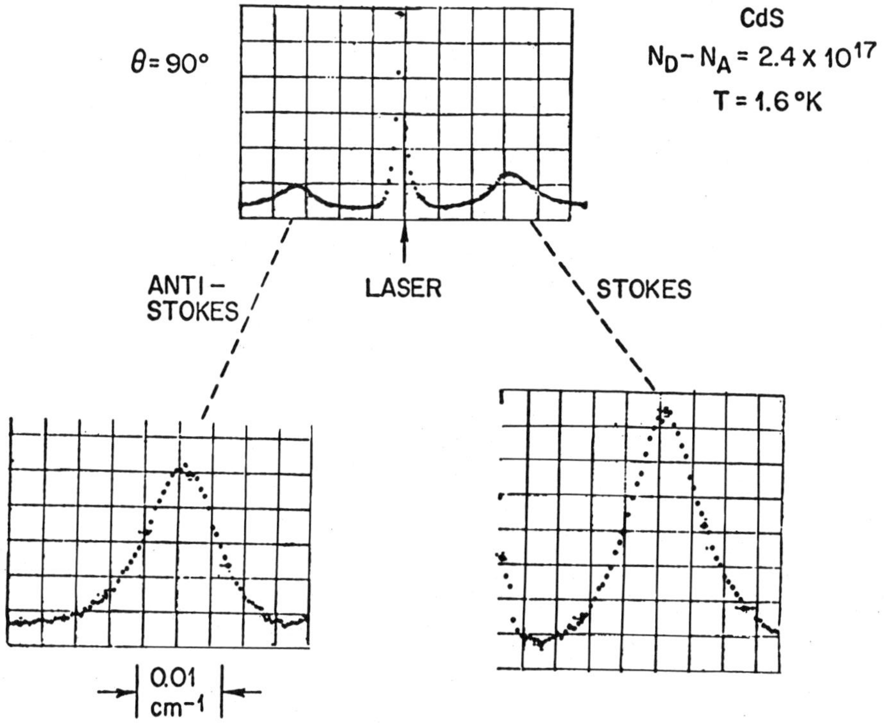

Fig. 5a

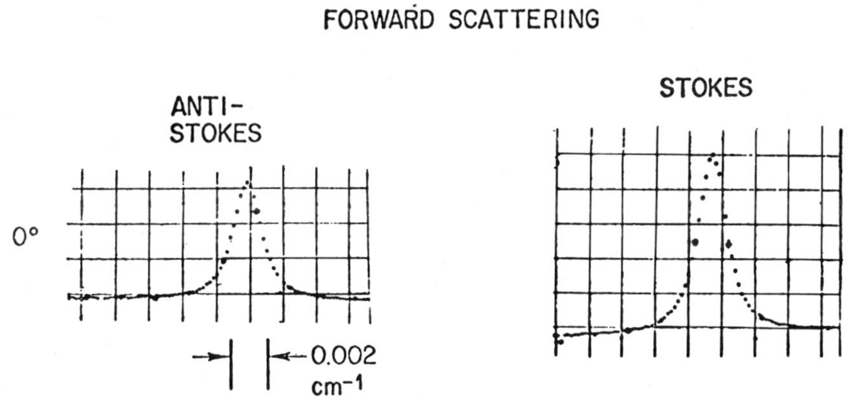

Fig. 5b

SFRS in a sample with $N = 2.4\ 10^{17}$. The lower portion of fig. 5a shows Stokes and Anti Stokes on an expanded scale. Fig. 5a is 90° scattering, Fig. 5b is 0° scattering. Both are taken with the same resolution

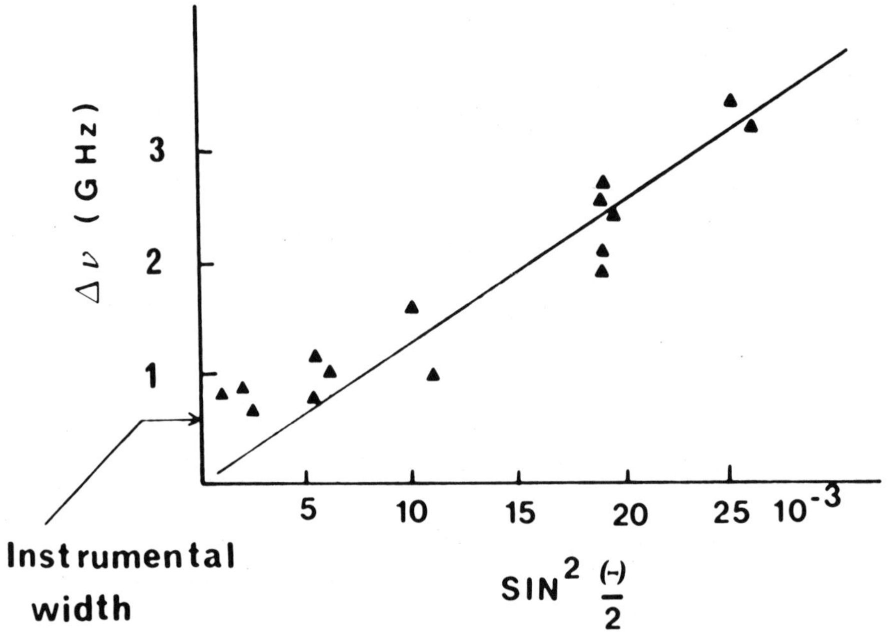

Fig. 6 : Linewidth of small angle forward scattering illustrating the diffusional motion of carriers via the q^2 dependence of the linewidth when N = 1.4 10^{18} and T = 1.6 K

The linewidth can be evaluated using relation (11) when the impurity band has merged into the conduction band so that v_F can be deduced from the electron concentration N, and effective mass m^* :

$$(18) \quad v_F = \frac{\hbar}{m^*} (3\pi^2 N)^{1/3}$$

τ_c can in turn be deduced from resistivity measurement (12) since

$$(19) \quad \rho = \frac{m^*}{Ne^2 \tau_c}$$

So equation (11) can be rewritten in that case

$$(20) \quad \Delta\nu = \frac{6.10^{15}}{\rho N^{1/3}} \text{ cm}^{-1}$$

ρ is in Ω cm and N in cm^{-3}.

The increased diffusional motion with increasing N and metallic character is illustrated in Table I where the results are compared with Eq (20). The agreement which is reasonable down to N = 2.4 10^{17} is much better than could be hoped. Below 10^{18} one does

	n	$\rho(2K)$ ohm-cm	$\Delta\nu_{eff}(calc)$ cm^{-1}	$\Delta\nu_{meas}(cm^{-1})$ $\theta = 90°$ $T = 2K$
A	1.0×10^{17}	$> 10^5$	-	0.003
B	2.4×10^{17}	20.0	0.016	0.008
C	5.0×10^{17}	1	0.25	0.125
D	1.4×10^{18}	0.1	1.9	2.0
E	4.0×10^{19}	0.0014	45	14

Table I : Comparison of measured SFRS linewidths with $\Delta\nu_{eff}$ calculated from eq (4), where ρ is in ohm-cm

not expect an effective mass formalism to be applicable since the impurity band has certainly a more complicated shape. Besides variation of linewidth by a factor of two are observed in different regions of the same sample. This is undoubtedly due to even slight inhomogeneities in concentration since $\Delta\nu$ varies so rapidly with N in this regime. It should be emphasized that the SFRS technique probes a region of the sample that in our case is of the order of 100 microns. Such variation of N are of course averaged in a typical transport measurement. Nevertheless there is clear evidence of increasing diffusion with increasing N though more detailed theoretical calculations are needed to understand the Mott transition region.

V. Measurement of χ_0 by Faraday rotation

In addition to the off diagonal component of the dipole $\vec{D}^{(2)}$ in eq (5) which gives rise to SFRS there is a diagonal part associated with σ_z. This corresponds to a dipole radiating at the same frequency ω_L but rotated with respect to the incident polarization. For instance

$$(21) \quad \langle a|D_y^{(2)}|a\rangle \sim \alpha \langle a|\sigma_z|a\rangle E_x e^{-i\omega_L t}$$

This rotated dipole gives rise to a Faraday rotation[14] per unit length along z given by

$$(22) \quad \frac{\phi}{l} = \frac{8\pi^2}{n} N \alpha \langle\sigma_z\rangle$$

Since the susceptibility is proportionnal to the derivative of the magnetization $\langle\sigma_z\rangle$ with respect to H, then the specific rotation $R = d/dH\ (\phi/l)$ is proportionnal to χ.

As α can be made extremely large for the donors by choosing near resonant light excitation as described for the SFRS, one has a way of measuring very selectively the χ of the donors independent of other magnetic impurity present in the sample. In figure 7a is shown the transmission of linearly polarized light through the sample as detected through a polaroïd. The separation between peaks is plotted versus H in fig. 7b. There is a small temperature independent rotation η, due to interband transitions which is substracted to give spin Faraday rotation $\phi - \eta$. This quantity follows a Brillouin function for S = 1/2 and g = 1.79 as expected for the donors in CdS. When R is studied as a function of temperature, a plot of 1/R versus T yields an antiferromagnetic Curie-Weiss θ of 0.3° K (the same results are found using 4965 Å excitation although the rotations are smaller). This would suggest searching for a spin glass at lower temperatures.

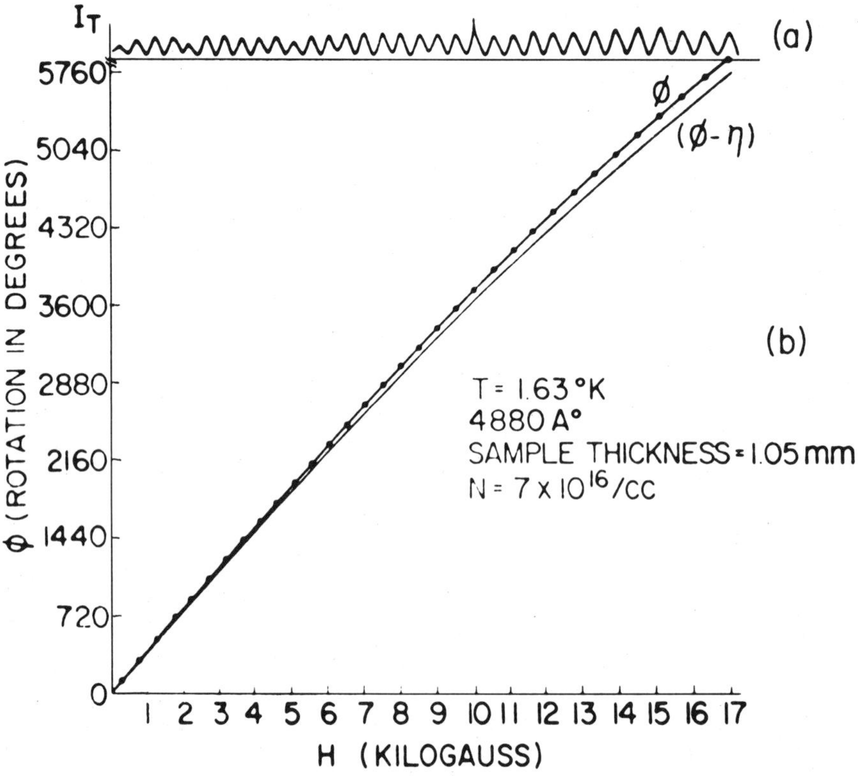

Fig. 7 : Faraday Rotation for localized centers $N = 7\ 10^{16}\ cm^{-3}$

In the very metallic region, for example with $N = 6 \times 10^{18}$, χ shows no saturation with H and is independent of T as one would expect from a Pauli susceptibility. However, in the metallic transition region, χ displays both temperature dependent as well as temperature independent components. This is illustrated in Fig. 8 for $N = 1.4 \times 10^{18}$. This temperature dependent component increases with decreasing concentration and is fairly dominant when $N = 2.4 \times 10^{17}$.

The variation of R with temperature, T, mirrors the variation of susceptibility with T in a given sample. To determine the absolute value of χ, however, one needs specific knowledge of the quantity α in eq (6). This may be easily calculated in the insulating limit where there is only one significant excited level corresponding to an exciton bound to the neutral donor (I_2) whose position and oscillator strength is well known[6]. Using this exciton level the calculated α agrees very well with the magnitude of the rotation observed in fig. (8). However, as one crosses to the metallic regions with increasing concentration, the bound exciton loses meaning and the question arises as to the changing spectrum of the electron-hole excitation. In the case of the very metallic samples, the lowest such excitation corresponds to the energy needed to promote an electron from the top of the valence band to the Fermi level now in the conduction band, and the oscillator strength is spread out among the unoccupied states. The most feasible way to determine α in this case would be experimentally by measuring the optical absorption spectrum. However, beyond the transition on the metallic side even up to $N \sim 2 \times 10^{18}$, it is observed that the luminescence feature corresponding to electron-hole recombination does not move and while it broadens somewhat, its width still remains small compared to the 5 meV separation from the laser line. Therefore, up to $N \sim 10^{18}$ we tentatively take α to be the same as in the dilute limit for our analysis of absolute χ in this region.

Thus if we compare the absolute magnitude of the temperature dependent component of the rotation for $N = 1.4 \times 10^{18}$ to that found per localized spin in the insulating region for $N = 7 \times 10^{16}$, we find that the temperature dependent part in fig. 8 corresponds to approximately 10^{16} electrons with free moments or 1% of the total. The same figure is arrived at by comparing the ratio of the temperature dependent and temperature independent parts assuming the latter to be Pauli like with a degeneracy temperature $T_F = 265°$ K corresponding to $N = 1.4 \times 10^{18}$. By contrast, for the $N = 2.4 \times 10^{17}$ sample which shows only a very small temperature independent susceptibility, an analysis of χ in terms $\chi_p + \chi_{c.w.}$ shows that the number of Curie-Weiss electrons is comparable to the number of Pauli like electrons. Very similar results for χ were obtained by Quirt and Marko[4] for Si : P.

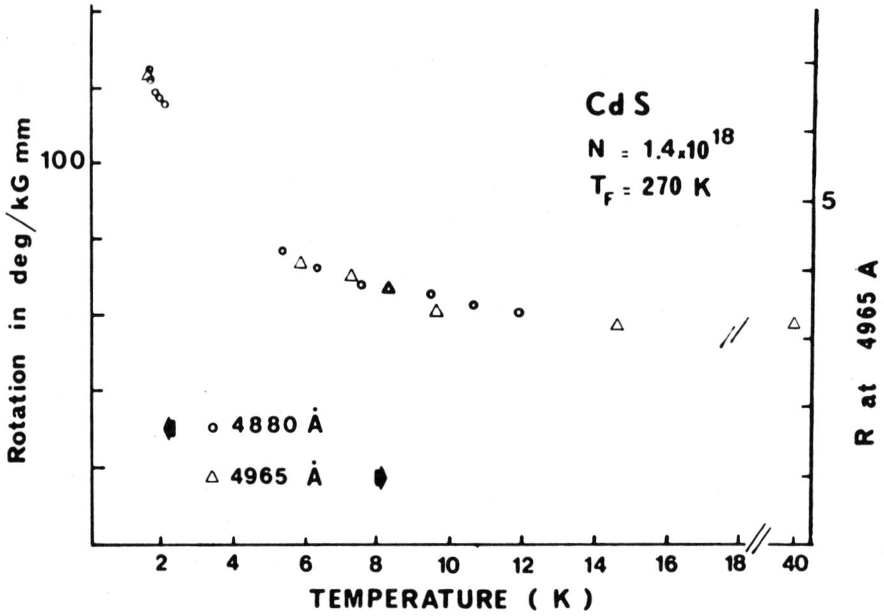

Fig. 8 : Susceptibility of donors in a concentrated sample
$N = 1.4\ 10^{18}$

VI. Discussion of results

Some may be inclined to interpret the two components χ_p and $\chi_{c.w.}$ that are observed as arising from well defined microscopic regions of metallic and insulating material with nondiffusing electrons respectively, associated with the clustering due to basic statistical randomness in donor positions. However, we will now present evidence that there are no nondiffusing spins beyond the metallic transition which seems to argue against the simultaneous coexistence of two such phases. If there were nondiffusing spins imbedded in a metallic matrix, we would certainly expect their number to be greatest close to the M-NM transition point as for example in the sample with $N = 2.4 \times 10^{17}$. If one looked at 90°

scattering in such a sample, one would expect to see a narrow line with no angular dependence, superimposed on the broader diffusional line with q^2 dependence. However, one always sees only a single line with q^2 dependence as shown in fig. 2a and no evidence for nondiffusing carriers. This is further illustrated in fig. 6 where again the narrow line observed at $\theta = 0$ broadens continuously as q^2 with increasing angle about $\theta = 0$. We are inclined to believe that the temperature component of χ is coming from electrons in singly occupied sites of the correlated electron motion[4,5]. Because of randomness some or all of these singly occupied sites may even have fixed positions in space. However, the electrons in these singly occupied sites are still in dynamic equilibrium with the rest of the electron sea and leave this site to be filled by another carrier replacing it. Equivalently stated, the electrons on singly occupied sites are Anderson local moments or Friedel virtual bound states which while giving rise to a $\chi_{c.w.}$ still have diffusive character in the SFRS.

References

1. N.F. Mott, Proc. Phys. Soc. (London) A 62, 416 (1949)
2. J. Hubbard, Proc. Roy. Soc. A (276), 238 (1963)
3. N.F. Mott, Journal de Physique C-4, 302 (1977)
4. J.D. Quirt, J.H. Marko, Phys. Rev. 7, 3842 (1973)
5. K.A. Chao, K.F. Berggren, Phys. Rev. Lett. 31, 880 (1975)
6. D.G. Thomas, J.J. Hopfield, Phys. Rev. B, vol. 128 n° 5, p.2135 (1962)
7. D.G. Thomas, J.J. Hopfield, Phys. Rev. 175 n° 3, 1021 (1968)
8. C.H. Henry, K. Nassau, Phys. Rev. B, vol. 1 n° 4, p. 1628 (1970)
9. J.F. Scott, T.C. Damen, P.A. Fleury, Phys. Rev. vol. 6, n° 10, 3856 (1972)
10. R. Romestain, S. Geschwind, G.E. Devlin, P.A. Wolff, Phys. Rev. Lett. vol. 33 n° 1, 10 (1974)
11. P.A. Wolff, S. Yven, P. Maldague, private communication
12. S. Toyotomi, K. Morigaki, J. Phys. Soc. Jap. 25, 807 (1968)
13. F.D. Adams, D.C. Look, L.C. Brown, D.R. Locker, Phys. Rev. B vol. 4 n° 7 (1971)
14. R. Romestain, S. Geschwind, G.E. Devlin, Phys. Rev. Lett. 75, 803 (1975)

ELECTRON PHONON INTERACTIONS AND CHARGE ORDERING IN INSULATORS

Y. Yamada

College of General Education

Osaka University, Toyonaka, Osaka, 560 JAPAN

I. CHARGE ORDERING IN INSULATORS

There are various kinds of crystals which undergo structural phase transitions upon change of any external circumstances, especially of temperature. These structural transitions are understood in terms of the instability of the lattice against some particular atomic displacement field, or phonon mode. Since it is obvious that a harmonic phonon system can not become unstable by change of temperature, some sort of interaction of harmonic phonons with some physical variables would be needed to trigger the phase transition.

In this lecture, let us take up localized electron system as the variable to trigger the phase transitions through the interaction with phonons. In this case, the ordering in the distributions of local charge density takes place below the transition temperature along with the condensation of a particular phonon mode. This is certainly in contrast to the topics concerning charge density waves in metals where the charge ordering is seen in itinerant electron system.

In the case of insulators, the electron associated with the ordering is more or less localized in space. Therefore, each electron can be specified by the site number rather than the wave vector K as is done in itinerant electron case.

The charge ordering process in insulators is generally considered to be as follows: In higher symmetry phase, there is a fluctuating part $\delta\rho$ in the charge density of the localized electrons, and, due to some sort of interactions between these fluctuating parts at different sites, a cooperative ordering of charge density will set in when the external circumstances such as temperature, pressure etc. are varied.

It would be convenient to expand the fluctuating part of the charge density localized at a site in spherical harmonics as follows:

$$\delta\rho = \sum_{\ell,m} R^m_\ell(r)\, Y^m_\ell(\theta,\phi) \tag{1}$$

The coeffiants of the ℓ'th terms are associate with the tensor components of 2^ℓ multipoles:

$$q^m_\ell = \sqrt{\frac{4\pi}{2\ell+1}} \int r^\ell R^m_{\ell(r)} r^2 dr \tag{2}$$

We may categorize the various types of charge order into "grades" by specifying the lowest order fluctuating term which 'freezes' in the low temperature phase as is shown in Table I. The most widely observed charge order is, of course, the case of 1st grade ordering, namely the dipolar ordering. Ferroelectric or antiferroelectric ordering fall into this class.

However, we can consider the case of even lower grade: the 0'th grade ordering. In this case, the total charge, integrated around each site, itself is fluctuating in the higher symmetry phase. Various kinds of mixed valence systems such as Fe_3O_4, SmS, etc. would be categorized as belonging to this class. Later, we will discuss Fe_3O_4 as an example of this class in some detail. On the other hand, we may also consider the cases of higher grades than dipolar ordering. In fact, from this stand point, various kinds of Jahn-Teller ordering is categorized as quadrapolar ordering. We will discuss this type charge ordering by taking up the case of $K_2PbCu(NO_2)_6$ as an example.

TABLE I

Classification of the charge ordering associated with local electronic state, with respect to the lowest multipoles appearing in the ordered phase.

grade	multipole	example
0th	monopole	mixed valence systems
1st	dipole	ferroelectrics, antiferroelectrics
2nd	quadrupole	Jahn-Teller systems
---	---	---

Whatever the grade of the charge order is, the most important characteristic of the ordering of localized electron, as compared with itinerant electron case, consists in that the energy eigenstates of electrons are consisted of well separated discrete levels. And the eigenstates relevant to the phase transition are restricted to the ones with excitation energies comparable to or less than kTc, Tc being the transition temperature. Usually only several eigenstates should be considered. In such the case, the pseudospin description of the electronic system is very useful.

We expand the electronic potential with respect to small atomic displacements, or more properly with respect to phonons. Then we can go through the procedure to obtain 'coupled pseudospin-phonon formalism' describing localized electron-phonon system. This procedure has been extensively discussed in the separate lectures in this seminar[1,2].

One of the advantage of introducing pseudospins to describe local electronic state is that the property of cooperative ordering of real spin system under various conditions are extensively investigated in connection with the study of magnetic materials. Hence, there are good chances that we can directly infer what will happen in our pseudospin system by making analogy with the corresponding real spin system. We will see later a good example where such analogy works well.

We concentrate ourselves to the lowest order bilinear coupling between pseudospins and phonons:

$$H_{int} = \frac{1}{\sqrt{N}} \sum_{\ell} \sum_{k,s} \xi_{k,s} \, \sigma_{\ell} \, Q_{k,s} \, e^{ik \cdot r_{\ell}} \quad (3)$$

where the spin operator σ_{ℓ} describes the state of electron localized at ℓ'site, $Q_{k,s}$ is the amplitude of phonon mode specified by the wave vector k and the branch number s. The property of the linearly coupled spin-phonon systems is discussed extensively in the separate lecture in this seminar[1]. The exchange of phonons gives rise to an effective spin-spin interaction causing spontaneous ordering of spin system below a critical temperature T_c, accompanied by a distortion of the lattice or a condensation of phonon mode.

In the following, we discuss two typical examples of the charge ordering of 2nd grade and 0'th grade. In both cases, it is shown that the pseudospin-phonon formalism can be worked out, though the physical contents for the 'pseudospin' is not the same. Thus, we try to view these phase transitions on a more or less unified stand point.

II. SECOND GRADE ORDERING : JAHN-TELLER ORDERING IN $K_2PbCu(NO_2)_6$

II·1 Introduction

The crystal of $K_2PbCu(NO_2)_6$ belongs to the isomorphous crystal system generally expressed as $R_2MCu(NO_2)_6$, where R=K, Rb, Cs, Tl and M=Ca, Sr, Ba, Pb. These crystals contain octahedral $Cu(NO_2)_6$ complex in common. As is well known, Cu^{2+} ion in a cubic crystal field is Jahn-Teller active, whence $Cu(NO_2)_6$ group has the tendency to become slightly distorted from the regular octahedron. Not only that, the crystal lattice as a whole would become distorted due to the interactions between local distortions. In fact, some of the crystals undergo phase transitions which is considered to be associated with the cooperative Jahn-Teller effect.

In particular, the system $R_2PbCu(NO_2)_6$ is very interesting in that it exhibits successive phase transitions,[3,4,5,6] $K_2PbCu(NO_2)_6$ being the typical example.[3,4] In this section, we try to understand the property of this sequencial cooperative Jahn-Teller ordering, as one of the interesting example of 2nd grade ordering, on the general basis discussed in the preveous section.

II.2 Successive Phase Transitions

The undistorted 'prototype' structure[7] is depicted in Fig. 1. It belongs to cubic system (space group F3m) and $Ni(NO_2)_6$ complex sitting at the corner of the cell forms the regular octahedron. It is well known that when Ni→Cu, the local 3d electrons of Cu^{2+} ion strongly couple to local distortions with Eg symmetry, conventionally specified as the Q_2-mode and the Q_3-mode.[8] The electronic state is represented in terms of Pauli spin operators σ^x and σ^z, and the stable ground state is expressed as a 'circle' in the two-dimensional σ^x-σ^z plane. In the pseudospin description, the cubic phase corresponds to the case when the spin direction is not fixed and σ_ℓ at the ℓ'th site is fluctuating around the circle.

The crystal of $K_2PbCu(NO_2)_6$ has the undistorted cubic structure at room temperature. This crystal undergoes two phase transitions at 280°K and 273°K successively.[3,4] The lower temperature phases are reported to have pseudotetragonal[9] unit cell with c/a<1. In order to understand overall feature of successive transitions, we show the precise measurement of the temperature variations of X-ray scattering angles of two fundamental reflections in Fig. 2. The intermediate phase (Phase II) is known to have tetragonal cell, while the lowest temperature phase, strictly speaking, has a lattice symmetry lower than orthothombic.

It is interesting to compare these with the temperature

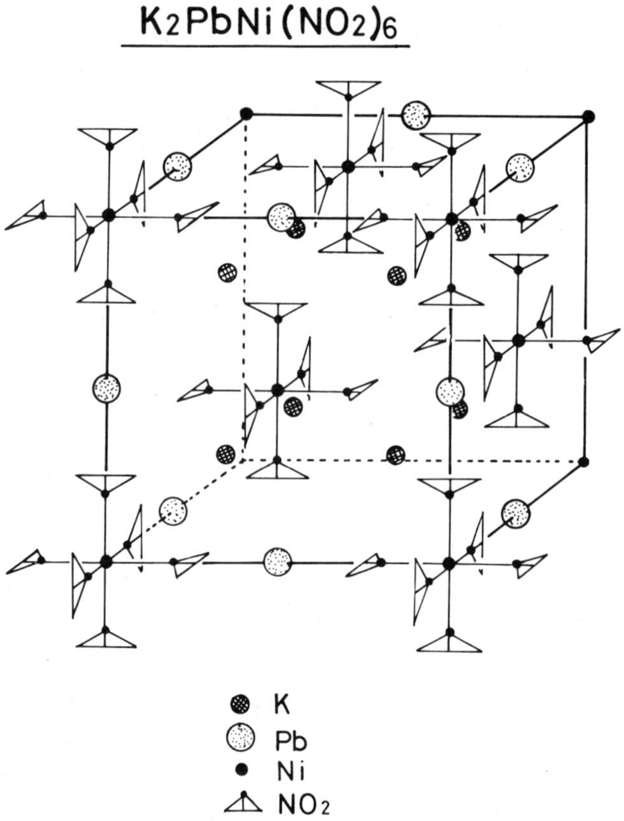

Fig. 1 The undistorted 'prototype' structure of $R_2MNi(NO_2)_6$. When Ni→Cu, the octahedral $Cu(NO_2)_6$ complex tends to be distorted due to Jahn-Teller effect.

dependence of ESR absorption lines of Cu^{2+} ions.[4,10] We see that there is a fairly good correspondence between the lattice distortion and the 3d electronic state of Cu ions, which indicates the important role of Jahn-Teller effect in these phase transitions.

Recently, we have carried out precise structural measurements in Phase II and Phase III[11,12] using X-rays and neutrons, which have revealed interesting features involved in these phase transitions.

II·3 Phase III : Canted Pseudospin Structure, Antiferrodistortive Phase

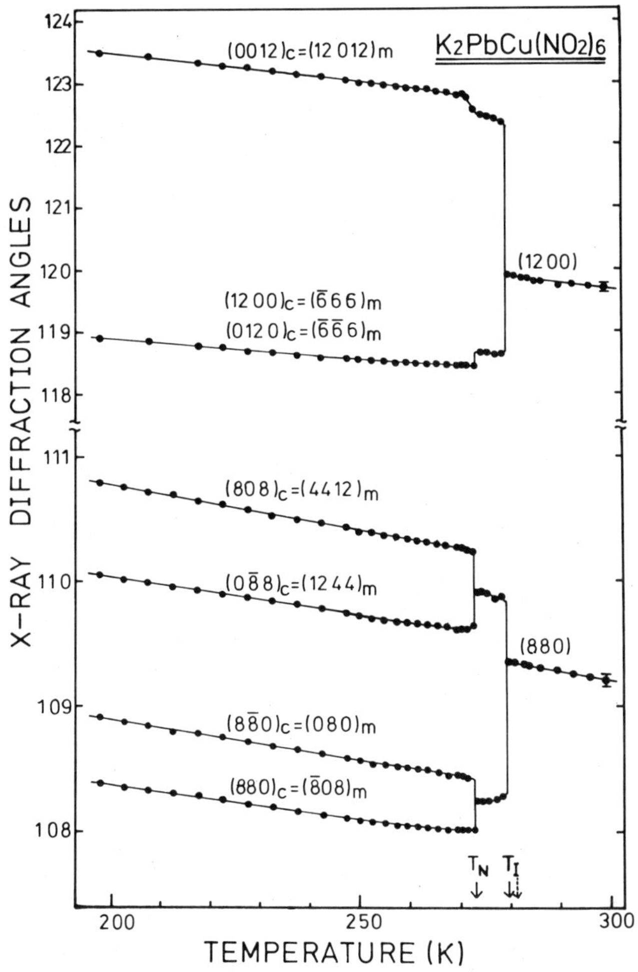

Fig. 2. Temperature dependences of the diffraction angle of the two fundamental Bragg reflections: $(1200)_c$ and $(880)_c$ in cubic indices. Phase II is indexed by a tetragonal lattice while Phase III by a monoclinic lattice.

In Phase III, we have observed the super lattice reflections at $(H\pm1/2,\ K\pm1/2,\ L\pm1/2)$ in cubic indices, namely at L-points of the Brillouin zone of fcc unit cell. This will imply that some of the L zone-boundary phonon modes condenses in Phase III. Therefore we look for the L zone-boundary modes which are compatible with the local Q_2- and the Q_3-mode. Following convensional symmetry analysis, we find that the L_3- and the L_5-mode are the possible

Jahn-Teller active phonons, whose displacement pattern are depicted in Fig. 3. We can easily see the local Q_3-mode is compatible with this L-mode. The structure analysis including super lattice reflections confirmed this displacement pattern.

This is an antiferrodistortive structure, where in one sub-lattice the a-axis is the direction of tetragonal elongation of the octahedron formed by $Cu(NO_2)_6$ complex while in the other the b-axis becomes the direction of elongation. In terms of the pseudospin description of local electronic configuration, this structure may be characterized as the 'canted spin' structure as is illustrated in Fig. 4.

Concerning the electronic state of Phase III, there have been very interesting debates between 2 groups based on ESR measurements. Reinen et al[6,13] assumes 'locally elongated' configuration implying that the wave function $\psi_g=2z^2-x^2-y^2$ should be taken as the ground state. The uniform contraction of the lattice ($c/a<1$) is attained by taking local unique axis along the a-axis or the b-axis alte-natively. (antiferrodistortive state).

On the other hand, Harrowfield[4,14] assumes 'locally contracted' configuration as the ground state. In this case, each ionic group contracts along the tetragonal unique axis, thus the uniform contraction of the lattice is attained simply by the ferrodistortive state. It is interesting that, as far as the 'averaged' property is concerned, these two models are equally good to explain the anisotropy of the g-tensor observed by ESR experiments. Our structural analysis is certainly consistent with Reinen's picture.

II·4 Phase III : 'Fan' Spin Structure, Incommensurate Phase

The property of Phase II is more interesting. We have performed neutron diffraction experiments in Phase II. The most remarkable point is that, in addition to the uniform bulk contraction, we have observed satellite reflections at $(H\pm0.425, K\pm0.425, 0)_c$ in cubic indices. That is, Phase II is considered as an 'incommensurate' phase. In Fig. 5, the diffraction pattern in (001) zone is schematically shown.

The model for the static structure in this phase would be that the local Jahn-Teller distortion is modulated along [110] with $|k_0|=0.425$. We look for phonons on Σ-line ($k=[\xi\xi0]$) which is compatible with the local Q_2- and Q_3-mode. The symmetry property of the low frequency acoustic phonons on Σ-line is summarized in Table II. We assumed that the most probable candidate of the phonon mode which is condensed in Phase II is TA_1-mode, propagating along [110] and polarized along [1$\bar{1}$0]. The structure based on this model is illustrated in Fig. 3. A preliminary structure analysis

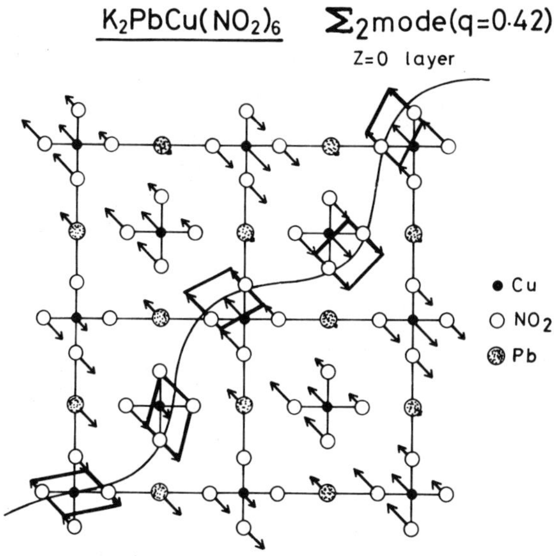

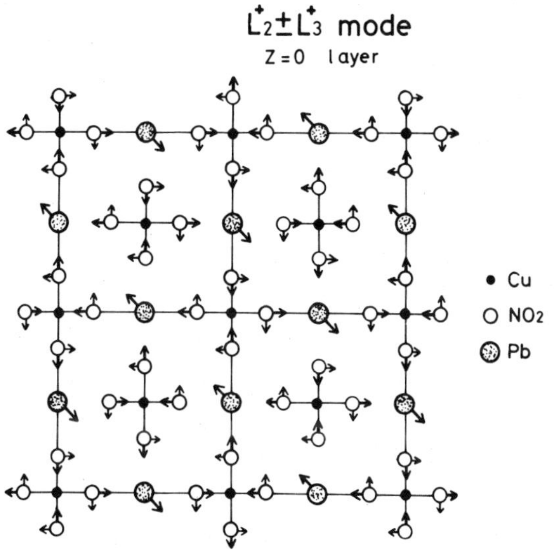

Fig. 3 Model structures in the low temperature phases.
Phase II (upper figure): Jahn-Teller active phonon mode (TA_1-mode) with wave vector k_0=(0.425, 0.425, 0) is 'freezed'. This is an incommensurate phase.
Phase III (lower figure): Jahn-Teller active zone boundary mode (k=1/2 1/2 1/2) is 'freezed'. This is an antiferrodistortive (commensurate) phase.

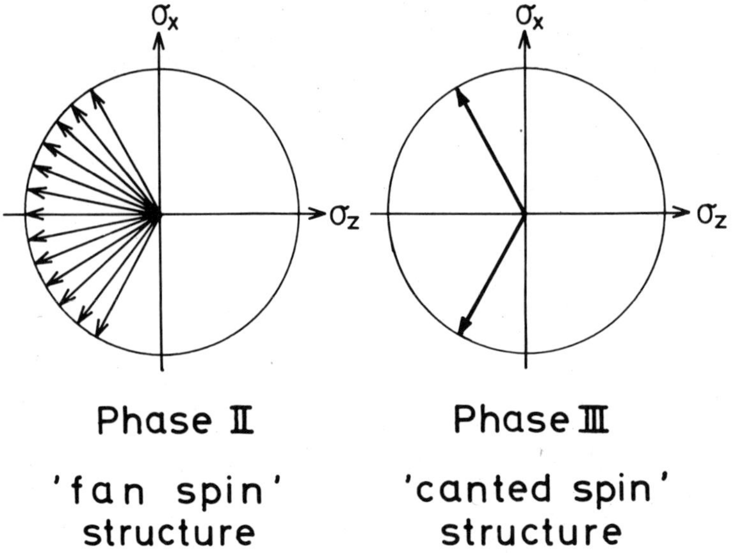

Fig. 4 Pseudospin descriptions of the low temperature phases. Phase II : a fan spin structure. The spin direction (the electronic state) in the σ^x-σ^z plane changes from site to site sinusoidally leaving a tetragonal contraction ($<\sigma^z><0$) as the average.
Phase III : a canted spin structure. In one sublattice, the octahedron is elongated along the a-axis while in the other it is elongated along the b-axis.

TABLE II

Symmetry property of the little group of $k=[\xi\xi 0]$ (Σ-line) of Fm3m. The compatibility of the local Jahn-Teller mode with these irreducible representations are also given in the third column.

$O_h^5(\Sigma)$=mm2		E	C_{2d}	m_z	m_d	local	phonon
A_1	Σ_1	1	1	1	1	Q_3	LA
A_2	Σ_3	1	1	-1	-1		
B_1	Σ_2	1	-1	1	-1	Q_2	TA_1
B_2	Σ_4	1	-1	-1	1		TA_2

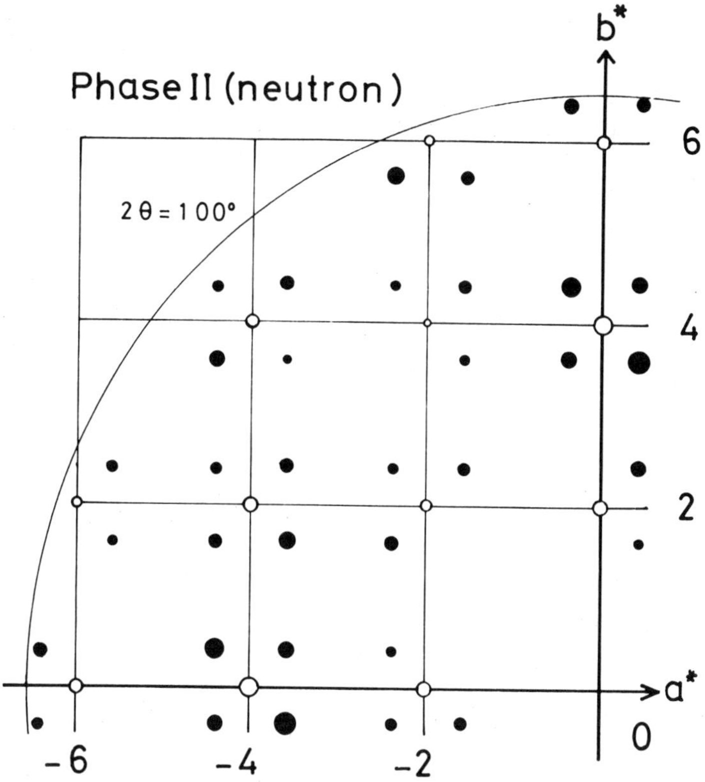

Fig. 5 The neutron diffraction pattern in (001) zone observed in Phase II. Many satellite reflections appeared around each fundamental reflections. Size of the solid circles roughly the observed intensity ratio between various satellites.

has shown that this is consistent with the observed intensity distribution of satellites.

It is worthwhile to point out that in terms of pseudospin description, this phase is expressed as a 'fan' spin structure. (See Fig. 4) In Phase II, it has the modulated local distortion of the Q_3-mode as well as the uniform distortion of the Q_2-mode. Therefore the orientation of the 'pseudospin' in the two dimensional σ^x-σ^z plane changes from site to site, which will result in the 'fan' spin structure. Such the structure has been studied in real spin system, and it is known that when a helical spin system is brought under the influence of external magnetic field perpendicular to the screw axis, this type structure is stabilized[15].

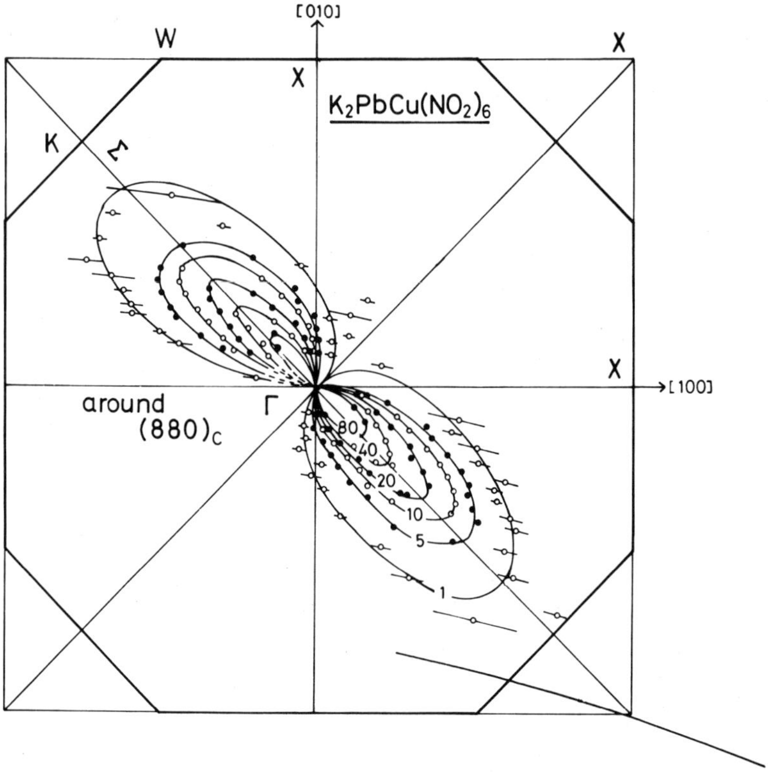

Fig. 6 Observed X-ray critical diffuse scattering around $(880)_c$ obtained in Phase I. Strong anisotropy at the Γ-point suggests that the interaction between local distortions are of long range.

Later, we will discuss the correspondence between our 'fan pseudospin' system and the fan structure in real spin systems.

From lattice dynamical point of view, this picture would imply the softening of the TA_1 branch particularly at $k \cong [0.4, 0.4, 0]$ avobe the transition temperature. We tried to observe softening of TA_1 branch along $k=[\xi\xi 0]$. However, we could not observe any well defined TA_1-mode above the upper transition point. Instead, as is often the case, strong quasielastic streaks were observed running along the <110> directions. Fig. 6 shows the intensity contours of observed X-ray diffuse scattering around (880) reflection. Particularly we notice the characteristic 'pinching' in the contour around Γ-point (the origin of the Brillouin zone). This pattern is very similar to that observed in ferroelectric substance such as KDP[16]. This correspondence is more than accidental. It is known that when the interaction between local distortions is via elastic strains, it falls off as $1/r^3$, exactly the same as electrostatic dipolar interaction. Such the long range forces, when Fouier transformed, always gives rise to the singlarity at the Γ-point. Therefore, we consider that the critical fluctuations is in fact mainly associated with the transverse acoustic mode.

II·5 Summary and Discussions

Summarizing, the successive phase transitions in $K_2PbCu(NO_2)_6$ have been studied by precise X-ray and neutron scattering measurements. The experimental results can be interpreted as sequencial cooperative Jahn-Teller phase transitions. In the intermediate phase, the local Jahn-Teller distortions are modulated with the wave vector $k_0=(0.42, 0.42, 0)$. In the lowest temperature phase, it has an antiferrodistortive phase where the axis of local tetragonal elongation alternates from site to site. From the structural view point the successive phase transitions in this crystal is characterized as normal→incommensurate→commensurate structural transitions which is accompanied by cubic→pseudo-tetragonal→pseudomonoclinic bulk distortion. On the other hand, the change of the electronic state of the Jahn-Teller active Cu^{2+} ion is characterized as para→fan→canted spin ordering.

The most conspicuous feature in the successive phase transitions of $K_2PbCu(NO_2)_6$ is the appearance of incommensurate phase where the local Jahn-Teller mode propagates with incommensurate wave vector k_0. It has been shown that in two dimensional Q_2-Q_3 plane, this structure corresponds to a 'fan structure' given in Fig. 4. It is known that in the magnetic system, this type of spin structure is realized when a helical spin system is brought into uniform magnetic field perpendicular to the screw axis. On the other hand, it is also well known that the 3d electronic state of Cu^{2+} ion in cubic crystalline field is completely described by Pauli spin operators, which is called 'pseudospins'. It would be interesting to look for formal correspondence between our 'pseudospin' system and real spin system and whence deduce the origin of stability of the 'fan pseudospin' structure by comparing with the real spins.

We begin with the well known Jahn-Teller hamiltonian,

$$H_{JT} = \sum_{\ell} g\,(\sigma_{\ell}^{x} Q_{2\ell} + \sigma_{\ell}^{z} Q_{3\ell}), \tag{4}$$

where σ_{ℓ}^{x} and σ_{ℓ}^{z} are Pauli spin operators to describle 3d electronic state of Cu^{2+} ions at the ℓ'site. On Σ-line ([ξξ0] in the Brillouin zone, these local modes are compatible with phonon modes belonging to $B_1(\Sigma_3)$ and $A_1(\Sigma_1)$ irreducible representations. Therefore, these local modes are expanded with respect to phonon modes as follows:

$$H = \sum_{\ell} g_0(\xi\sigma_\ell^x + \eta\sigma_\ell^z) + \sum_{\ell}\sum_{k,\Sigma_3} \frac{g_{k,\Sigma_3}}{\sqrt{N}} (Q_{k,\Sigma_3}\sigma_\ell^x e^{ik\cdot r_\ell} + c.c.) + \sum_{\ell}\sum_{k,\Sigma_1} \frac{g_{k,\Sigma_1}}{\sqrt{N}} (Q_{k,\Sigma_1}\sigma_\ell^z e^{ik\cdot r_\ell} + c.c.) \tag{5}$$

where, we have included the coupling to uniform strains: $\xi = e_{xx} - e_{yy}$, $\eta = 2e_{zz} - e_{xx} - e_{yy}$. The importance of the role of strains has been pointed out by Kanamori[17] and Elliott et al[18]. Among these, we only retain the most important terms which become condensed in the low temperature phase to obtain

$$H = \sum_{\ell} g_0(\sigma_0^x\xi + \sigma_\ell^z\eta) + \sum_{\ell} \frac{g_{k_0 2}}{\sqrt{N}} (Q_{k_0,TA_1}\sigma_\ell^x\, e^{ik_0\cdot r_\ell} + c.\ c.) + \frac{g_{k_0 3}}{\sqrt{N}} (Q_{k_0 LA}\sigma_\ell^z\, e^{ik_0 r_\ell} + c.\ c.) \tag{6}$$

In the molecular field approximation, we may introduce the following effective 'pseudomagnetic fields'

$$H_0^z \equiv -g_0\langle\eta\rangle \tag{7}$$

$$H_{k_0,\ell}^x \equiv -g_{k,TA_1}\langle Q_{k_0,TA_1}\rangle \cdot \cos k_0\cdot r_\ell\ , \tag{8}$$

$$H_{k_0,\ell}^z \equiv -g_{k,LA}\langle Q\ k_0 LA\rangle \cdot \sin k_0\cdot r_\ell\ . \tag{8'}$$

H_0^z is a uniform 'pseudofield' while $H_{k_0,\ell}$ is a staggered pseudo-field modulated with wave vector k_0. Making use of these expressions we have

$$H_{JT} = -\sigma_\ell^z H_0^z - \sum_{\ell}(\sigma_\ell^x H_{k_0\ell}^x + \sigma_\ell^z H_{k_0\ell}^z) \tag{9}$$

which corresponds to the hamiltonian studied by Nagamiya[15] to investigate fan structure in the real spin system, where H_0^z and $H_{k_0\ell}$ mean uniform external magnetic field and the staggered exchange field. The only difference is that in Eq. (9), H_0^z stands for the spontaneous strain, not the applied field. To make this point consistent, we should introduce a higher order anharmonic interaction :

$$H_{anh} = A\sum_{\ell} Q_{k_0 TA_1} Q_{-k_0 TA_1}\cdot\sigma_\ell^z\ . \tag{10}$$

By the suitable choice of parameters g_0, g_{k_0}, and A, we should be able to obtain the self consistent solution to stabilize the fan pseudospin structure.

III. ZEROTH GRADE ORDERING IN Fe_3O_4

III·1 Introduction

The 123 K phase transition of magnetite is one of the 'time honored' problemes which has been the subject of extensive studies since Verwey[19] proposed the existance of 'charge ordering' below the transition in 1941. The 'charge order' means that the total charge integrated around each particular site alternates from site to site. Thus, in the framework given in §1, this falls into the case of 0'th grade ordering.

The crystal structure of Fe_3O_4 in the high temperature phase is so called 'inverse spinel' type. The tetrahedrally coordinated A-sites are solely occupied by Fe^{3+} ions, while the octahedral B-sites accomodate Fe^{2+} ions and the remaining Fe^{3+} ions at random with equal probability. Large electrical conductivity of this substance at room temperature is considered to be due to hopping of d-electrons within the disordered B-sites. It is this disorder that causes the 123 K phase transition.

The important characteristic of the 123 K phase transition is that the electrical conductivity decreases by factor of 10^{-2} at the transition.[20] Based on this fact, Verwey proposed the ordering of charges within the octahedral site below the transition temperature which eliminates the disorder of Fe^{2+} and Fe^{3+}, whence explains the drastic decrease of the conductivity. The 'frozen' charge ordering scheme proposed by Verwey is explained in Fig. 7. In this ordering, the unit cell remains unchanged.

However, recent observations on this phase transition by electron diffraction[21], neutron diffraction[22], Mössbauer effect[23,24] and magnetic resonance[24,25] have casted doubt on the validity of the Verwey type charge ordering scheme. The details of historic aspects of the experimental progress have been discussed in the seperate lecture.[26]

As is illustrated in there, the electron diffraction pattern[21] exhibits some super structure lines indicating that the unit cell in the low temperature phase is doubled along the cubic principal axes. Moreover the neutron diffraction experiments[22] have shown that the doubling of the unit cell is mainly caused by small displacements of atoms. This strongly suggests that some internal lattice mode is associated with this phase transition.

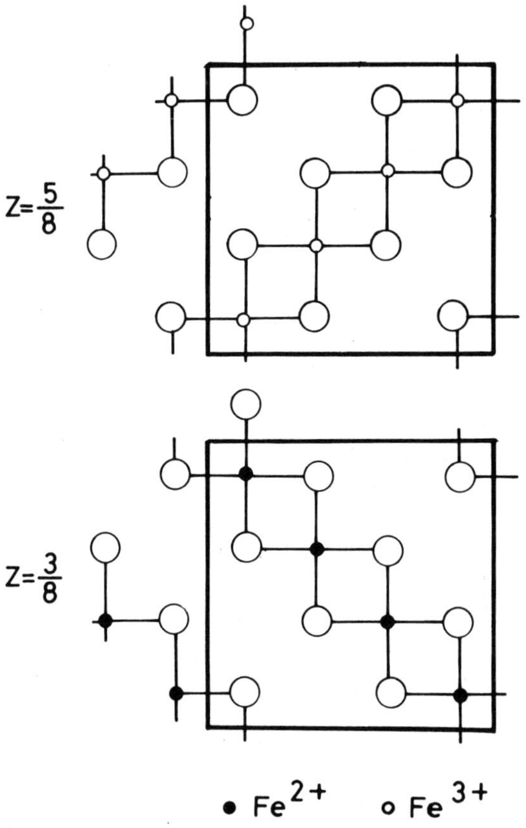

Fig. 7. The charge ordering scheme proposed by Verwey[1]. Two atomic planes occupied by Fe^{2+} and Fe^{3} alternates successively along the z-direction. Cubic unit cell is not doubled in this ordering scheme.

When combined with the existence of the charge ordering below T_V, the phase transition temperature, these considerations naturally lead to the assumption that the interaction between electrons and phonons is playing the crucial role in this phase transition. The important implication of the above assumption is that the charge density fluctuation which is coupled to the unstable phonon mode is also an 'internal' mode modulated by the same wave length (twice the cubic unit cell length).

In the following, we try to understand the 123 K transition in Fe_3O_4 on the basis that this phase transition is viewed as the instability of a coupled charge density-phonon mode driven by the coupling between local electrons and phonons.[27]

III·2 Symmetry Property of The Phonon Field and The Charge Density Field

The observed neutron critical scattering has shown strong peaks at (H, 0, L±1/2] reciprocal lattice points accompanied by ridge-like streaks running along [001] direction[28]. This suggests that the relevant phonon modes which become unstable at the phase transition have the wave vector $k=(00\xi)$ (specified as Δ-points in the reciprocal space). Especially, the mode with $k=(00\ 1/2)$ will become condensed in the low temperature phase as is confirmed by the electron diffraction and the neutron diffraction measurements. Therefore we begin with examining the symmetry property of the phonon field as well as the charge density field at Δ-points in the cubic reciprocal space.

Phonon Field

The little group of $k=(00\xi)$ of space group O_h^7 has five irreducible representations denoted by Δ_1, Δ_2, Δ_3, Δ_4, and Δ_5. Among these, Δ_5 is a two dimensional representation and the others are all one dimensional, Δ_1 being the identity representation. The 42 phonon branches are characterized by the following symmetry properties,

$$\Delta = 7\Delta_1 + 3\Delta_2 + 7\Delta_3 + 3\Delta_4 + 11\Delta_5. \qquad (11)$$

Among these phonon branches, the Δ_5-modes are particularly important for the later discussions. It should be pointed out that all the singlet modes share the common feature in the pattern of displacement in that the atom pair which is related by two fold axis along [001] moves out of phase, while in the Δ_5-mode, they move in phase as is seen in Fig, 8. It should be also mentioned that for an arbitrary value of $k=(00\xi)$, the atomic displacement is modulated along [001] direction with the corresponding wave length a/ξ. Thus, for the particular case of $k=(00\ 1/2)$, the wavelength is twice the cubic unit cell dimension along the c-axis.

Charge Density Field

The electrons which are associated with the 123 K phase transition are a pair of excess 3d-electrons coming from two Fe^{2+} ions per primitive cell located at octahedral sites. Since the magnetic ordering has been attained at this temperature, the spin degeneracy is already lifted. Therefore, there are four orbitals per primitive cell to accommodate these two electrons.

As for the electronic states of these electrons above T_V, there are evidences[29,30,31] that these electrons are localized at

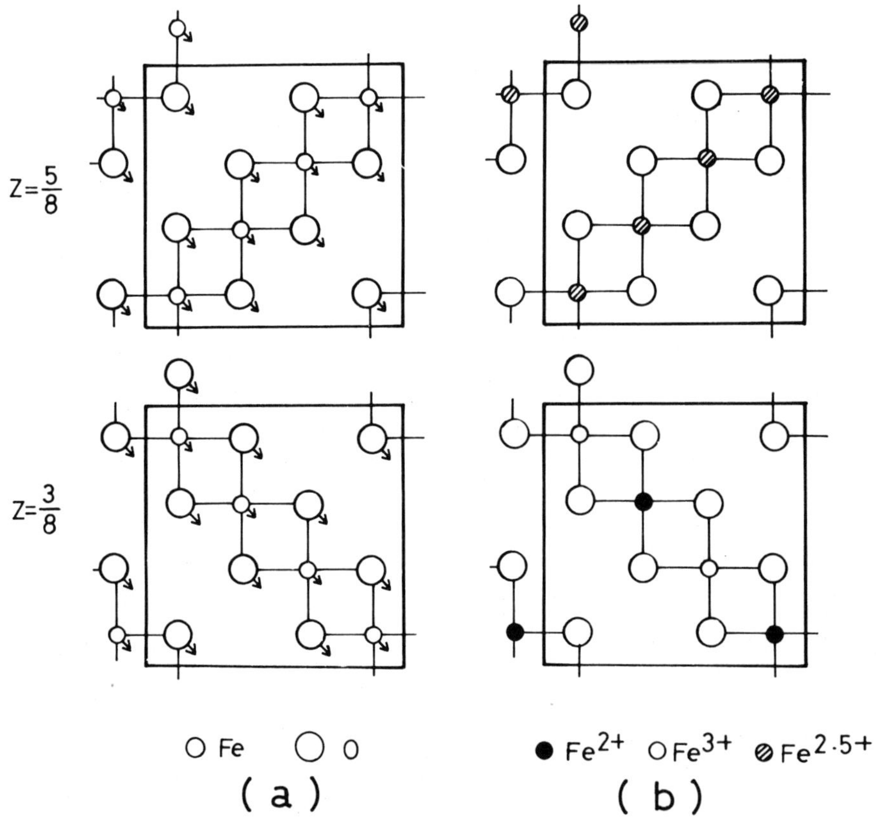

Fig. 8. Schematic discription of the displacement pattern as well as the charge density pattern with the symmetry Δ_5^1. This pattern is modulated along the z-axis with an arbitrary wave vector.

each site rather than forming bands, thus giving rise to a hopping disorder in charge density at the octahedral sites. We take this standpoint and consider the symmetry property of the localized charge density fluctuations. First we assume the condition of local charge neutrality within the unit cell (Anderson's condition)[32] is satisfied. We have then six configurations of two electrons in the unit cell as is given in Fig. 9. The pattern of charge density waves with an arbitrary wave vector k is obtained by multiplying the phase factor $e^{ik\cdot r_i}$. We are especially interested in the modes with k=[00ξ]. Along this particular line, the allowed charge density patterns are further limited[33] to cases (c)~(f) in Fig. 9.

We may define the charge density fluctuation operator as follows:

$$\delta\rho = C_m^+ C_m - 1/2 \qquad m = 1, 2, 3, 4 \tag{12}$$

where $C_m^+ C_m$ is the number operator of the 3d electron localized at m'th site in the unit cell. The relevant two electron states (c)~(f) in Fig. 9, will be given by: $|1010\rangle$, $|1001\rangle$, $|0110\rangle$ and $|0101\rangle$, in terms of the eigen functions for these number operators as is illustrated in Fig. 9. We then symmetrize this charge density fluctuation field (a scaler field), so that the pattern of the charge density with wave vector k will belong to one of the irreducible representations of the same little group Δ as phonons does.

It is shown that we can construct the following symmetrized 'charge density modes'

$$\begin{aligned}
\delta\rho(\Delta_1) &= \delta\rho_1 + \delta\rho_2 + \delta\rho_3 + \delta\rho_4 - 2 \\
\delta\rho(\Delta_4) &= \delta\rho_1 + \delta\rho_2 - (\delta\rho_3+\delta\rho_4) \\
\delta\rho(\Delta_5^1) &= \delta\rho_1 - \delta\rho_2 \\
\delta\rho(\Delta_5^2) &= \delta\rho_3 - \delta\rho_4
\end{aligned} \tag{13}$$

Among these, only the Δ_5-modes have non-zero eigenvalues when operated on the above defined four eigenstates. By taking these four eigenfunctions as basis vectors, we have

$$\delta\rho(\Delta_5^1) = \begin{pmatrix} 1 & 0 & 0 & 0 \\ 0 & 1 & 0 & 0 \\ 0 & 0 & -1 & 0 \\ 0 & 0 & 0 & -1 \end{pmatrix} \equiv \sigma^z(\Delta_5^1) \tag{14}$$

$$\delta\rho(\Delta_5^2) = \begin{pmatrix} 1 & 0 & 0 & 0 \\ 0 & -1 & 0 & 0 \\ 0 & 0 & 1 & 0 \\ 0 & 0 & 0 & -1 \end{pmatrix} \equiv \sigma^z(\Delta_5^2) \tag{15}$$

Namely, we have deduced (4-dimensional) pseudospin representations to express charge density fluctuation field.

It is worthwhile to point out that as $k \to 0$, the Δ_4-and the Δ_5-modes become degenerate to form Γ_{25} (triplly degenerate) mode, and the charge density pattern proposed by Verwey is exactly one of these triplly degenerate modes. It should be also noted that in the case of the Δ_5-mode condensation, there is a particular complication involved. Since the Δ_5 is doublly degenerate any linear combination of these two modes is equally stable. In order to eliminate this ambiguity, we have to take into account the higher order interaction.

III·3 Pseudospin-phonon Formalism and Neutron Scattering Cross Sections

We assume that the bilinear coupling between the phonon field and the charge density field is important to trigger the phase transition. The interaction hamiltonian is then generally given by

$$H_{int} = \sum_{\ell, m} \sum_{\ell' m' \alpha} g^{\alpha}_{\ell m \ell' m'} \delta\rho_{\ell m} u^{\alpha}_{\ell' m'} , \quad \alpha = x, y, z, \tag{16}$$

where $u^{\alpha}_{\ell' m}$ is the α-component of the displacement vector at site $(\ell' m')$.

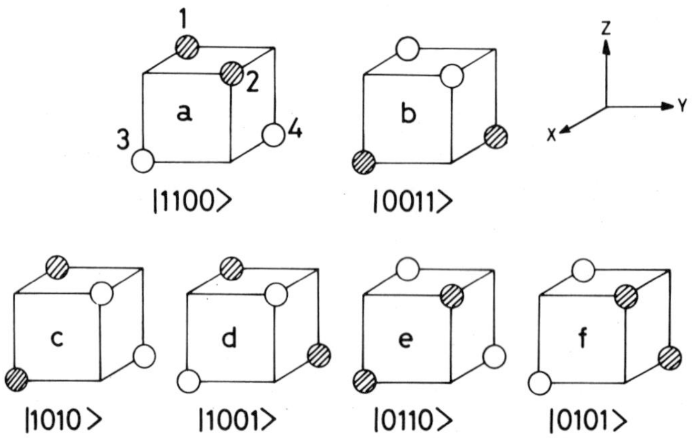

Fig. 9. Possible charge configurations satisfying the local charge neutrality condition. The state vectors corresponding to each configuration are given below each cubes.

Making use of the pseudospin representations of the charge density operator, we rewrite the above expression into the familiar pseudospin-phonon formalism,

$$H_{int} = \sum_k \zeta_{(k,\Delta_5)} \{\sigma^z_{(k,\Delta_5^1)} Q_{(k,\Delta_5^1)} + \sigma_{(k,\Delta_5^2)} Q_{(k,\Delta_5^2)}\} \tag{17}$$

Here,

$$\sigma^z_{(k,\Delta_5^1)} = \frac{1}{\sqrt{N}} \sum_\ell \sigma^z_{\ell(\Delta_5^1)} e^{ik\cdot r_\ell} . \tag{18}$$

The general property of such the coupled system has been discussed extensively elsewhere. Namely, by a suitable choice of the coupling parameter, the system undergoes the phase transition, below which the ordering of 'pseudospins' (the charge density) and the condensation of the phonon take place at the same time. It is predicted that the symmetry of both charge density mode and the phonon mode should belong Δ_5-type. In particular the pattern of the charge density below T_V should be as given in Fig. 8.

Let us consider the experiments to verify the above model. The neutron scattering cross sections at $T \lesssim T_V$ is particularly interesting from the following reason. As has been stated, the spin degeneracy is already lifted in the temperature region in question. Hence, the charge density ordering is always accompanied by the same ordering of magnetic spin density distribution. Therefore, both the atomic displacement and the charge ordering are detectable by neutrons as nuclear and magnetic scattering respectively. These scattering cross sections are expressed as

$$\frac{d^2\sigma}{d\Omega d\omega}_{nuc} = NkT \sum |F^s_d (k)|^2 \phi^s_{QQ(k,\omega)} , \tag{19}$$

$$\frac{d^2\sigma}{d\Omega d\omega}_{mag} = N \left(\frac{e^2\gamma}{mc}\right)^2 \sum_s |F^s_{m(k)}|^2 \phi_{\sigma\sigma(k,\omega)} . \tag{20}$$

Here, ϕ^s_{QQ} and $\phi_{\sigma\sigma}$ stand for the spectral densities of the correlation functions of phonons and pseudospins;

$$\phi^s_{QQ(k,\omega)} = \int < Q_{(k,\ s,\ o)} Q_{(k,\ s,\ t)} > e^{i\omega t} dt \tag{21}$$

$$\phi^s_{\sigma\sigma(k,\omega)} = \int < \sigma^z_{(k,\ s,\ 0)} \sigma^z_{(k,\ s,\ t)} > e^{i\omega t} dt \tag{22}$$

Below T_V, these simply give $\sum_{kn} \delta(k-k_h \pm k_0)\delta(\omega)$. $F^s_d(k)$ and $F^s_m(k)$ are the structure factors due to nuclear displacements and charge (magnetic spin) distributions belonging to the symmetry s respectively. They are explicitly given by

$$F_d^S(k) = \sum_j \frac{b_j e^{-W_j}}{\sqrt{M_j}} (k \cdot e^j_{k,s})\, e^{ik\cdot r_j} \tag{23}$$

where b_m is the neutron scattering amplitude, M_m the mass and $e^m_{(k,s)}$ is the polarization vector of the m'th atom. And

$$F_m^S(k) = \sum_{j'} S_m^S \, f_{(k)} e^{ik\cdot r_{j'}} \tag{24}$$

where S_m^S is the magnetic moment of Fe^{2+} ion perpendicular to k, f(k) is the magnetic form factor.

Therefore, in principle, we can determine, by analizing the intensity distribution of the super lattice reflections, what is the pattern of the charge ordering as well as that of the atomic displacements in the actual crystal.

III·4 Summary

(1) We have seen that in Fe_3O_4, a charge ordering and lattice distortion take place at the same time. The atomic displacement field as seen by neutron scattering is associated with the phonon mode at the middle of the Σ-line (K=00 1/2), not the uniform (zone center) mode.

(2) Assuming electron-phonon coupling scheme similar to polaron (namely, linear coupling between charge density and local atomic displacement), the effective interaction hamiltonian is formally reduced to pseudospin-phonon coupling, whose property is extensively investigated.

(3) Starting from this hamiltonian, we have predicted a possible charge ordering pattern, the Δ_5-type ordering.

(4) It has been also pointed out that in order to investigate such the coupled charge-displacement field, neutron scattering technique is particularly powerful, because neutron wave can 'sense' atomic displacements through nuclear interaction, and charge density (magnetic spin density) through magnetic interaction.

The detailes of the experimental results recently performed on this material will be discussed by Shirane in the separate lecture[26].

References

[1] Stinchcomb, this conference
[2] H. Thomas, this conference
The relevant references are sited in the above lectures
[3] B. V. Harrowfield and R. Weber, Phys. Lett. 38A, 27 (1972).
[4] B. V. Harrowfield, A. J. Dempster, T. E. Freeman and J. R. Pilbrrow, J. Phys. C. 6, 2058 (1973).
[5] D. Reinen, C. Friebel and K. P. Reetz, J. Solid State Chem. 4 103 (1972).
[6] D. Mullen, G. Heger and D. Reinen, Solid State Comm. 17 1249. (1975)
[7] S. Takagi and M. D. Joesten, Acta Cryst B31, 1968 (1975) : The prototype structure is seen in the crystals containing Jahn-Teller inactive complex such as $K_2PbNi(NO_2)_6$.
[8] M. D. Sturge, Solid State Physics, 20, 91 (1967). Ed. F. Seitz, D. Turn bull H. Ehrenreich Academic Press, New York.
[9] S. Takagi, P. G. Lemhert and M. D. Joesten, J. Am. Chem. Soc. 96' 21, 6606 (1974).
[10] C. Friebel, Z. anorg. allg. Chem. 417, 197 (1975).
[11] Y. Noda, M. Mori and Y. Yamada, Solid State Commun. 19, 1071 (1976).
[12] Y. Noda, M. Mori and Y. Yamada, to be published in Solid State Commun.
[13] D. Reinen, Solid State Commun. 21, 137 (1977).
[14] B. V. Harrowfield, Solid State Commun. 19, 983 (1976).
[15] T. Nagamiya, Solid State Physics 20, 306 (1967) edited by F. Seitz, D. Turnbull and H. Ehrenreich Academic Press, New York
[16] J. Skalyo Jr., B. C. Frazer and G. Shirane, Phys, Rev. B1, 278 (1970).
[17] J. Kanamori, J. appl. Phys. Suppl. 31, 14 (1960),
[18] R. J. Elliott, R. T. Harley, W. Hayes and S. R. P. Smith, Proc. Roy. Soc. A328, 217 (1972).
[19] E. J. W. Verwey and P. W. Haayman, Physica 8, 979 (1941). E. J. W. Verwey, P. W. Haayman and N. C. Romeijin, J. Chem. Phys. 15. 181 (1947).
[20] S. Chikazumi, AIP Proc. 29, 382 (1976).
[21] T. Yamada, K. Suzuki and S. Chikazumi, Appl. Phys. Lett. 13, 172 (1968).
[22] J. Samuelson, E. J. Bleeker, L. Dobrzynski, and T. Riste, J. Appl. Phys. 39, 1114 (1968).
[23] R. S. Hargrove and W. Kundig, Solid State Commun. 8, 303 (1970)
[24] M. Rubinstein and D. W. Forester, Solid State Commun. 9, 1675 (1971).
[25] M. Rubinstein, G. H. Strauss and F. J. Bruni, AIP. Conf. Proc. No.10, 1384 (1972).
[26] G. Shirane, this conference
[27] Y. Yamada, AIP. Conf. Proc. No. 24, 79 (1975).
[28] Y. Fujii, G. Shirane and Y. Yamada, Phys. Rev. B11 2036 (1975).

[29] W. Kündig and R. S. Hargrove, Solid State Commun. 7, 223 (1969).

[30] A. A. Samokhvalov, N. M. Tutikov and G. P. Skovnyakov Soviet Phys, Solid State 10, 2172 (1969).

[31] D. L. Camphausen, Solid State Commun. 11, 99 (1972).

[32] P. W. Anderson, Phys. Rev. 102, 1008 (1956)

[33] T. Matsuura and A. Yoshimori, Prog. Theor. Phys. 36, 679 (1966).

THE VERWEY TRANSITION IN MAGNETITE*

G. Shirane

Brookhaven National Laboratory

Upton, New York 11973 U. S. A.

We will first review the characteristics of the Verwey transition in magnetite at 123 K. An ordering takes place among Fe^{2+} and Fe^{3+} ions on the octahedral sites of the inverse spinel structure, accompanied by a sudden change of electrical resistivity. We shall discuss in some detail recent neutron scattering measurements at Brookhaven in collaboration with Chikazumi's group at the University of Tokyo.

I. INTRODUCTION

The Verwey transition in magnetite is one of the oldest problems in magnetism as well as in phase transitions.[1] We shall review a part of this complex problem from the viewpoint of electron-phonon interactions, the topic of this conference. In the first lecture, we shall survey the historical background up to 1973. The second lecture concerns recent neutron scattering experiments carried out at Brookhaven in collaboration with Professor Chikazumi's group at the University of Tokyo.

Let us start by discussing the major characteristics of this phase transition at 123 K. At higher temperatures, magnetite has the inverse spinel structure, a structure which many technically important magnetic materials possess. The unit cell contains 8 Fe_3O_4 units, 8 Fe^{3+} on the tetrahedral sites and the random distribution of 8 Fe^{2+} and 8 Fe^{3+} ions on the octahedral sites. A

* Work performed under the auspices of the U. S. Energy Research and Development Administration.

very pronounced phase transition has been observed around T_V=123 K. The most dramatic anomaly is in resistivity, as shown in the most recent data (Fig. 1) in Chikazumi's review article.[2] It has a high conductivity for $T>T_V$, but it becomes a reasonably good insulator below T_V.

Verwey[1] proposed a model, thirty years ago, that this phase transition is caused by an electronic charge ordering among the Fe^{2+} and Fe^{3+} ions on the octahedral sites. This ordering scheme is shown in Fig. 2 together with the concomitant orthorhombic symmetry. This is a very simple and attractive model which implies that alternate layers (or strings) of Fe^{2+} and Fe^{3+} are stacked along the c axis. Subsequent magnetic[3], dilatometric, and x-ray[4] measurements all appeared to support the orthorhombic symmetry required by the Verwey model. Then in 1958, Hamilton[5] presented the most convincing evidence in his celebrated neutron scattering paper. A magnetic cross section was observed at (002) below T_V

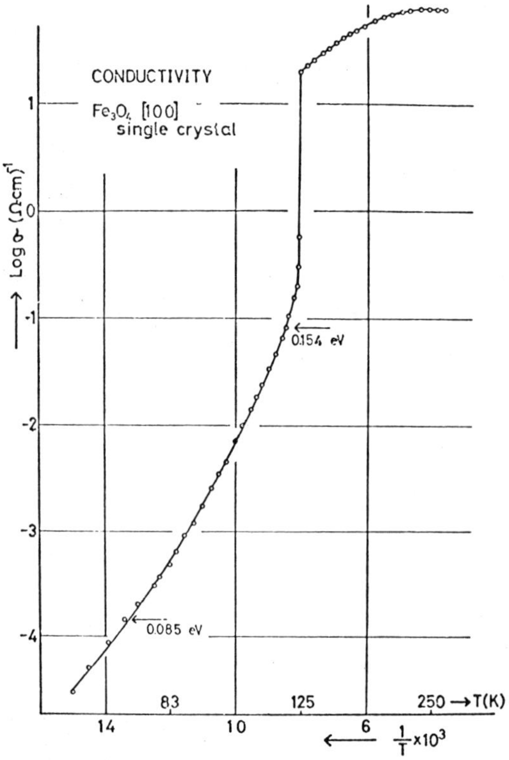

Fig. 1. Temperature dependence of conductivity for Fe_3O_4 along the cubic [100]. (Ref. 2).

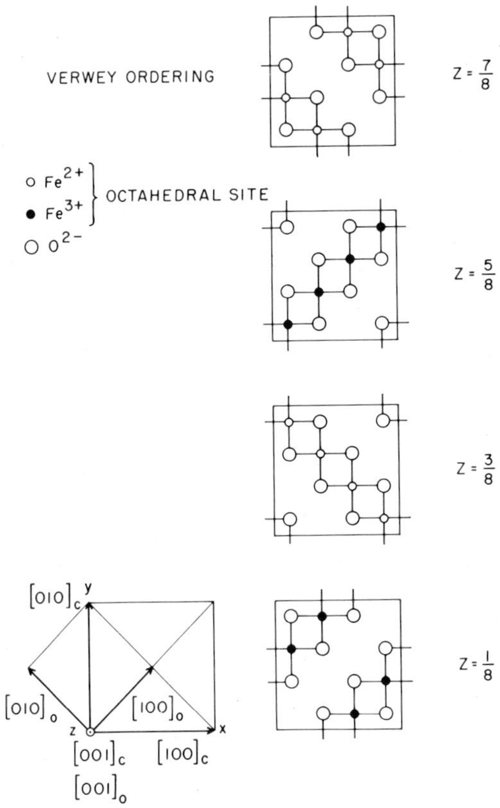

Fig. 2. The Verwey ordering scheme together with the resulting orthorhombic symmetry.

where none is allowed above T_V. At that point, the problem appeared to be completely solved.

Since then, many new experimental results have been reported which indicate that some basic modification is needed to the original Verwey model. The most important discovery was reported by Samuelsen et al[6] in 1968. In their neutron diffraction experiment below T_V, they observed satellites at reciprocal-lattice points with half integer such as (4 0 ½). This study, as well as independent electron diffraction work by Yamada et al[7] both reported that such satellites appear only along the orthorhombic c axis. In particular, Samuelsen et al noted that the satellite intensities are proportional to Q^2, where $\vec{Q}$ is a reciprocal lattice vector. Based on this fact they concluded that the satellites result mainly from atomic displacements and they are not directly due to the magnetic ordering of Fe^{2+} and Fe^{3+}. Other experiments such as NMR and Mössbauer measurements showed the existence of more than two nonequivalent octahedral sites below T_V. There was other experimental evidence to indicate that the low temperature symmetry is lower than orthorhombic.

II. CRYSTAL AND SYMMETRY

Let us now examine more closely the crystallographic aspect of this problem. Figs. 3 and 4 are taken from Chikazumi's review article[2] and demonstrate the crystallographic notation used in this talk. The unit cell below T_V has a nearly rhombohedral shape as shown in Fig. 4, though the true symmetry is monoclinic or lower. We may use two types of cell description: (A) Pseudo-rhombohedral cell with [001] axis as the unique magnetic c axis and [110] axis as the monoclinic b axis. In this, $\alpha = 90° - 0.16°$. (B) Pseudomonoclinic cell with the common c axis with the rhombohedral cell.

There are two types of twins: i) c axis zig-zag (c* common) as shown in Fig. 3 and ii) monoclinic a and b axes rotated. One can view these domains as 4 different <111> directions of the rhombohedral cell. These twinnings have caused considerable difficulty and confusion in the interpretation of experimental results. A truly untwinned crystal has been obtained recently by Chikazumi's group by a combined field cooling and a squeezing operation.[2,8] It has been known for some time that the magnetic

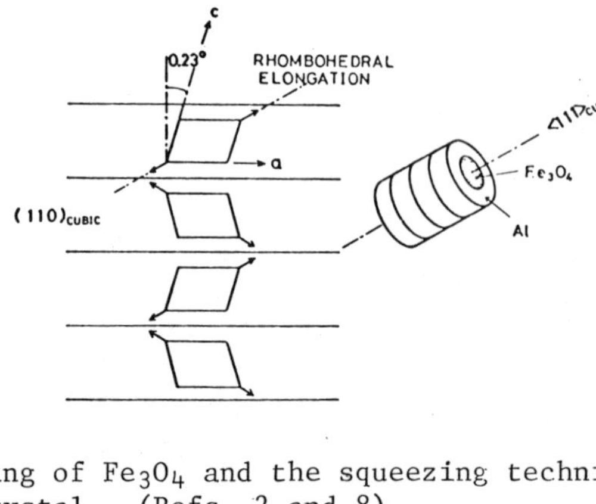

Fig. 3. Twinning of Fe_3O_4 and the squeezing technique to create an untwinned crystal. (Refs. 2 and 8).

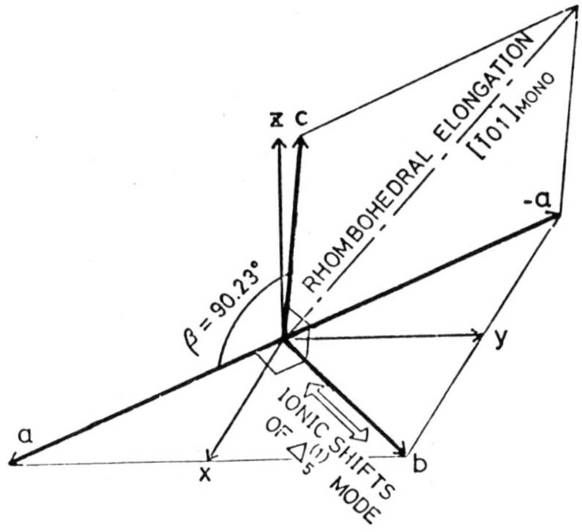

Fig. 4. Low temperature unit cell of magnetite. abc refers to the monoclinic axes. (Ref. 2).

c axis can be aligned by an <100> magnetic field through T_V. If one puts tight aluminum rings aroung the <111> axis, then this particular rhombohedral elongation will be established. For neutron scattering purposes, a cylindrical crystal (3mm in diameter and 6mm in length) was successfully made into a 99% untwinned crystal.

There is an additional advantage to these squeezed crystals. In order to separate out magnetic components from nuclear scattering,one has to apply a magnetic field below T_V. In a field cooled crystal (without squeezing) the c axis tends to follow the field direction. The desired experimental condition is to reorient the spin direction without changing the c axis. This is realized in the squeezed monocrystal and thus we can experimentally separate out small magnetic components.

The rhombohedral cell shape was properly identified by early x-ray studies[9] but was disregarded in favor of the orthorhombic assignment by the single crystal studies.[4] More recent x-ray[10] as well as neutron diffraction studies[11,12] have now clearly established the rhombohedral cell shape.

A series of neutron scattering experiments have been carried out as a joint project between the University of Tokyo and Brookhaven. In the following we shall discuss in some detail the results of these measurements.

III. CRITICAL SCATTERING

When our neutron scattering work was initiated, the c-axis doubling was already established. We, however, assumed that the basic magnetic ordering was already firmly established by Hamilton's neutron experiment. It appeared that the transition is accompanied by atomic displacements as clearly established by Samuelsen et al.[6] The key experiment was to look for critical scattering at both reflections, the (40½) type due to atomic displacements and the magnetic (002) reflection. This could give us a clue to the true order parameter of this complex phase transition. As we will see later, the (002) reflection has been absent at all temperatures! But this is the story after.

One crystallographic piece of information. The space group above T_V is Fd3m; reflections such as (200), (600), and (420) are missing because of space group requirements. The Verwey ordering scheme requires (002) type magnetic reflections appear as one can easily see in Fig. 2.

Fig. 5 depicts a typical example of critical scattering reported by Fujii et al.[13] The sharp increase of the (40½) type reflection is limited to a narrow temperature range above T_V. It shows the typical divergence toward T_c, which is a few degrees below T_V (see Fig. 10). The critical scattering was measured at several Brillouin zones and shown, in general, to be quite similar to those of the satellite below T_V.

Yamada[14] proposed a model to explain this critical scattering based upon an electron phonon coupling. This model involves the Δ_5 phonon mode as shown in Fig. 6. Ionic shifts of oxygen and Fe ions are such that they give a larger space for Fe^{2+} ions and a smaller space for Fe^{3+}. The most important result of the Yamada model is the charge ordering scheme which differs from that predicted by the Verwey model. The Fe^{3+} ions are not forming chains in c planes but both Fe^{2+} and Fe^{3+} form alternate chains as shown in Fig. 6. Since the Yamada model modulates a charge density wave along the c axis, it creates grey (disordered) layers as well.

There is a unique and simple prediction of the Yamada model for the magnetic cross section due to the spin ordering below T_V. The (002) type should be missing and the (20½) type reflection should show the main magnetic scattering. This prediction, however, is only partially fulfilled. Before discussing the low

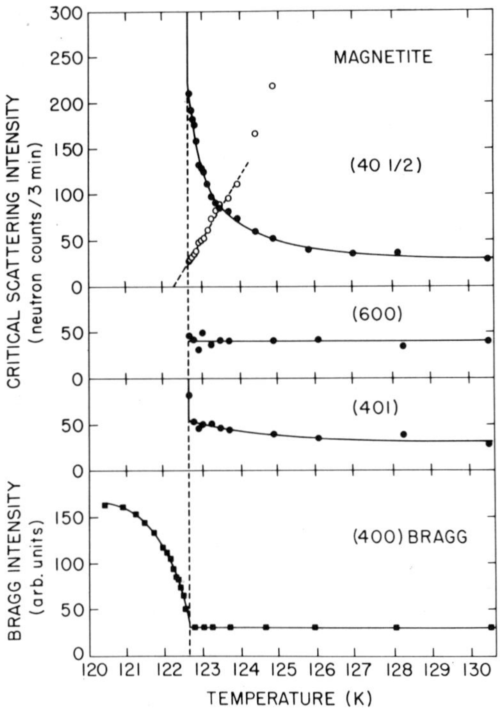

Fig. 5. Critical scattering just above T_V. Open circles for (40½) is T/I. (Ref. 13).

temperature study, we describe an additional type of diffuse scattering which is widely distributed in q space over a large temperature range, as shown in Fig. 7 and 8. This type of diffuse streak was first recognized by Chiba et al[15] in their electron diffraction study. A detailed study was carried out by Shapiro et al[16] on a large magnetite crystal. Somewhat surprisingly, this scattering showed one-dimensional nature along the <001> direction. It shows a gradual temperature dependence (Fig. 9) with extrapolated divergence at 106 K.

At present, we can offer only a qualitative picture for this diffuse scattering and its relation to the sharp critical peak described above. A major feature of the profile can be explained by an elongated correlation range based upon the Yamada model. The shape of the elliptical cross section is determined by the shape of the 1-D "clusters." This correlation grows with little change of shape for wide temperature range (see Fig. 10). Only at

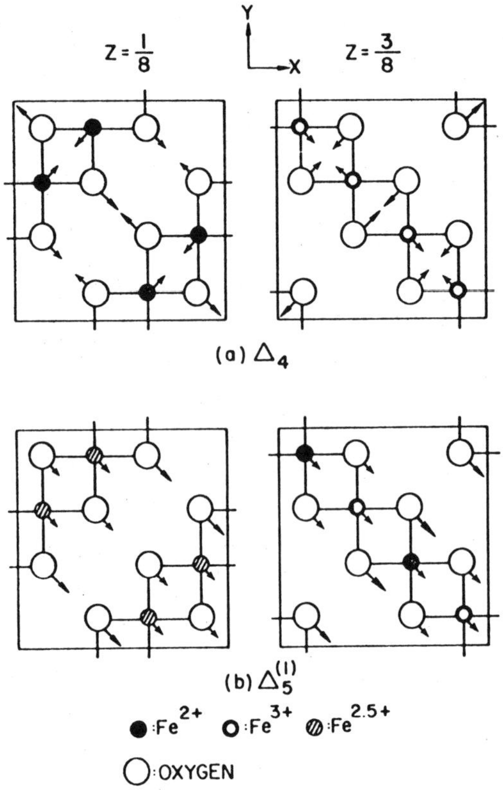

Fig. 6. The Yamada model based upon the Δ_5 phonon mode. Δ_4 corresponds to the Verwey model. (Ref. 14).

a few degrees above T_V, the 3-D critical scattering sets in, peaking at positions corresponding to low temperature satellites. The latter behavior is shown for (40½) by the steep broken line in Fig. 10.

IV. STRUCTURE BELOW T_V

It is essential to have a reliable knowledge of the low temperature structure to establish the charge ordering scheme in magnetite. As it turned out, this is a difficult and complex problem. Moreover, there is one important aspect of the analysis we did not fully appreciate at the outset.[12] In ordinary magnetic structure determination, one uses the magnetic structure factor

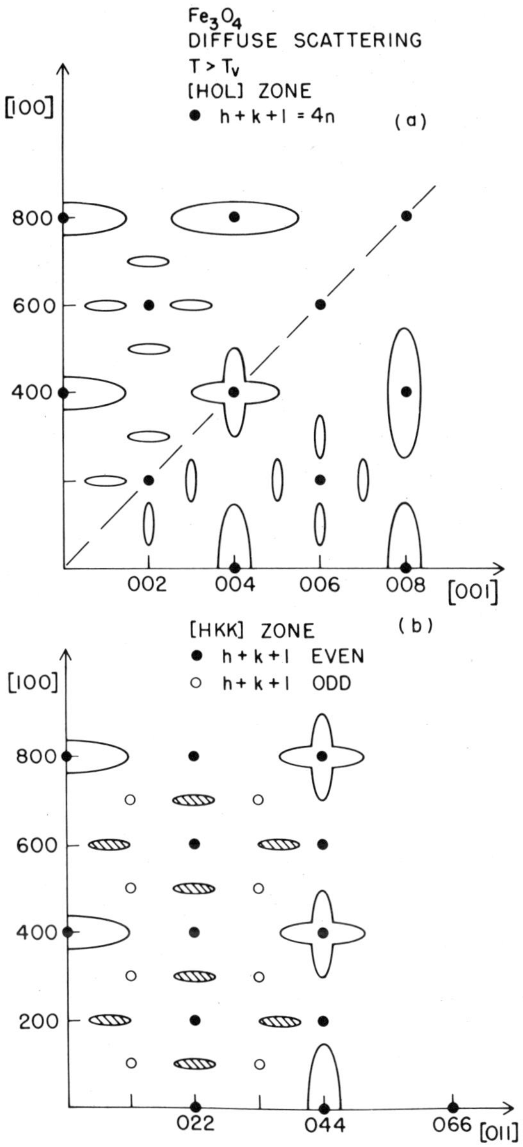

Fig. 7. Intensity distribution of diffuse scattering in two different zones. (Ref. 16).

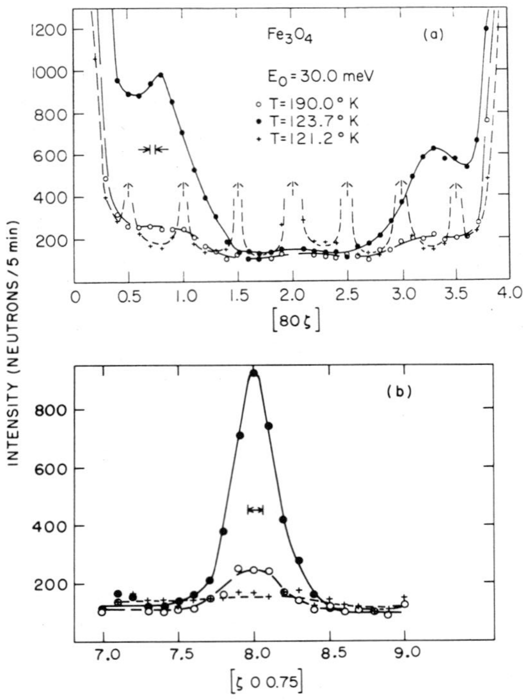

Fig. 8. Cross sections of diffuse streak shown in Fig. 7. (Ref. 16). Broad peaks around ζ=0.75 and 3.25 result from 1-D correlation.

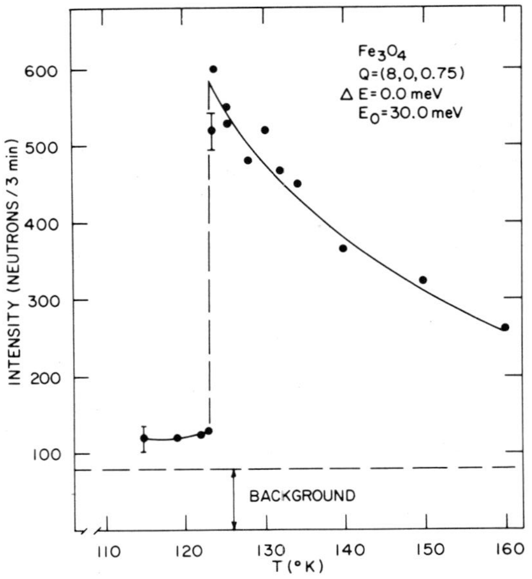

Fig. 9. Temperature dependence of diffuse scattering in Fe_3O_4. (Ref. 16).

$$F_M = \sum_j \Delta P_j \, e^{i(\vec{Q}\cdot\vec{r}_j)} \tag{1}$$

where $Q = a^*h + b^*k + c^*\ell$, and $\vec{r}_j$ atomic position and ΔP_j is due to magnetic modulation, namely $\pm 0.5\mu_B$ for the difference between the Fe^{2+} and Fe^{3+} magnetic moments. Usually we can use the cubic parameter $\vec{r}_j$ and neglect the higher order effect of atomic shift Δ_j at T_V.

For magnetite we have to use a more complete formula. This originates from the unusual situation that the spin modulation ΔP_j is superposed on a much larger average component $\bar{P}$, corresponding to $4.5\mu_B$. Now (1) becomes

$$F_M = \sum_j (\bar{P} + \Delta P_j) \, e^{i\vec{Q}\cdot(\vec{r}_j+\vec{\Delta}_j)} \tag{2}$$

Expanding in Δ_j we obtain

$$F(Q) \cong \sum_j \bar{P} \, e^{i\vec{Q}\cdot\vec{r}_j} + \sum_j \Delta P_j \, e^{i\vec{Q}\cdot\vec{r}_j} + i \sum_j \bar{P} \, (\vec{Q}\cdot\vec{\Delta}_j) \, e^{i\vec{Q}\cdot\vec{r}_j} \tag{3}$$

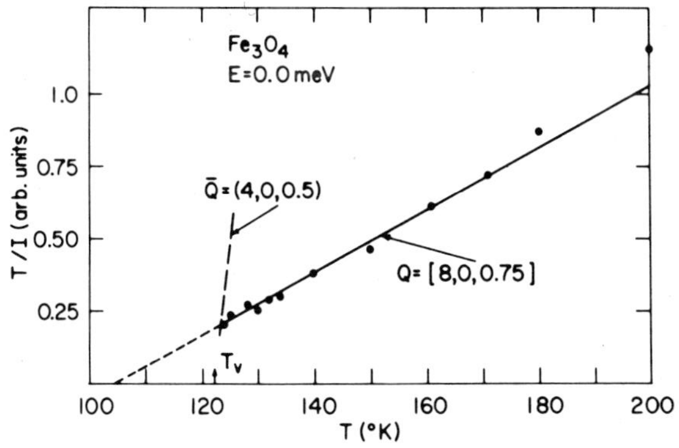

Fig. 10. Comparison of temperature dependence of 1-D diffuse peak at (8,0,0.75) and 3-D critical scattering at (40½). (Ref. 16).

The last term cannot be ignored for magnetite because $\bar{P}$ is considerably larger than ΔP_j. Thus we can establish ΔP_j only if we know Δ_j accurately enough.

Three types of neutron scattering experiments have been carried out so far for this purpose: (1) High resolution data were obtained for selected reflections, in particular (002) and (20½) to establish magnetic scattering below T_V. Measurements were carried out on a "regular" field cooled[12] as well as "squeezed" field cooled crystals.[18] (2) A polarized beam study on selected reflections from squeezed monocrystal.[17] Structure analysis at 78 K utilizing 1400 reflections.[18] Only the part (1) has been published and the other two are now in the process of final data analysis.

Table I lists some examples of the high resolution scattering experiment. The attempt here is to establish reliably a weak scattering cross section in the presence of much stronger reflections. The most crucial part of the experiment was to eliminate simultaneous reflections (see Fig. 11). Table I includes two simple charge ordering schemes, the Verwey model and the AB model. The latter is calculated for the double c axis. The Yamada model is a more general AB type modulation based upon the specific phonon mode coupling.

We have not yet established a satisfactory charge ordering scheme below T_V. We can, however, rule out some models unambiguously because of very low limits established in this experiment. First of all, the original Verwey ordering scheme is ruled out because of extremely low limits set for the magnetic cross sections for (402) and (002). The Chikazumi-Chiba model[19] is also ruled out. Previous neutron data must have been severely distorted by simultaneous reflections.

In Table I, $F_{\vec{M}}^2$(cal) for AB model (Fig. 12) involves only ΔP_j terms and not $i\bar{P}(\vec{Q}\cdot\vec{\Delta}_j)$ terms. Agreement with observed is less than satisfactory and, in particular, (20½) poses the vital disagreement. In fact, we have yet to prove that any part of the magnetic cross sections is due to the spin modulation ΔP_j, and not to the magneto-distortive term $\bar{P}(\vec{Q}\cdot\vec{\Delta}_j)$. Table I demonstrates that the observed magnetic cross sections are quite weak. It is conceivable, though very unlikely, that large magnetic peaks exist outside of (h0ℓ) and (hhℓ) zones. If the actual spin modulation is less than $0.15\mu_B$, compared with full $0.5\mu_B$, then the corresponding magnetic intensity ($\sim\Delta P^2$) would be difficult to detect.

Preliminary results of the current structure analysis[18] show several promising aspects. (1) The Δ_j's have been determined by least square analysis without any preassumed model. Yet they

Table I

Comparison of F_M^2(obs) with two model calculations. These are put into absolute units of millibarns per cubic unit cell (Ref. 12). F_M^2(cal) assumes full Fe^{2+} - Fe^{3+} modulations of $\pm 0.5 \mu_B$. Total F^2(obs), nuclear plus magnetic, are given for (400) and (40½) for comparison.

	F_M^2(obs)	F_M^2(cal)	
h0ℓ		Verwey model	AB model
402	< 16	10,500	---
002	< 26	17,980	---
20½	< 72	---	3,500
201½	∿ 300	---	3,100
202½	∿ 400	---	2,400
	F^2(obs)		
400	4×10^6		
40½	1.3×10^4		

possess essential features of the Yamada model. (2) The magnetic cross sections based upon this structure give good agreement with the recent polarized beam experiment.[17] (3) The Fe atoms on the octahedral sites may be divided into different kinds with respect to their distances to surrounding oxygens, in accord with NMR and Mössbauer measurements.[20-22] The final step to the spin modulation is still missing.

ACKNOWLEDGMENTS

I would like to thank S. Chikazumi, M. Iizumi, T. Koetzle, S. M. Shapiro, and Y. Yamada for many illuminating discussions during the course of more recent collaboration.

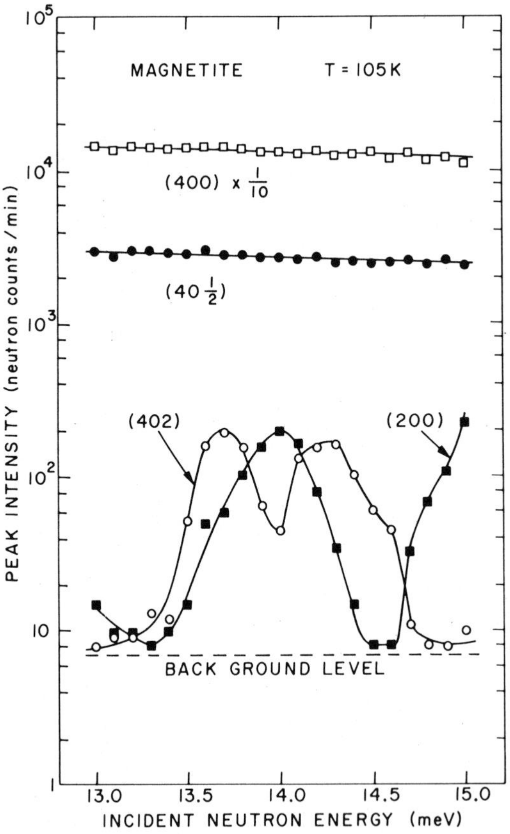

Fig. 11. Test of simultaneous reflection for selected peaks in Fe_3O_4. (Ref. 12).

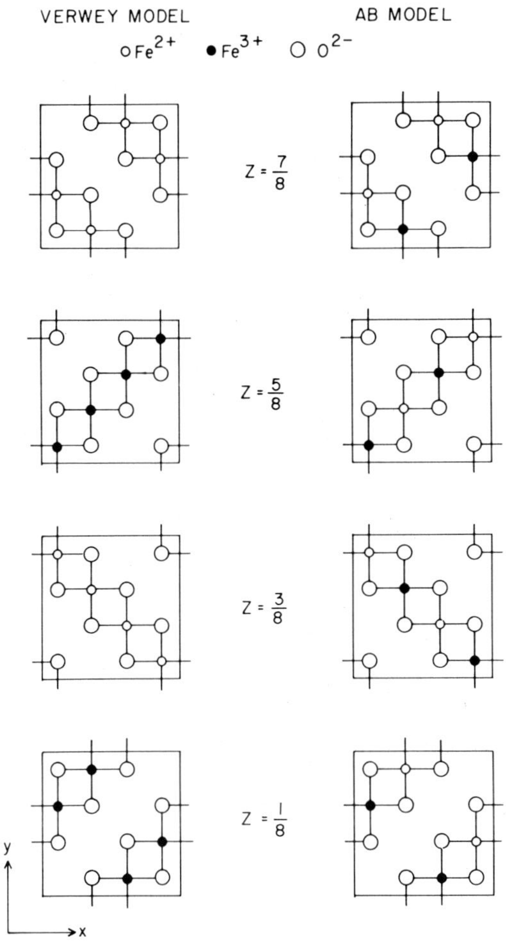

Fig. 12. Two charge ordering schemes in Fe_3O_4.

REFERENCES

1. E. J. W. Verwey and P. W. Haayman, Physica 8, 979 (1941); E. J. W. Verwey, P. W. Haayman, and N. C. Romeijin, J. Chem. Phys. 15, 181 (1947).
2. S. Chikazumi, AIP Proc. 29, 382 (1976).
3. B. A. Calhoun, Phys. Rev. 94, 1577 (1954).
4. S. C. Abrahams and B. A. Calhoun, Acta Crystallogr. 6, 105 (1953).
5. W. C. Hamilton, Phys. Rev. 110, 1050 (1958).
6. E. J. Samuelsen, E. J. Bleeker, L. Dobrzynski, and T. Riste, J. Appl. Phys. 39, 1114(1968); Kjeller Rept. KR-122 (1967).
7. T. Yamada, K. Suzuki, and S. Chikazumi, Appl. Phys. Lett. 13, 172 (1968).
8. K. Abe, Y. Yamamoto, and S. Chikazumi, J. Phys. Soc. Japan 41, 1894 (1976).
9. N. C. Toombs and H. P. Rooksby, Acta Crystallogr. 4, 474 (1951); H. P. Rooksby and B. T. M. Willis, Acta Crystallogr. 6, 565 (1953).
10. J. Yoshida and S. Iida, J. Phys. Soc. Japan 42, 230 (1977).
11. M. Iizumi and G. Shirane, Solid State Commun. 17,433 (1975).
12. G. Shirane, S. Chikazumi, J. Akimitsu, K. Chiba, M. Matsui, and Y. Fujii, J. Phys. Soc. Japan 39, 947 (1975).
13. Y. Fujii, G. Shirane, and Y. Yamada, Phys. Rev. B 11, 2036 (1975).
14. Y. Yamada, AIP Proc. 24, 79 (1975).
15. K. Chiba, K. Suzuki, and S. Chikazumi, J. Phys. Soc. Japan 39, 839 (1975).
16. S. M. Shapiro, M. Iizumi, and G. Shirane, Phys. Rev. 14, 200 (1976).
17. G. Shirane, M. Iizumi, J. Schweitzer, and S. Chikazumi, to be published.
18. M. Iizumi, T. Koetzle, G. Shirane, S. Chikazumi, M. Matsui, and S. Todo, to be published.
19. S. Chikazumi, K. Chiba, M. Matsui, J. Akimitsu, and S. Todo, Proc. Intern. Conf. Magnetism, Moscow (NAUKA, Moscow, 1974) Vol. 1-(1) 137.
20. M. Rubinstein and D. W. Forester, Solid State Commun. 9, 1675 (1971).
21. R. S. Hargrove and W. Kundig, Solid State Commun. 8, 303 (1970).
22. S. Iida, K. Mizushima, M. Mizoguchi, J. Mada, S. Umemura, K. Nakao, and J. Yoshida, AIP Proc. 29, 388 (1976).

PARTICIPANTS

ANDRESEN, A.F.	Institutt for Atomenergi, P.O.B. 40, 2007 Kjeller, Norway
AUSLOOS, M.R.	Dept. of Physics, University of Liege, 4000 Sart Tilman, Belgium
AXE, J.D.	Dept. of Physics, Brookhaven National Lab., p.t. Risø Research Establishment, 4000 Roskilde, Denmark
BAK, P.	NORDITA, Blegdamsvej 17, 2100 Copenhagen, Denmark
BERRE, B.	Dept. of Physics, Agricultural University of Norway, 1432 Ås-NLH, Norway
BLEIF, H.J.	Hahn-Meitner Institut für Kernforschung, Postfach 390128, 1 Berlin 39, Germany
DESJARDINS, S.	Dept. of Physics, University of Rhode Island, Kingston, R.I. 02881, USA
DULTZ, W.	Dept. of Physics, University of Regensburg, Universitätsstrasse 31, 8400 Regensburg, W-Germany
EVANS, B.J.	Dept. of Chemistry, University of Michigan, Ann Arbor, Michigan 48109, USA
FEDER, J.	Dept. of Physics, University of Oslo, Blindern, Oslo 3, Norway
FEINER, L.F.	Philips Research Laboratorium, Eindhoven, The Netherlands
FITZGERALD, W.J.	Institut Laue-Langevin, B.P. 156, Centre de Tri, 38042 Grenoble-Cedex, France

FLEURY, P.A.	Bell Laboratories, 600 Mountain Ave., Murray Hill, N.J. 07974, USA
FORTUIN, C.M.	Inst. of Theoret. Physics, University of Nijmegen, The Netherlands
FOSSHEIM, K.	Dept. of Physics, The Technical Univerisity of Norway, 1734 Trondheim-NTH, Norway
FRIEDEL, J.	Lab. de Physique des Solides, Université Paris-Sud, Batiment 510, 91405 Orsay, France
HAFNER, J.	Max-Planck-Institut für Festkörperforschung, Büsnauer Strasse 171, 7000 Stuttgart 80, W-Germany
HARLEY, R.T.	Clarendon Laboratory, University of Oxford, Parks Road, Oxford OX1 3PU, England
HEMMER, P.C.	Dept. of Theoretical Physics, The Technical University of Norway, 7034 Trondheim-NTH Norway
HÖCK, K.H.	Dpet. of Physics and Astronomy, Ruhr-Universität Bochum, 463 Bochum-Querenburg, W-Germany
JANSSEN, T.	Inst. of Theoretical Physics, Catholic University, Toernooiveld, Nijmegen, The Netherlands
JØSSANG, T.	Dept. of Physics, University of Oslo, Blindern, Oslo 3, Norway
KJEMS, J.K.	Dept. of Physics, Research Establishment Risø, 4000 Roskilde, Denmark
LEROUX HUGON, P.	C.N.R.S. Lab. de Physique des Solides, 1 Place A. Briand, 92190 Meudon, France
LUTHER, A.	NORDITA, Blegdamsvej 17, 2100 Copenhagen, Denmark
MAASKANT, W.J.A.	Gorlaeus Laboratoria, Rijksuniversiteit Leiden, Postbus 75, Leiden, The Netherlands
McMILLAN, W.L.	Dept. of Physics, University of Illinois, Urbana, Ill. 61801, USA

MØLLENBACH, K.	Dept. of Physics, Research Establishment Risø, 4000 Roskilde, Denmark
NEDELLEC, P.	Lab. de Physique des Solides, Université Paris-Sud, Batiment 510, 91405 Orsay, France
OTTINSEN, T.	Dept. of Physics, University of Oslo, Blindern, Oslo 3, Norway
PAQUET, D.	C.N.E.T., Groupt P.E.C., 196 rue de Paris, 92220 Bagneux, France
de Pater, H.	Lab. voor Technische Natuurkunde, Technische Hogeschool, Delft, The Netherlands
POPPE, U.	Inst. für Festkörperforschung der KFA, Postfach 1913, 517 Jülich 1, W-Germany
PYNN, R.	Institut Laue-Langevin, B.P. 156 Centre de Tri, 38042 Grenoble-Cedex, France
RICCA, A.M.	C.I.S.E., P.B. 3986, Milan, Italy
ROMESTAIN, R.	Université Scientifique et Médicale, Lab. de Spectrométric Physique, B.P. 53, 38041 Grenoble-Cedex, France
RYAN, J.F.	Clarendon Laboratory, University of Oxford, Parks Road, Oxford OX1 3PU, England
Di SALVO, F.J.	Bell Laboratories, 600 Mountain Avenue, Murray Hill, N.J. 07974, USA
SHIRANE, G.	Dept. of Physics, Brookhaven National Laboratory, Upton, L.I., N.Y. 11973, USA
SHULTZ, M.J.	DEAP/Pierce Hall, Harvard University, Cambridge, Mass. 02139, USA
SIEGEL, E.	Molecular Energy Research Company, 23 Bergenline Ave., Westwood, N.J. 07675, USA
SLAGSVOLD, B.J.	Dept. of Physics, The Technical University of Norway, 7034 Trondheim-NTH, Norway
SMITH, S.R.P.	Dept. of Physics, University of Essex, Wivenhoe Park, Colchester CO4 3SQ, England

STEIGMEIER, E.F. Laboratories RCA, Badenerstrasse 569, 8048 Zürich, Switzerland

STEINSVOLL, O. Institutt for atomenergi, 2007 Kjeller, Norway

STINCHCOMBE, R.B. Dept. of Theoretical Physics, University of Oxford, 12 Parks Road, Oxford OX1 3PQ, England

TAYLOR, D.R. Clarendon Laboratory, University of Oxford, Parks Road, Oxford OX1 3PU, England

TESTARDI, L.R. Bell Laboratories, 600 Mountain Avenue, Murray Hill, N.J. 07974, USA

THOMAS, H. Dept. of Physics, University of Basel, 4056 Basel, Switzerland

de WOLFF, P.M. Lab. voor Technische Natuurkunde, Technische Hogeschool, Delft, The Netherlands

YAMADA, Y. College of General Education, Osaka University, 1-1 Machikaneyama, Toyonaka-sho, Osaka 560, Japan

YOUNGBLOOD, R. Dept. of Physics, Research Establishment Risø, 4000 Roskilde, Denmark

ØSTGAARD, E. Dept. of Physics, NLHT, 7000 Trondheim, Norway

ORGANIZING COMMITTEE:

ANDERSEN, E.
JARRETT, G.
RISTE, T.
} Institutt for atomenergi, P.O.B. 40, 2007 Kjeller, Norway

SUBJECT INDEX